SPRINGER
LAB MANUAL

Springer-Verlag Berlin Heidelberg GmbH

Erwin E. Sterchi · Walter Stöcker

Proteolytic Enzymes
Tools and Targets

With 57 Figures, including one Color Plate

Springer

PROF. DR. ERWIN E. STERCHI

Institut für Biochemie und Molekularbiologie
Universität Bern
Bühlstrasse 28
CH-3012 Bern

PROF. DR. WALTER STÖCKER

Westfälische Wilhelms-Universität Münster
Institut für Zoophysiologie
Hindenburgplatz 55
D-48143 Münster

ISBN 978-3-642-47807-9   ISBN 978-3-642-59816-6 (eBook)
DOI 10.1007/978-3-642-59816-6

Cover design: design & production GmbH, D-69121 Heidelberg
Typesetting: Mitterweger Werksatz GmbH, D-68723 Plankstadt
SPIN: 10086278 27/3137 5 4 3 2 1 0 - Printed on acid free paper

Preface

Proteolytic enzymes play significant roles in numerous cellular and extracellular processes in health and disease. Our knowledge on this group of enzymes has increased considerably in recent years. For example, more than 500 entries providing a comprehensive bibliography have recently been documented in the Handbook of Proteolytic Enzymes (Barrett, Rawlings, Woessner 1998). Further, an outstandingly large number of publications have been dedicated to the aspartic proteinases, especially to the enzyme of the human immuno deficiency virus (HIV), which is a key target in the treatment of AIDS. Other new developments led to the recognition of the caspases, a new family of cysteine proteinases involved in apoptosis. Another milestone was placed by the structural and functional analysis of the proteasome, whose catalytic principle relies on an amino terminal threonine residue, and thus appears to be a variation of the mechanism originally observed in the serine proteinases. The recognition of the importance of metallo proteases in developmental processes, arthritis and cancer have added another highlight on recent achievements.

In addition to these arbitrarily selected new discoveries of proteolytic principles in a variety of biological systems, proteases have gained importance as laboratory tools in the experimental investigation of peptides and proteins. Many researchers working in a wide range of specialist fields are thus faced with proteolytic enzymes in one way or the other. They may find that proteolytic activity plays an integral part in the particular system they are studying and wish to characterize the enzyme or enzymes involved. They may be faced with the problem of peptidases present in their experimental set-up that interfere with the particular system under investigation. Or, they may wish to use proteolytic enzymes as laborarory tools in their study of a particular protein. Invariably, investigators not primarily involved in protease research may often lack the specialist know how for the characterization, or the inhibition of proteolytic activities or the use of these proteolytic enzymes as tools.

The aim of this book is to provide methodological support for both the use and the analysis of peptidases. The topics covered in the first part give an overview of assay-methods for proteolytic enzymes using natural and artificial substrates. The inhibition of proteolytic enzymes is addressed in a chapter with methods to suppress unknown proteolytic activites. The second part of the book deals with the investigation of proteolytic enzymes per se and covers expression, purification and characterization of proteolytic enzymes. Finally, the third part is a selection of chapters on the use of proteolytic enzymes as laboratory tools.

It is with clear intent that the addresses of the authors who have contributed to this book are given in full detail. This enables readers to approach the individual authors directly if they encounter difficulties in investigating a particular proteolytic activity in their system.

We thank all the authors who have contributed to this book. We also thank Dr. Rolf Lange and Dr. Jutta Lindenborn from Springer-Verlag for their advice and guidance in editing this book.

Erwin E. Sterchi
Walter Stöker

Contents

Chapter 18
Protease-Catalyzed Peptide Synthesis

Appendix
Supplement to Chapter 9: Crystallization Conditions for Proteases

Chapter 1

Introduction: Nomenclature and Classes of Peptidases

A. J. KENNY

Introduction

Proteolytic enzymes - you either love them or you hate them. To some researchers the only good proteinase is a dead one, no longer able to degrade the delicate protein they are striving to isolate: their interest in this book will focus on the best ways to irreversibly inhibit any contaminating proteinases. For the rest of us proteolytic enzymes provide a never ending source of pleasure and interest. Although they catalyse but a single reaction, the hydrolysis of a peptide bond, the various ways they achieve this, their ubiquitous distribution among all life forms, their multiplicity of locations inside, outside and at the surface of cells and, above all, their enormous diversity of function make them one of the most fascinating group of enzymes.

Proteolytic enzymes frequently exist as zymogens in contact with potential substrates which are hydrolysed only when the zymogen is activated by another proteinase, an arrangement that ensures that the pancreas does not digest itself and that, when required, our blood coagulates and our complement system can be brought into action. These particular proteinases have their function outside cells and have been well studied for very many years. Those whose actions are within cells are more numerous, much more difficult to study and only relatively recently has our level of knowledge become comparable to that of the extracellular proteinases (for review, see Bohley 1995). The cell-surface peptidases are a much smaller group, specializing in the hydrolysis of relatively simple peptides rather than proteins. As a group they do not require to be activated, but lie in wait for the arrival of a susceptible peptide substrate, sometimes activating it, as in the conversion of angiotensin I to angiotensin II, but more usually inactivating the peptide, thereby terminating a hormonal or neuropeptide signal (for review, see Kenny and Hooper 1991 and Kenny and Boustead 1997).

A. J. Kenny, Le Presbytère, St. Etienne d'Albagnan, 34390, FRANCE (*phone* +33 (0)467 97 15 37; *fax* +33 (0)467 97 15 37)

Nomenclature and Terminology

The terminology of a group of enzymes that has been studied for more than a century is inevitably complicated and there is a need to bring order and consistency. All these enzymes are correctly designated as **peptidases**, indicating that they hydrolyse peptide bonds (which, in all but a few cases, are α-peptide bonds). Those enzymes that require the presence of an unsubstituted N- or C-terminus in the substrate are **exopeptidases**, those that do not are **endopeptidases**. The terms proteinase and protease (both implying the ability to hydrolyse macromolecular substrates) are roughly equivalent to the term endopeptidase. The secondary and tertiary structure of protein substrates usually prevents attack by exopeptidases. The action of an endopeptidase is generally not favoured by the presence of a free N- or C-terminus close to the scissile bond. Exopeptidases remove a single amino acid, a dipeptide or a tripeptide from one or other terminus, actions which are the basis for the classification of the exopeptidases. Similar considerations of specificity cannot be applied to the endopeptidases, which are best distinguished by their active sites yielding the four main classes: **serine, cysteine, aspartic** and **metallo**-endopeptidases. The same catalytic mechanisms operate among the exopeptidases, so that some of them can be further subdivided on the same basis (Fig. 1).

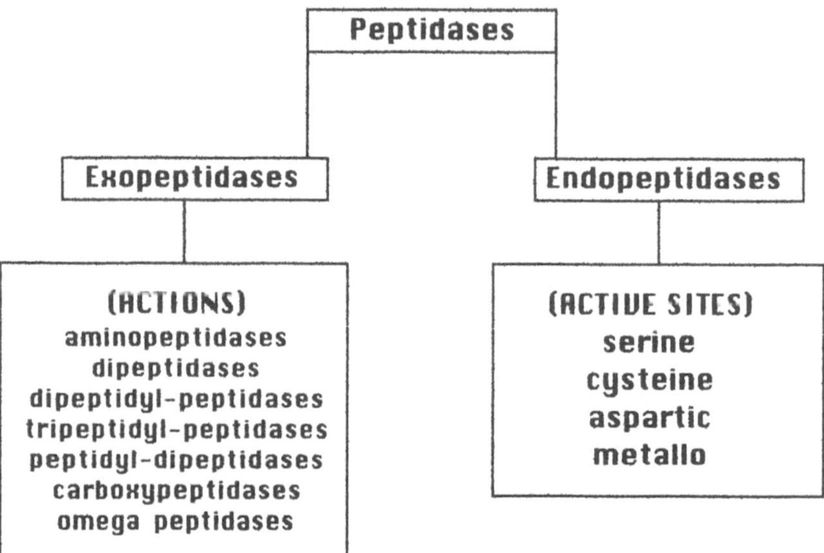

Fig. 1. The main classes of peptidases

The four types of active site were first recognized by the use of some group-specific inhibitors. Serine peptidases (in common with other serine hydrolases, e.g. acetylcholinesterase) react with organophosphate compounds, such as di-isopropyl phosphofluoridate (DFP or DipF) in such a way that they catalyse their own death by acylating a single serine residue at the active site, essentially an irreversible step. The toxicity of this reagent has led to the use of other reagents, e.g. phenylmethylsulphonyl fluoride (PmsF) and 3,4-dichloroisocoumarin (3,4-DCI) in its place. In the catalytic mechanism of cysteine peptidases the thiol group of a single cysteine residue plays an essential role. This group is susceptible to oxidation and can react with a variety of reagents: heavy metals (e.g. Hg), iodoacetate, N-ethyl-maleimide and a highly selective inhibitor, effective for many but not all in this group, E-64, a peptide epoxide, N-(L-3-*trans*-carboxyoxiran-2-carbonyl)-L-leucyl-amido(4-guanidino)butane. The aspartic peptidases were first recognized by their highly acidic pH optima, only later did a specific inhibitor, pepstatin A (derived from a strain of *Streptomyces*) become available. The metallopeptidases are usually recognized by their susceptibility to inhibition by chelating agents such as EDTA and 1,10-phenanthroline. When knowledge of the amino acid sequences of peptidases became commonplace and particularly when 3-dimensional protein structures began to emerge, this functional division of peptidases was placed on a firmer basis and detailed mapping of the active sites became possible. The designation of a peptidase on the basis of single amino acid residue at the active site, though useful, is somewhat simplistic. In the case of endopeptidases it was already clear from kinetic studies that the binding of substrates (and inhibitors) involved interaction at a number of subsites on either side of the pair of residues containing the scissile bond. In the terminology of Schechter and Berger (1967) these two residues are P1 and P1', binding to peptidase sites S1 and S1' (Fig. 2) The better understanding of the detailed interaction of substrate and enzyme from classical kinetics and from modelling of the 3-dimensional structure has permitted the design and synthesis of highly specific inhibitors. Nature is no less inventive. There are many low molecular weight potent inhibitors of microbial origin, some of broad specificity, e.g. amastatin for a range of aminopeptidases, others of high specifity, e.g. phosphoramidon for endopeptidase-24.11. Inhibitors of high molecular weight include some nonspecific proteins e.g. α-2-macroglobulin, which can entrap endopeptidases, sequestering them from their protein substrates. Some are group specific, e.g. the cystatins which bind tightly but reversibly to many cysteine endopeptidases and others are specific to a small group of related peptidases, such as TIMP (tissue inhibitor of metallo-proteinases) which binds tightly to collagenase and other matrix metallo-endopeptidases.

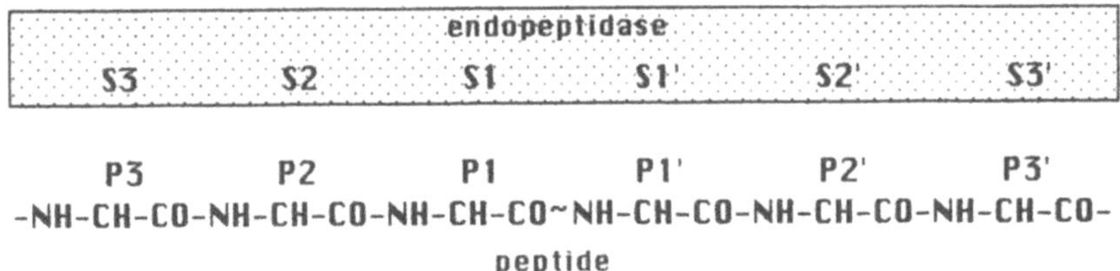

Fig. 2. Terminology of the subsite interactions between peptide substrate and endopeptidase active site. ~ indicates the scissile bond

The Enzyme Commission (E.C.) Classification

All the hydrolases are designated by the International Union of Biochemistry and Molecular Biology (1992) as E.C. 3.- and the peptidases as E.C. 3.4. The main classes of peptidases are defined by a third numeral (11 to 24) as listed in Table 1. The exopeptidases are classified mainly on the basis of their actions. Only peptides with an unsubstituted terminus are attacked, with the exception of a very small number, grouped as **omega peptidases** (3.4.19.-) which can release certain modified terminal residues. Examples acting at the N-terminus are acylaminoacyl peptidase which will release an acetyl or formyl N-terminal residue; pyroglutamyl peptidase, able to release the cyclic residue; β-aspartyl peptidase, able to split an isopeptide bond. Others are directed to the C-terminus, e.g. peptidyl glycinamidase which releases a C-terminal glycinamide and γ-glutamyl carboxypeptidase able to release a C-terminal glutamic acid linked by an isopeptide bond.

Table 1. Types of peptidase defined in the Enzyme Nomenclature list of the International Union of Biochemistry and Molecular Biology (1992).

E.C. number	Peptidase type	Action
EXOPEPTIDASES		
3.4.11.-	Aminopeptidase	N-terminal residue released
3.4.13.-	Dipeptidase	Acts only on dipeptides
3.4.14.-	Dipeptidyl peptidase	N-terminal dipeptide released
	Tripeptidyl peptidase	N-terminal tripeptide released
3.4.15.-	Peptidyl dipeptidase	C-terminal dipeptide released
3.4.16.-	Carboxypeptidase (serine)	C-terminal residue released
3.4.17.-	Carboxypeptidase (metallo)	C-terminal residue released
3.4.18.-	Carboxypeptidase (cysteine)	C-terminal residue released

Table 1. Continous

E.C. number	Peptidase type	Action
3.4.19.-	Omega peptidase	Releases modified residues from N- or C-termini
ENDOPEPTIDASES		
3.4.21.-	Serine endopeptidase	
3.4.22.-	Cysteine endopeptidase	
3.4.23.-	Aspartic endopeptidase	
3.4.24.-	Metallo-endopeptidase	
3.4.99.-	Endopeptidase of unknown catalytic mechanism	

Aminopeptidases (3.4.11.-)

The actions of aminopeptidases depend on various factors: the identity of the P1 residue, the adjacent residues at P1' and P2' and on peptide chain length. In naming these enzymes the E.C. classification gives major emphasis to the identity of the P1 or P1' residue, yielding names such as membrane alanyl aminopeptidase (3.4.11.2), glutamyl aminopeptidase (3.4.11.7) or X-Pro aminopeptidase (3.4.11.9). In the case of the last named, also known as aminopeptidase P, the name accurately describes the restricted specificity of the peptidase, but for the other two the term is misleading. The striking feature of alanyl aminopeptidase is its very broad specificity which is in no way limited to alanyl peptides, hence its other name, aminopeptidase N, from its preference for neutral (uncharged) sidechains; similarly glutamyl aminopeptidase also releases aspartyl residues, hence its other name, aminopeptidase A (acidic sidechains). It is therefore important to avoid taking the approved names too literally or too restrictively. Chain length may also be important: alanyl aminopeptidase prefers oligopeptide substrates, while aminopeptidase W (X-Trp aminopeptidase, 3.4.11.16) is most effective with dipeptides. Most of the aminopeptidases that have been characterized are metallo-enzymes.

Dipeptidases

This very small group of peptidases are distinguished from the aminopeptidases by their inability to hydrolyse peptides containing more than two residues. In other respects they may have few specificity requirements, e.g. membrane dipeptidase (3.4.13.19) will even hydrolyse Gly-D-Phe.

Dipeptidyl Peptidases and Tripeptidyl Peptidases

These release di- or tripeptides from the N-termini of their substrates. In assigning a peptidase to this group, it is important to characterize the released fragment as a di- or tripeptide. The sequential action of other peptidases may confuse if the assay depends on the release of a terminal chromogenic or fluorogenic moiety (see e.g. Kenny and Ingram 1988). Serine peptidases predominate in this group.

Peptidyl Dipeptidases

The best known example is peptidyl dipeptidase A (3.4.15.1, angiotensin converting enzyme) a Zn-metallopeptidase found at the cell surface. With most substrates it behaves 'correctly', releasing the C-terminal dipeptide, but curiously, it can attack the amidated peptide, substance P, to release the C-terminal tripeptide.

Carboxypeptidases

They are classified according to the class of active site, three types being recognized: serine (3.4.16.-), metallo (3.4.17.-) and cysteine (3.4.18.-), with most of them characterized as Zn-metallopeptidases.

Endopeptidases

Enzymes that attack "interior" rather than terminal peptide bonds defy classification on the principles that are appropriate for the exopeptidases, where the properties of the substrate are paramount. It is rare to find an endopeptidase that is specific for a bond involving a single type of side chain and so there is no logic in attempting to devise a nomenclature or classification relating to specificity. Hence the current division into the four groups indicating the catalytic mechanism: **serine** endopeptidases (3.4.21.), **cysteine** endopeptidases (3.4.22.), **aspartate** endopeptidases (3.4.23.) and **metallo**-endopeptidases (3.4.24.) while a final group (3.4.99.) contains a handful of enzymes for which the mechanism remains undefined.

Peptidase Families and Clans

The relative ease by which cDNA-derived sequences can now be obtained has added another level of sophistication to the classification of peptidases, one based on evolutionary considerations. At an early stage the pentapeptide motif, HEXXH, was observed to be a characteristic of Zn-binding in a number of metallopeptidases, the motif providing two of the three ligands for the Zn atom. The process has advanced rapidly, enabling Rawlings and Barrett (1993) to examine over 600 sequences of peptidases, distinguishing between **families** and groups of families, designated as **clans**. They recognized 84 families, some of which were related to each other, thus creating family clans. The smallest class was that of the aspartic peptidases, 20 enzymes in two families and a single clan, depending on a pair of aspartate residues for the catalytic mechanism. There were 14 families among the cysteine peptidases, two of which were assigned to clan A and three to clan B distinguished by the active site residues. The largest class was that of the metallopeptidases, 25 families, of which 13 were in clan A, possessing the common Zn-binding motif, HEXXH. Other families had HXXE and HXXEH binding motifs and some others bound two Zn atoms. Families may include both prokaryotic and eukaryotic enzymes. On the other hand most of the viral proteinases show no relationship to non-viral proteinases. The families also exhibit a variety of functional differences, some even lacking peptidase activity, while in others carboxypeptidases, aminopeptidases or endopeptidases coexist. At the time of going to print the number of families had grown considerably, reflecting the immense research activity in the field (see Table 2).

Table 2. Evolutionary classification of peptidases based on amino acid sequences (from Barrett et al., 1998, and MEROPS – the peptidase database
url: http://www.bi.bbsrc.ac.uk/Merops/merops.htm)

Class	Families	Clans	Active site residues
Serine (EC 3.4.21.)	S1 - S44	PA (S1-3, 6, 7, 29-32, 35, 43)	His, Asp, Ser
		SB (S8)	Asp, His, Ser
		SC (S9, 10, 15, 28, 33, 37)	Ser, Aso, His
		SE (S11-13)	Ser, Lys, (SXXK)
		SF (S24, 26, 41, 44)	Ser, Lys, (His)
		SH (S21)	His, Ser, His
		SK (S14)	Ser, His
		SX (S16, 18, 19, 38)	
Cysteine (EC 3.4.22)	C1 - C47	CA (C1, 2, 6-10, 12, 16, 19, 21, 23, 27-29, 31-36, 39, 41, 42, 47)	Cys, His, Asp (Asn)
		PA (C3, 4, 24, 30, 37, 38)	His, Asp, Cys
		CD (C11-14, 25)	His, Cys
		CE (C5)	His, Glu (Asp), Cys
		CF (C15)	Glu, Cys, His
		CG (C17)	Cys, Cys
		CH (C46)	Cys, Thr, His
		CX (C22, 26, 40)	

Table 2. Continous

Class	Families	Clans	Active site residues
Aspartic (EC 3.4.23)	A1 - A21	AA (A1-3, 9, 11-13, 16-8) AB (A6, 21)	Asp, Asp Asp, Asn
Metallo (EC 3.4.24)	M1 - M51	MA (M1, 2, 4-13, 30, 36, 48) MC (M14) MD (M15, 45) ME (M16, 44) MF (M17) MG (M24) MH (M18, 20, 25, 48, 40, 42 MJ (M38) MX (M3, 26, 27, 32, 34, 35, 41, 43, 50, 51)	His, Glu, His (HEXXH) His, Glu, His His, Asp, His His, Glu, His (HXXEH) Lys, Asp, Asp, Asp, Glu Asp, Asp, His, Glu, Glu His (Asp), Asp, Glu, Asp (Glu), His His, His, Lys, His, His, Asp
Threonine	T1, T3	PB (T1, 3)	Thr, Ser or Cys

References

Bohley P (1995) The fates of proteins in cells. Naturwissenschaften 82:544-550

International Union of Biochemistry and Molecular Biology (1992) Enzyme Nomenclature. Academic Press Inc. San Diego, New York, Boston, London, Sydney, Tokyo and Toronto.

Kenny AJ, Hooper NM (1991) Peptidases involved in the metabolism of bioactive peptides. In: Henriksen JH, Degradation of bioactive substances, physiology and pathology. CRC Press Inc 47-79

Kenny AJ, Ingram J (1988) Is there a tripeptidyl peptidase in the renal brush border membrane? Biochem J 255: 373-376

Rawlings ND, Barrett AJ (1993) Evolutionary families of peptidases. Biochem J 290:205-218

Schechter I, Berger A (1967) On the size of the active site in proteases. I Papain. Biochem Biophys Res Commun. 27: 157-162

The following books may be useful sources of more detailed information:

Barrett AJ (ed) (1994) Proteolytic enzymes: serine and cysteine peptidases. Methods Enzymol vol 244 Academic Press Inc. San Diego, New York, Boston, London, Sydney, Tokyo and Toronto pp. 765

Barrett AJ (ed) 1995 Proteolytic enzymes: aspartic and metallopeptidases. Methods Enzymol vol 248 Academic Press Inc. San Diego, New York, Boston, London, Sydney, Tokyo and Toronto

Barrett AJ, Rawlings ND, Woessner JF (eds) (1998). Handbook of Proteolytic Enzymes. Academic Press.

Beynon RJ, Bond JS (eds) (1989) Proteolytic enzymes, a practical approach. IRL Press at Oxford University Press Oxford New York Tokyo pp 259

Kenny AJ, Boustead CM (eds) (1997) Cell-Surface Peptidases in health and disease. Bios Scientific Publishers, Oxford. pp 384

Part I

The Assay and Inhibition of Proteolytic Enzymes

Chapter 2

Chromogenic Peptide Substrates

H. KIRSCHKE AND B. WIEDERANDERS

Introduction

This chapter deals with spectrophotometric methods for the determination of proteolytic enzymes by their chromogenic substrates. The choice of sensitive substrates for several enzymes is facilitated by a summary of kinetic constants included in additional tables.

Synthetic substrates are indispensable to kinetic measurements, testing of inhibitors, routine assays e.g. during isolation, and further characterization of proteolytic enzymes. Synthetic peptide or amino acid substrates used for kinetic measurements contain only one cleavable bond. Enzymatic hydrolysis liberates chromogenic or chromophoric residues which can be determined by simple spectrophotometric methods (Kirschke and Wiederanders 1984).

The peptidyl or amino acyl residues determine selectivity and specificity. The sensitivity is usually dependent on the chromogenic or chromophoric group. But there are also examples where the detector group has an influence on k_{cat} or K_m values (Barrett and Kirschke 1981).

The thioester bond is extremely sensitive to enzymatic hydrolysis so that e.g. lower concentrations of elastase have been determined using a peptidyl thioester substrate than with a fluorogenic substrate, although fluorescence detection is several times more sensitive than spectrophotometric measurements (Powers and Kam 1995).

Nowadays, chromogenic or chromophoric substrates are employed if the substrate has the same sensitivity as a fluorogenic one, or for measurements

Correspondence to: H. Kirschke, Martin-Luther-University Halle-Wittenberg, Department of Medicine, Institute of Physiological Chemistry, Hollystrasse 1, Halle, 06097, Germany (*phone* +49-(0)3461-721960; *fax* +49-(0)0345-5573811; *e-mail* 0345530320-0001@t-online.de)

B. Wiederanders, Clinic of the Friedrich-Schiller-University Jena, Institute of Biochemistry I, Jena, 07740, Germany

where very high sensitivity is not required e.g. during the isolation procedure of an enzyme to avoid considerable dilution of the enzyme samples, or if the spectrophotometer is better equipped for kinetic measurements than the fluorimeter.

Subprotocol 1
4-Nitroanilide Substrates

Principle   Free or *N*-protected aminoacyl or peptidyl 4-nitroanilides are hydrolyzed by proteases to result free 4-nitroaniline which is measured spectrophotometrically. The measurement (absorbance at 410 nm) can be performed continuously or after inactivation of the enzyme in stopped assays. At pH <3.5 A_{410} is decreasing. The released 4-nitroaniline can also be diazotized and coupled to *N*-(1-naphthyl)-ethylene diamine to result a red azo dye with absorbance at 546 nm (see Subprotocol 7). Here, we describe the continuous spectrophotometric assay.

Materials

Equipment   – 1-cm matched cuvettes
– spectrophotometer equipped with a thermo-regulated cuvette holder and linked to either a chart recorder or a computer

Substrate   Dissolve substrate in the appropriate concentration (for K_m see Table 1 – 4) as a stock solution in water or in an organic solvent. A stock solution in Me$_2$SO can be stored usually at 4 °C for several weeks. Many nitroanilide substrates autolyse in solutions with pH >7.

Buffer   100 mM buffers are recommended (for pH see Table 1 – 4). Activate cysteine peptidases by 5 mM thiol compounds and EDTA in the buffer (activated buffer).

Enzyme   Dilute enzyme solutions with buffer or 0.01% Brij-35 in buffer.

Table 1. Metallopeptidases

Enzyme	Substrate	pH	$k_{cat}(s^{-1})$	$K_m(\mu M)$	k_{cat}/K_m (mM^{-1}s^{-1})	Ref.
Carboxypeptidase N EC 3.4.17.3 human	FA-Ala-Lys-OH	7.5	97	0.34	285	a
Meprin A EC 3.4.24.18	H-Arg-Pro-Pro-Gly-Phe(NO$_2$)-Ser-Pro-Phe-Arg-OH	8.7	40.9	0.29	141	b
Astacin EC 3.4.24.21	H-Arg-Pro-Pro-Gly-Phe(NO$_2$)-Ser-Pro-Phe-Arg-OH	8.7	78.7	0.085	922	b
	H-Arg-Pro-Pro-Gly-Phe(NO$_2$)-Ala-Pro-Phe-Arg-OH	8.7	834	0.306	2730	b
	Suc-Ala-Ala-Ala-NHPhNO$_2$	8.0	0.25	0.97	0.258	c
Pseudolysin EC 3.4.24.26	FA-Ala-Leu-Ala-OH	8.0	144	0.16	900	d
	FA-Gly-Leu-Gly-OH	8.0	7	0.23	30	d
Gelatinase A EC 3.4.24.35 human	Ac-Pro-Leu-Ala-^{s}Nva-Trp-NH$_2$	6.0	7.5	0.045	167	e
Thermolysin EC 3.4.24.27	Ac-Pro-Leu-Ala-^{s}Nva-Trp-NH$_2$	6.0	1.7	0.36	4.7	e
Stromelysin 1 EC 3.4.24.17	Ac-Pro-Leu-Ala-^{s}Nva-Trp-NH$_2$	6.0	15	0.59	25	e
Leucine aminopeptidase EC 3.4.11.1 bovine	H-Leu-NHNap	8.0	0.1	0.25	0.4	f
	H-Leu-NHPhNO$_2$	8.0	0.75	1.1	0.7	f
	H-Leu-hydrazide	8.0	3280	1.7	1929	f
Peptidyl dipeptidase A (ACE) EC 3.4.15.1	FA-Phe-Gly-Gly-OH	7.5	317	0.3	1057	g
	FA-Phe-Ala-Phe-OH	7.5	128	0.093	1380	g

a Skidgel 1995; *b* Wolz and Bond 1995; *c* Stöcker and Zwilling 1995; *d* Saulnier et al. 1989; *e* Stein and Izquierdo-Martin 1994; *f* Hanson and Frohne 1976; *g* Bünning et al. 1983

Table 2. Aspartic peptidases

Enzyme	Substrate	pH	$k_{cat}(s^{-1})$	$K_m(\mu M)$	k_{cat}/K_m $(mM^{-1}s^{-1})$	Ref.
Cathepsin E EC 3.4.23.34	H-Pro-Pro-Thr-Ile-Phe-Phe(NO_2)-Arg-Leu-OH	3.5	75	0.03	2500	a
Cathepsin D EC 3.4.23.5	H-Phe-Gly-His-Phe(NO_2)-Phe-Val-Leu-OMe	4.0	0.87	0.09	9.7	b
	H-Pro-Thr-Glu-Phe-Phe(NO_2)-Arg-Leu-OH	3.1	173	0.55	315	b
Pepsin EC 3.4.23.1 human	H-Pro-Thr-Glu-Phe-Phe(NO_2)-Arg-Leu-OH	3.1	72	0.17	424	b
Gastricsin EC 3.4.23.3 human	H-Pro-Thr-Glu-Phe-Phe(NO_2)-Arg-Leu-OH	3.1	9	2.0	4.5	b
Retropepsin EC 3.4.23.16 HI-virus	H-Lys-Ala-Arg-Val-Leu-Phe(NO_2)-Glu-Ala-Met-OH	4.7	20	0.02	1000	c
Penicillopepsin EC 3.4.23.20	H-Pro-Thr-Glu-Phe-Phe(NO_2)-Arg-Leu-OH	3.1	24	0.005	4800	b

a Kageyama 1995; b Dunn et al. 1986; c Richards et al. 1990

Table 3. Serine peptidases

Enzyme	Substrate	pH	$k_{cat}(s^{-1})$	$K_m(\mu M)$	k_{cat}/K_m $(mM^{-1}s^{-1})$	Ref.
Carboxypeptidase C EC 3.4.16.5 yeast	FA-Phe-Leu-OH	6.5	82	21	3900	b
Trypsin EC 3.4.21.4	Z-Lys-SBzl	8.0	75	50	1500	c
	Z-Arg-SBzl	7.5	94	5.3	18000	c
	Tos-Gly-Pro-Arg-NHPhNO$_2$	7.5	69a	17	4000	d
Thrombin EC 3.4.21.5	Z-Lys-SBzl	8.0	35	40	880	c
	Z-Arg-SBzl	7.5	7.0	1.9	3700	c
Elastase, Pancreatic EC 3.4.21.36 porcine	Boc-Ala-Ala-Nva-SBzl	7.5	570	84	6800	c
Elastase, Leucocyte EC 3.4.21.37 human	Boc-Ala-Pro-Nva-SBzl(4-Cl)	7.5	10	0.08	130000	c
Chymotrypsin EC 3.4.21.1	Boc-Ala-Ala-Phe-SBzl	7.5	135	9.1	15000	c
Cathepsin G, Leucocyte EC 3.4.21.20 human	Suc-Val-Pro-Phe-SBzl Suc-Ala-Ala-Pro-Phe-NHPhNO$_2$	7.5	22	19	1200	c
Coagulation Factor Xa EC 3.4.21.6	Bz-Ile-Glu-Gly-Arg-NHPhNO$_2$	8.4	140a	83	1700	d
Plasmin EC 3.4.21.7	Z-Lys-SBzl	8.0	50a	24	2080	d
	Tos-Gly-Pro-Lys-NHPhNO$_2$	8.0	34a	150	227	d
u-Plasminogen activator EC 3.4.21.73	Z-Lys-SBzl	8.0	60a	30	2000	d
Tissue kallikrein EC 3.4.21.35 human	Z-Tyr-OPhNO$_2$	8.8	-	7	-	e
Tryptase EC 3.4.21.59 human	Suc-Val-Pro-Phe-NHPhNO$_2$	8.0	100	75	1330	f
Acrosin EC 3.4.21.10	Bz-Arg-OEt	8.0	11500	101	114000	g
Granzyme A EC 3.4.21.78 mouse	Z-Lys-SBzl	7.5	22	130	170	h
	Z-Gly-Arg-SBzl	7.5	59	160	370	h
Granzyme B EC 3.4.21.79 mouse	Boc-Ala-Ala-Asp-SBzl	7.5	116	500	230	h

a k_{cat} values are based on molarities determined by titration; *b* Mortensen et al. 1994; *c* Powers and Kam 1995; *d* Lottenberg et al. 1981; *e* Geiger et al. 1977; *f* Tanaka et al. 1983; *g* Akama et al. 1994; *h* Odake et al. 1991

Table 4. Cysteine peptidases

Enzyme	Substrate	pH	k_{cat} (s^{-1})	K_m (μM)	k_{cat}/K_m (mM^{-1}s^{-1})	Ref.
Cathepsin B	Z-Arg-Arg-NHNap	6.0	178a	0.19	937	b
EC 3.4.22.1	Bz-Arg-NHNap	6.0	25a	4.3	6	c
human						
rat	Bz-Arg-NHNap	6.0	24a	4.0	6	d
Cathepsin H	H-Arg-NHNap	6.5	16.6a	0.071	234	d
EC 3.4.22.16	H-Arg-NHPhNO$_2$	6.8	13.0a	0.2	65	e
rat	H-Lys-NHNap	6.5	12.6a	0.092	137	d
	Suc-Ala-Ala-Ala-Ala-NHPhNO$_2$	6.5	0.9a	0.7	1	f
porcine	Bz-Arg-NHNap	6.8	8.8a	0.46	19	g
Cathepsin L	Z-Lys-OPhNO$_2$	5.5	19.8a	0.01	1980	c
EC 3.4.22.15 rat						
Dipeptidyl	H-Ala-Ala-NHNap	6.0	248	0.19	1300	h
peptidase I	H-Gly-Arg-NHNap	6.0	1300	0.1	13000	h
= Cathepsin C	H-Gly-Phe-NHNap	6.0	79	0.17	465	h
EC 3.4.14.1	H-Ser-Met-NHNap	6.0	510	0.17	3000	h
rat						
bovine	H-Gly-Phe-NHPhNO$_2$	5.0	-	2.51	-	i
Chymopapain	Boc-Ala-Ala-Gly-NHPhNO$_2$	6.8	0.5a	4.2	0.1	k
EC 3.4.22.6						
Glycyl endopeptidase	Boc-Ala-Ala-Gly-NHPhNO$_2$	6.8	22a	5.2	4.2	k
EC 3.4.22.25						
Papain	Boc-Ala-Ala-Gly-NHPhNO$_2$	6.8	1.6a	1.4	1.1	k
EC 3.4.22.2	Z-Phe-Cit-NHPhNO$_2$	6.2	42.6	2.2	19.1	l
	Bz-Arg-NHPhNO$_2$	7.2	0.74	2.9	0.3	l

a k_{cat} values are based on molarities determined by titration; b Knight 1980; c Barrett and Kirschke 1981; d Kirschke et al. 1986; e Tchoupe et al. 1991; f Brömme et al. 1987; g Takahashi et al. 1984; h McDonald et al. 1969; i Ohsawa et al. 1993; k Buttle et al. 1990; l Gray et al. 1984

Procedure

1. Set wavelength at 410 nm, scale expansion at 0.1 absorbance and chart speed 1–2 cm min^{-1}.

2. Place 50–475 µl enzyme solution in the cuvette. Add 500 µl of buffer and effector. The volume is made up to 975 µl with water. Cysteine peptidases have to be activated by preincubation for 5–10 min with the activated buffer.

3. Add 25 µl of substrate stock solution, place a small piece of Parafilm over the cuvette and invert it several times. Begin recording immediately. If the reaction is too fast, either dilute the enzyme or adjust the spectrophotometer to greater scale expansion, or faster chart speed.

4. Check the autolysis of substrate by replacing the enzyme solution with buffer or 0.01% Brij-35 and run also an enzyme blank.

5. Read the $\Delta A_{410\,nm}$ with respect to time and calculate the released nmoles min^{-1}. 1 unit of enzyme is defined as the amount required to release 1 µmol of product min^{-1}.
   4-nitroaniline $\epsilon_{410} = 8800$ M^{-1} cm^{-1} (Erlanger et al. 1961)

Continuous spectrophotometric assay

Subprotocol 2
2-Naphthylamide Substrates

Free or *N*-protected aminoacyl or peptidyl 2-naphthylamides and 4-methoxy-2-naphthylamides are hydrolyzed by peptidases to result free 2-naphthylamine which is coupled to Fast Garnet Base. The red azo dye is kept in solution by Brij-35 (or other detergents) and the colour is measured spectrophotometrically. 2-Naphthylamide substrates can also be used in continuous spectrophotometric assays at 340 nm. But this assay is not very sensitive ($\epsilon_{340} = 1780$ M^{-1} cm^{-1}) and can only be used in neutral or alkaline buffers. Very sensitive is the continuous fluorimetric assay.

Principle

Note: (Warning!) 2-naphthylamine and the naphthylamides can cause cancer. The solutions should be treated as very poisonous.

Materials

Equipment
- 1-ml test tubes
- spectrophotometer, wavelength 520 nm, 1 cm-microcuvette
- Fast Garnet Base

Substrate Dissolve substrate in the appropriate concentration (for K_m see Table 1 – 4) in a buffer of the desired pH. In case of poor solubility, the substrate should be dissolved first in methanol or Me_2SO and then diluted with buffer (final concentration of the solvent should not exceed 10%). Most of the 2-naphthylamide substrates are stable in solution at room temperature for several days.

Buffer 50 mM – 100 mM buffers are recommended. Activate cysteine peptidases by thiol compounds (cysteine, mercaptoethanol, dithiothreitol etc.) in the presence of EDTA. The final concentration of these compounds in the buffer should not exceed 6 mM (activated buffer).

Enzyme Dilute enzyme solutions with buffer or 0.01% Brij-35 in buffer.

Colour reagents
- Diazonium salt:
  Dissolve 225 mg Fast Garnet Base in 50 ml of ethanol. Add 30 ml of 1 M HCl with stirring and dilute to 100 ml with water and store at 4 °C for several months.
- 0.2 M $NaNO_2$ solution:
  The solution of 1.4 g $NaNO_2$ in 100 ml of water is stable for 1 day.
- Brij-35 solution:
  The 0.4% solution of Brij-35 in water should also contain an inhibitor of the respective enzyme to stop the enzymatic reaction. In assays of cysteine peptidases the inclusion of mersalyl acid or p-chloromercuribenzoate is necessary not only to inactivate the enzyme but also to bind the essential thiol compounds (1mM in the final assay) which destroy the azo dye.
- Colour reagent:
  Mix 1 ml of the diazonium salt with 0.1 ml of $NaNO_2$ on ice. After 5 min, add 0.4% Brij-35 to 100 ml. This reagent is stable at 4°C for 1 day.

Standard Store the 10 mM 2-naphthylamine stock solution (14.3 mg/10 ml methanol) at 4 °C and dilute 1:50 (0.2 mM), 1:100 (0.1 mM), etc. with buffer prior to use.

▪ Procedure

1. Incubate 50 µl of enzyme solution (preincubate 25 µl solution of cysteine proteases and 25 µl activated buffer at 37 °C for 5 – 10 min) in a 1-ml-test tube with 100 µl of substrate in buffer at 37 °C for 10 min.

2. Stop the enzymatic reaction by addition of 500 µl of the colour reagent. Blanks of substrate and enzyme solutions have to be incubated, as well.

3. Add 100 µl standard solutions (0.2 mM, 0.1 mM, etc.) to 50 µl of buffer and 500 µl of colour reagent. As a control for compounds in the enzyme solution which disturb the colour development, incubate 100 µl of standard solution with 50 µl of enzyme at 37 °C.

4. After 15 min (up to 15 h) of colour development read the assay and standard tubes in a spectrophotometer at 520 nm. Use a microcuvette (1 cm) or, otherwise, increase the volumes given above by the same factor.   **Measurement**

2-naphthylamine ϵ_{520} = 28000 M^{-1} cm^{-1} (Knight 1980)

The nominal unit is release of 1 µmol of product per min under the reaction conditions. The 0.1 mM standard solution containing 10 nmole of 2-naphthylamine in 100 µl gives ΔA = 0.430 and corresponds in a 10-min-assay to 1 munit of enzyme in the tube.   **Calculation**

Subprotocol 3
Thioester Substrates

Free or *N*-protected aminoacyl or peptidyl thiobenzyl esters are hydrolyzed by peptidases to result free benzene thiols which react with either 4,4'-dithiopyridine (Grassetti and Murray 1967) or 5,5'-dithiobis(2-nitrobenzoic acid) (Ellman 1959).   **Principle**
Another type of thioester substrates are penta- or hexapeptides with the scissile thioester bond in the middle of the peptide: -Leu-Gly-^{s}Leu-Leu-, where the NH group of leucine has been replaced by an S atom. Reaction product is SH-Leu-Leu- which also react with 4,4'-dithiopyridine or dithiobis(2-nitrobenzoic acid).
The enzymatic reaction can be followed continuously by the increasing absorbance at 324 nm and 412 nm, respectively. The concentration of the yellow reaction products can also be measured in stopped assays.

Peptidyl thioesters are not suitable substrates for cysteine and other peptidases which need SH-compounds for optimum activity, because the activating thiol compounds react with the reagents.

Materials

Equipment
- 1-cm cuvettes
- spectrophotometer equipped with a thermo-regulated cuvette holder and linked to either a chart recorder or a computer
- wavelength either 324 nm or 412 nm
- 5,5'-dithiobis(2-nitrobenzoic acid) (Ellman's reagent) or 4,4'-dithiopyridine

Substrate   Peptidyl thioesters are usually dissolved in Me_2SO and stored at $-20\,°C$. The concentration in the stock solution (12 mM) should be 40 times higher than in the final assay (0.3 mM).

Buffer   100 mM buffers are recommended (for pH see Table 1 – 4).

Enzyme   Dilute enzyme solutions with buffer or 0.01% Brij-35 in buffer.

Reagent   Dissolve 19.8 mg dithiobis(2-nitrobenzoic acid) or 14.7 mg 4,4'-dithiopyridine x 2 HCl in 10 ml of water. The 5 mM stock solutions are diluted prior to use 1:5 with buffer.

Procedure

For the procedure see the continuous spectrophotometric assay (Subprotocol 1). Set wavelength at either 324 nm for the reaction with 4,4'-dithiopyridine or 412 nm with dithiobis(2 nitrobenzoic acid).

620 µl buffer, 330 µl reagent in buffer (1 mM) and 25 µl substrate stock solution (12 mM) were mixed in a 1-cm cuvette. 25 µl enzyme solution are added to initiate the reaction. The nonenzymatic rate of absorbance increase should be subtracted from the enzyme-catalyzed rate.

4,4'-dithiopyridine, pH 7.5

$\epsilon_{324} = 19800$ M^{-1} cm^{-1} (Grassetti and Murray 1967)

dithiobis(2-nitrobenzoic acid), pH 8.0

$\epsilon_{412} = 13600$ M^{-1} cm^{-1} (Ellman 1959) $\epsilon_{405} = 13260$ M^{-1} cm^{-1} (Ellman 1959)

Subprotocol 4
4-Nitrophenylalanine Containing Peptides

Oligopeptides containing the sequence -Xaa-Phe(4-nitro)-Xaa- in the middle of the molecule are sensitive substrates to aspartic and metallopeptidases.

Enzymatic hydrolysis at the *N*- or *C*-terminal side of Phe(NO_2) causes a shift of the absorbance maximum at 279 nm (-Phe(NO_2-)- in the intact peptide) towards longer wavelengths, if the negatively charged carboxyl group of Phe(NO_2) is generated and towards shorter wavelength, if the positively charged α-amino group of Phe(NO_2) is released. Monitoring the absorbance at 300 – 310 nm, cleavages of the substrate at the *C*-terminal side of Phe(NO_2) result in an increase and at the *N*-terminal side in a decrease of the absorbance.

Principle

Materials

- 1-cm cuvettes
- spectrophotometer equipped with a thermo-regulated cuvette holder and linked to either a chart recorder or a computer
- wavelength 300 – 310 nm

Equipment

Dissolve substrate in water, buffer or organic solvent.

Substrate

100 mM buffer of the desired pH are recommended (see Table 1 – 2).

Buffer

Dilute enzyme solutions with buffer or 0.01% Brij in buffer.

Enzyme

Procedure

For the procedure see the continuous spectrophotometric assay (Subprotocol 1). The wavelength is set at 300 or 310 nm and the decreased or increased absorbance is monitored. In this case it is recommended to mix first substrate, buffer and effector in the cuvette and then start the reaction by the addition of enzyme.

Subprotocol 5
4-Nitrophenyl Ester Substrates

Principle *N*-Protected aminoacyl 4-nitrophenyl esters are hydrolyzed by several enzymes - but because of their spontaneous autolysis at pH $\geq$ 7.0, they can be used as substrates only at acidic pH values. None of these substrates have been found to be specific for one enzyme. Therefore, the application of the aminoacyl nitrophenyl esters are restricted to indicate the activity of purified enzymes rather than to detect one enzyme in a crude extract. The specificity of these substrates to special enzymes can be enhanced by *N*-protected **peptidyl** nitrophenyl esters. 4-Nitrophenol liberated by enzymes can be monitored continuously by the increase in A_{405} in a spectrophotometric assay. The absorbance is pH dependent.

Materials

Equipment
- 1-cm cuvettes
- spectrophotometer equipped with a thermo-regulated cuvette holder and linked to either a chart recorder or a computer
- wavelength 400 – 420 nm

Substrate Dissolve the substrate in 95% acetonitrile or Me_2SO in concentrations 40 times higher than in the final assay. The stock solutions can be stored at – 20 °C.

Buffer 100 mM buffers, pH < 7.0, are recommended.

Enzyme Dilute enzyme solutions with buffer or 0.01% Brij-35 in buffer.

Procedure

For the procedure see the continuous spectrophotometric assay (Subprotocol 1). The wavelength is set at 405 nm and the increased absorbance is monitored. The nonenzymatic rate of absorbance increase should be subtracted from the enzyme catalyzed rate.

Subprotocol 6
N-Protected Dipeptides

Aminoacylated dipeptides are hydrolyzed by carboxypeptidases to result free and *N*-protected amino acids. The intact *N*-protected dipeptides absorb light of short wavelength dependent on the acyl group. So, Bz- and Z-groups require for monitoring wavelengths <300 nm, the region where also proteins absorb. For 3-(2-furyl)acryloyl-dipeptides maximal absorbance was determined at 337 nm. Therefore, FA-dipeptides can be used as substrates for carboxypeptidases during purification procedures in the presence of large amounts of proteins (Remington and Breddam 1994). The enzymatic hydrolysis can be followed continuously by the decreasing absorbance.

Principle

Materials

- 1-cm cuvettes
- spectrophotometer equipped with a thermo-regulated cuvette holder and linked to either a chart recorder or a computer
- wavelength 337 nm

Equipment

FA-dipeptides are usually dissolved in methanol in concentrations about 40 times higher than in the final assay (about 0.5 mM). See Table 1 – 4 for K_m values. Several FA-dipeptides are water soluble and thus e.g. 20 mM solutions of the substrates in water can be stored frozen in aliquots for several months and diluted 1:40 in the final assay.

Substrate

100 mM buffers are recommended (for pH see Table 1 – 4).

Buffer

Dilute enzyme solutions with buffer or 0.01% Brij-35 in buffer.

Enzyme

Procedure

For the procedure see the continuous spectrophotometric assay (Subprotocol 1). The wavelength is set at 337 nm and the decreased absorbance is monitored. In this case, it is recommended to mix first substrate, buffer and effector in the cuvette and then start the reaction by the addition of enzyme.

Calculation The ΔA for intact substrates and split products is different for each substrate. Therefore, ΔA for total hydrolysis of a substrate at a precise concentration in the assay should be used to calculate units of the enzyme (Skidgel 1995).

Subprotocol 7
Aromatic Amide Substrates

Principle Amides of amino acids or peptides and aromatic amines such as p-aminobenzoic acid, aniline or 4-nitroaniline are hydrolyzed by peptidases to result the aromatic amine, which can be diazotized and coupled to N-(1-naphthyl)-ethylene diamine. The red azo dye is measured spectrophotometrically at 546 nm (Bratton and Marshall 1939, Ikehara et al. 1994). The enzymatic reaction can be stopped by specific inhibitors or by trichloro acetic acid.

Materials

Equipment
- 1-ml test tubes
- spectrophotometer, wavelength 546 nm
- microcentrifuge
- 10% trichloro acetic acid
- Ammonium sulfamate ($NH_4SO_3NH_2$), N-(1-naphthyl)-ethylene diamine

Substrate, buffer Dissolve substrate in methanol or Me_2SO to give a 50 times higher concentration than in the final assay. Dilute the substrate stock solution prior to use 1:40 with the appropriate buffer.

Enzyme Dilute enzyme solutions with buffer or 0.01% Brij-35 in buffer.

Colour reagents
- 0.1% $NaNO_2$:
  Solution in water is stable for 1 day.
- 0.5% $NH_4SO_3NH_2$:
  Solution in water is stable at 4 °C for several months. This reagent destroys the excess of nitrite, otherwise the colour development is disturbed.
- Colour reagent:
  Solution of 50 mg N-(1-naphthyl)-ethylene diamine in 100 ml of methanol is stable at 4 °C for several months.

1 mM stock solution of the aromatic amine in water (methanol, Me$_2$SO) is **Standard**
diluted 1:10 (100 µM), 1:20 (50 µM), 1:25 (40 µM) etc. with water and subsequently 1:2 with 10% trichloro acetic acid.

Procedure

1. Incubate 50 µl of enzyme solution (add effector, if necessary) and 200 µl substrate-buffer at 37 °C for exactly 10 min.

2. Stop the enzymatic reaction by addition of 250 µl 10% trichloro acetic acid. Spin the tubes in a microcentrifuge for 10 min.

3. The resulting supernatant (TCA-supernatant) is used for the colour **Colour** development. The TCA precipitation can be omitted, if the protein con- **development** centration of the sample does not disturb the colour development.

4. 100 µl TCA-supernatant (or 100 µl standard-TCA), 100 µl 0.5 M HCl and 50 µl 0.1% NaNO$_2$ were mixed and at intervals each of 5 min 50 µl 0.5% NH$_4$SO$_3$NH$_2$ and 100 µl 0.05% colour reagent were added and carefully mixed after the additions.

5. After 1.5 h (up to several hours in the dark) of colour development read **Measurement** the assay and standard tubes in a spectrophotometer at 546 nm. Use a microcuvette, or otherwise, increase the volumes given above by the same factor.
   $\epsilon_{546} = 57600$ M^{-1} cm^{-1} (Bratton and Marshall 1939)

The nominal unit is release of 1 µmol of product per min under the **Calculation** reaction conditions. 100 µl of the 40 µM standard 1:2 diluted with TCA, contain 2 nmole of the amine (10 nmole in the 250 µl assay) and give $\Delta A = 0{,}288$. This corresponds in a 10-min-assay to 1 munit of enzyme in the tube.

Subprotocol 8
Hydrazide Substrates

Amino acyl hydrazides have been used as substrates of aminopeptidases, **Principle** which are frequently capable to cleave the amide bond between the carboxyl terminus of the amino acid and the hydrazine residue. The liberated hydrazine reacts with 4'-dimethylaminobenzaldehyde to give an azo dye which can be measured spectrophotometrically at 436 nm (maximum of a broad peak).

Materials

Equipment
- 1-ml and 5-ml test tubes
- spectrophotometer, wavelength 436 nm
- 4'-dimethylaminobenzaldehyde

Substrate The amino acyl hydrazides are water soluble. For the determination of leucine aminopeptidase dissolve 54.5 mg L-Leu-hydrazide x 2 HCl in 1 ml of water (0.25 M solution). The substrate solution is stable at $-20\,°C$ for several weeks and at room temperature for one day.

Enzyme Dilute enzyme solutions with buffer.

Buffer Most aminopeptidases are active at alkaline pH. For leucine aminopeptidase, 0.5 M TRIS-HCl buffer pH 8.5 is recommended. Some metallo peptidases need preincubation with metal ions for complete activation.

Colour reagent Dissolve 4 g 4'-dimethylaminobenzaldehyde in 100 ml ethanol. Mix 1 part of this solution with 17 parts of 0.1 N HCl (Haschen et al. 1968). The solution is stable for one day.

Procedure

1. Mix 200 µl 0.5 M buffer, 100 µl enzyme, 100 µl substrate solution (0.25 M), and 100 µl water.

2. Between 5 and 30 min incubation at 25 °C 10 µl of the solution are withdrawn and added to 5 ml freshly prepared colour reagent. The sample is left in the dark for 1 h at room temperature to allow colour development.

3. Read absorbance at 436 nm and calculate the released nmoles min^{-1}. 4,4'-dimethylbenzilideneazine ϵ_{436} = 44000 M^{-1} cm^{-1} (Hanson and Frohne 1976)

References

Akama K, Terao K, Tanaka Y, Noguchi A, Yonezawa N, Nakano M, Tobita T (1994) Purification and characterization of a novel acrosin-like enzyme from boar cauda epididymal sperm. J Biochem Tokyo 116:464-470

Barrett AJ, Kirschke H (1981) Cathepsin B, cathepsin H, and cathepsin L. Methods Enzymol 80:535-561

Bünning P, Holmquist B, Riordan JF (1983) Substrate specificity and kinetic characteristics of angiotensin converting enzyme. Biochem 22:103-110

Bratton AC, Marshall EK Jr (1939) A new coupling component for sulfanilamide determination. J Biol Chem 128:537-550

Brömme D, Bescherer K, Kirschke H, Fittkau S (1987) Enzyme-substrate interactions in the hydrolysis of peptides by cathepsins B and H from rat liver. Biochem J 245:381-385

Buttle DJ, Ritonja A, Pearl LH, Turk V, Barrett AJ (1990) Selective cleavage of glycyl bonds by papaya proteinase IV. FEBS Lett 260:195-197

Dunn BM, Jimenez M, Parten BF, Valler MJ, Rolph CE, Kay J (1986) A systematic series of synthetic chromophoric substrates for aspartic proteinases. Biochem J 237:899-906

Ellman GL (1959) Tissue sulfhydryl groups. Arch Biochem Biophys 82:70-77

Erlanger BF, Kokowsky N, Cohen W (1961) The preparation and properties of two new chromogenic substrates of trypsin. Arch Biochem Biophys 95:271-278

Geiger R, Mann K, Bettels T (1977) Isolation of human urinary kallikrein by affinity chromatography. Determination of human urinary kallikrein, I. J Clin Chem Clin Biochem 15:479-483

Grassetti DR, Murray JF (1967) Determination of sulfhydryl groups with 2,2'- or 4,4'-dithiopyridine. Arch Biochem Biophys 119:41-49

Gray CJ, Boukouvalas J, Szawelski RJ, Wharton CW (1984) Benzyloxycarbonyl-phenylalanyl-citrulline p-nitroanilide as a substrate for papain and other plant cysteine proteinases. Biochem J 219:325-328

Hanson H, Frohne M (1976) Crystalline leucine aminopeptidase from lens (α-aminoacyl-peptide hydrolase; EC 3.4.11.1). Methods Enzymol 45:504-521

Haschen RJ, Farr W, Reichelt D (1968) Photometrische Bestimmung der "klassischen" Leucinaminopeptidase im Blutplasma und -serum. Z Klin Chem Klin Biochem 6:11-18

Ikehara Y, Ogata S, Misumi Y (1994) Dipeptidyl-peptidase IV from rat liver. Methods Enzymol 244:215-227

Kageyama T (1995) Procathepsin E and cathepsin E. Methods Enzymol 248: 120-136

Kirschke H, Schmidt I, Wiederanders B (1986) Cathepsin S. The cysteine proteinase from bovine lymphoid tissue is distinct from cathepsin L (EC 3.4.22.15). Biochem J 240:455-459

Kirschke H, Wiederanders B (1984) Methoden zur Aktivitätsbestimmung von Proteinasen. Martin-Luther-Univ. Halle-Wittenberg, Halle (Saale)

Knight CG (1980) Human cathepsin B. Biochem J 189:447-453

Lottenberg R, Christensen U, Jackson CM, Coleman PL (1981) Assay of coagulation proteases using peptide chromogenic and fluorogenic substrates. Methods Enzymol 80:341-361

McDonald JK, Zeitman BB, Reilly TJ, Ellis S (1969) New observations on the substrate specificity of cathepsin C (dipeptidyl aminopeptidase I). Including the degradation of β-corticotropin and other peptide hormones. J Biol Chem 244:2693-2709

Mortensen UH, Remington SJ, Breddam K (1994) Site-directed mutagenesis on (serine) carboxypeptidase Y. Biochem 33:508-517

Odake S, Kam CM, Narasimhan L, Poe M, Blake JT, Krahenbuhl O, Tschopp J, Powers JC (1991) Human and murine cytotoxic T lymphocyte serine proteases: subsite mapping with peptide thioester substrates and inhibition of enzyme activity and cytolysis by isocoumarins. Biochem 30:2217-2227

Ohsawa Y, Nitatori T, Higuchi S, Kominami E, Uchiyama Y (1993) Lysosomal cysteine and aspartic proteinases, acid phosphatase, and an endogenous cysteine proteinase inhibitor, cystatin-β, in rat osteoclasts. J Histochem Cytochem 41:1075-1083

Powers JC, Kam CM (1995) Peptide thioester substrates for serine peptidases and metalloendopeptidases. Methods Enzymol 248:3-18

Remington SJ, Breddam K (1994) Carboxypeptidases C and D. Methods Enzymol 244:231-248

Richards AD, Phylip LH, Farmerie WG, Scarborough PE, Alvarez A, Dunn BM, Hirel PH, Konvalinka J, Strop P, Pavlickova L, Kostka V, Kay J (1990) Sensitive, soluble chromogenic substrates for HIV-1 proteinase. J Biol Chem 265:7733-7736

Saulnier JM, Curtil FM, Duclos MC, Wallach JM (1989) Elastolytic activity of *Pseudomonas aeruginosa* elastase. Biochim Biophys Acta 995:285-290

Skidgel RA (1995) Human carboxypeptidase N: lysine carboxypeptidase. Methods Enzymol 248:653-663

Stein RL, Izquierdo-Martin M (1994) Thioester hydrolysis by matrix metalloproteinases. Arch Biochem Biophys 308:274-277

Stöcker W, Zwilling R (1995) Astacin. Methods Enzymol 248:305-325

Takahashi T, Dehdarani AH, Schmidt PG, Tang J (1984) Cathepsins B and H from porcine spleen. Purification, polypeptide chain arrangements, and carbohydrate content. J Biol Chem 259:9874-9882

Tanaka T, McRae BJ, Cho K, Cook R, Fraki JE, Johnson DA, Powers JC (1983) Mammalian tissue trypsin-like enzymes. Comparative reactivities of human skin tryptase, human lung tryptase, and bovine trypsin with peptide 4-nitroanilide and thioester substrates. J Biol Chem 258:13552-13557

Tchoupe JR, Moreau T, Gauthier F, Bieth JG (1991) Photometric or fluorometric assay of cathepsin B, L and H and papain using substrates with an aminotrifluoromethylcoumarin leaving group. Biochim Biophys Acta 1076: 149-151

Wolz RL, Bond JS (1995) Meprins A and B. Methods Enzymol 248:325-345

Abbreviations

Ac-	=	acetyl-
Boc-	=	t-butyloxycarbonyl-
Bz-	=	benzoyl-
Cit	=	citrulline
FA-	=	β-2-furylacryloyl-
-^{s}Gly-	=	-SCH_2CO-
Me_2SO	=	dimethylsulfoxide
-NHNap	=	-2-naphthylamide
-NHNapOMe	=	-2-(4-methoxy)naphthylamide
-$NHPhNO_2$	=	-4-nitroanilide
Nva	=	norvaline
-OEt	=	-ethylester
-OMe	=	-methylester
-$OPhNO_2$	=	-4-nitrophenylester
$Phe(NO_2)$	=	4-nitro-phenylalanine
-SBzl	=	-thiobenzylester
Suc-	=	succinyl-
Tos-	=	tosyl-
Z-	=	benzyloxycarbonyl-

Chapter 3

Fluorometric Assays

DAVID S. AULD

Introduction

The introduction of a fluorescent group into a proteinase substrate can lead to several useful assays. Miniature thin layer chromatography based assays can rapidly detect enzymatic activity during eg purification procedures. High performance liquid chromatography, HPLC, based assays can also determine the peptide bonds cleaved during catalysis. If the fluorophore is attached at the C-terminal through an ester or amide bond it can provide a direct assay of the activity of some endoproteases. Incorporation of the fluorophore and a quencher of its fluorescence into the substrate yields intramolecular quenching assays for both exo- and endopeptidases. Intermolecular quenching between a fluorophore in the enzyme and a quenching group in the substrate allows investigation of the catalytic mechanism of the enzyme through direct inspection of the formation and breakdown of ES complexes. All of these assays will be covered in the following sections. The dansyl, Dns, group and the Trp/Dns energy donor/acceptor pair will be used in demonstrating the various assays.

Peptide Substrate Synthesis and Purification

If the fluorescently labeled peptide substrate for the protease under investigation can not be bought it will need to be synthesized. Solid phase synthesis is the best method available for synthesis of peptide substrates. The two current solid phase strategies are the Boc/Bzl strategy for weak acid

David S. Auld, Harvard Medical School, Center for Biochemical and Biophysical Sciences and Medicine and Department of Pathology, 250 Longwood Avenue, BostonMassachusetts, 02115, USA (*phone* 617-432-4009; *fax* 617-566-3137; *e-mail* David_Auld@hms.harvard.edu or Auld@fas.harvard.edu)

(trifluoroacetic acid, TFA) removal of the α-amino protecting group and strong acid (hydrogen fluoride, HF) global removal of amino acid side chain protecting groups and the peptide from the resin (Barany and Merrifield, 1980) and the Fmoc/t-butyl strategy for weak base (piperidine) removal of the α-amino protecting group and weak acid (TFA) removal of side chain protecting groups and the peptide from the resin (Atherton and Sheppard, 1989). Some of the cleaved side chain protecting groups can be removed by washing the cleaved, deprotected peptide with an organic solvent such as ethyl acetate. The crude, cleaved peptides can be extracted from their resins by 1 to 25 % solutions of acetic acid, lyophilized and purified by HPLC techniques. These methods allow the facile incorporation of naturally fluorescent amino acids such as Trp or Tyr as well as those containing ionic side chain groups such as Arg, Lys, His, Glu, Asp and Cys that may be needed for specificity. It is now possible to introduce other fluorescent or quenching groups by this approach. Thus a fluorescent label such as a dansyl group can be introduced easily by using dansyl chloride under mild basic (triethylamine) conditions to modify the last alpha amino group while the peptide is still attached to the resin (Ng and Auld, 1989). Reagents for the incorporation of several fluorophores and quenching groups are given in Table 1.

Table 1. Incorporation of Fluorescent Donor and Acceptor Groups into Peptides in Solid Phase Synthesis

Fluorescent Donor	Acceptor/ Quencher	Reagent(s)	Reference
Abz		Boc-Abz-ODhbt	Meldal et al, 1991
Lys(Abz)		Fmoc-Lys(BocAbz)-ODhbt	Meldal et al, 1991
	Tyr(NO2)	Fmoc-Tyr(NO2)	Meldal et al, 1991
Trp		Boc-Trp(CHO)	Ng and Auld, 1989
	Dns	Dansyl Chloride	Ng and Auld, 1989
Mca		7-methoxycoumarin-4-acetic acid	Nagase et al, 1994
	Lys(Dnp)	Fmoc-Lys-Dnp)	Nagase et al, 1994
	Dnp(N-term)	2,4-dinitrofluorobenzene	Welch et al, 1995
	Dpa	Fmoc-Asn, DnpF	Knight et al, 1992
EDANS		Boc-γAbu-Glu(EDANS)-OH	Maggiora et al, 1992
	DABCYL	DABCYL-BOP	Maggiora et al, 1992

Subprotocol 1
Thin Layer Chromatography Based Assay

A fluorescent label at either the C- or N-terminal end of the peptide can be very useful for designing a miniature assay using thin layer chromatography as the means of separating the products of the proteolytic cleavage of the peptide. The assay volume can be as small as 10 µl (about 0.2 µl applied to the TLC plate) provided the fluorescent label emits in the visible region and has a good quantum yield on the TLC medium. The dansyl group and micropolyamide on Mylar are a good choice since even 20 dansylated aminoacids can be separated by two dimensional chromatography on a 3 by 3 inch plate of this material.

Materials

Reagents
- micropolyamide (Schleicher & Schuell, F1700) or silica gel (Eastman 6060)
- micro centrifuge tubes (e.g. Fisherbrand 0.5 ml polypropylene)
- 1.1-1.2 ID by 100 mm melting point tubes, open both ends (Pyrex Corning)
- TLC solvent (Table 2)

Table 2. Chromatography Solvents for TLC Assays

Micropolyamide	
Solvent	Solvent Ratio
Water/ Formic Acid	200/3
Water/HCl	40/1
Methanol/Water/Glacial Acetic Acid	30/20/1
Chloroform/Methanol/Glacial Acetic Acid	17/2/1
Carbon tetrachloride/Dioxane/Glacial Acetic Acid	9/9/2
Benzene/Glacial Acetic Acid	9/1
Silica Gel	
Butanol/Glacial Acetic Acid/Water	4/1/1 or 8/1/1
Ethanol/30% Ammonia	25/1

- propane torch (eg 14 oz Turbo Torch)   **Equipment**
- constant temperature Bath (eg. Julabo Exotherm P5)
- fluorescent light source (eg. Blak-Ray Lamp Longwave UV 366 nm)
- pipeting device (Pipetman P20)
- vortex (eg. Fisher Vortex Genie 2)
- Cap style stopper 40 x 100 cm weighing bottle (Kimax)
- Hair Dryer (Milwaukee Heating Tool, Coddington Mfg.)

Procedure

1. Pipet applicators are prepared by drawing out capillaries in a propane flame. They should have a 1-3 cm tip length so that about 0.2 µl of solution is applied to the TLC medium.

2. Micropolyamide on Mylar is the best TLC medium for separating dansylated peptides. In some cases a silica gel plate will work well enough. The substrate and fluorescent product of the protease catalyzed reaction must be tested in different solvent systems in order to find one that will separate them. Table 2 lists some solvents that have worked for dansylated peptides on micropolyamide and silica gel.

3. The 15 cm square sheet of micropolyamide should be cut to a size just sufficient for the assay. A 2 x 5 cm piece can be used for one assay having about 6 time points.

4. A 9 to 45 µl solution of buffered fluorescent substrate is placed in a 0.5 ml   **Substrate**
microfuge tube in a constant temperature bath. One to five µl of enzyme solution is added to start the reaction. The solution is gently vortexed and placed back in the temperature bath. Aliquots are removed at specified times using the flame extruded glass capillary and deposited on the micropolyamide plate. The application of the aliquot to the sheet can be followed by using the fluorescent light to illuminate the sample spot. The concentration of fluorescent substrate in the assay can be as low as about 25-50 µM. A typical assay in our lab uses about 100 µM substrate. For a peptide having a MW of 1000, 1 mg of peptide is enough for one thousand 10 µl assays having a substrate concentration of 100 µM. Since the enzyme concentration will frequently be in the range 10^{-10} to 10^{-7} M, the assay only uses fM to pM amounts of enzyme per assay. This procedure can be used to follow the purification of an enzyme (Fig. 1). In this case one aliquot of enzyme is removed from each fraction tube from the purification and added to an assay solution. One aliquot is removed from each assay tube at a specified time and applied to the chromatogram.

5. After all the aliquots are applied the TLC plate is placed in a weighing bottle that has enough solvent (few ml) for developing the chromatogram. The height of this solvent must not exceed the line of aliquot application. The top is placed on the bottle. If the concentration of peptide is $\geq 100\ \mu M$, the development of the chromatogram can be followed by placing the fluorescent light source next to the bottle.

6. After drying the TLC plate with a hot air blower, the reaction can be visualized using the fluorescent lamp (Fig. 1). Alternatively quantitation of the reaction can be accomplished using devices that scan fluorescent plates.

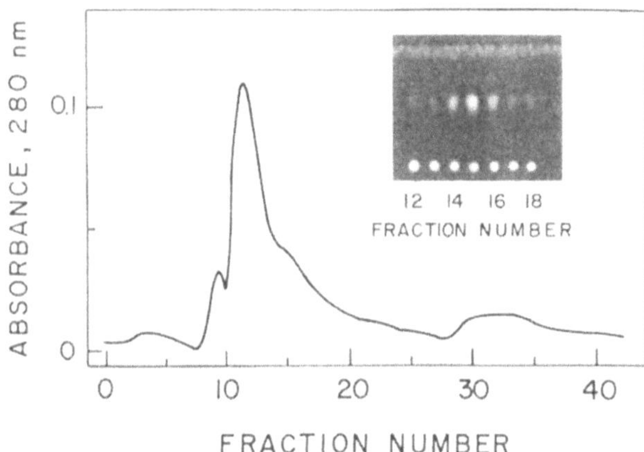

Fig. 1. Gel filtration of partially purified human skin fibroblast collagenase (Bond et al, 1986). The insert shows the fractions containing collagenase activity using Dns-PQGIAG-D-R as the substrate using a TLC assay. The product Dns-PQGI has an R_f of 0.61 using 9/9/2 carbon tetrachloride/dioxane/acetic acid on micropolyamide.

Subprotocol 2
High Performance Liquid Chromatography (HPLC) Based Assay

A miniature assay can also be based on the use of HPLC as the means to separate the substrate and product of the reaction. If a fluorescent label is present in the substrate, the fluorescence or absorption properties of the label can be used to follow the reaction. Wavelengths of 220 and 260 nm are particularly useful to monitor the absorbance of the peptide bond and the dansyl group, respectively. The fluorescence of the dansyl group can be monitored at wavelengths above 400 nm. This type of assay like the TLC assay is also quite conservative in the amount of enzyme consumed per assay point (fM quantities, see below).

Materials

- micro centrifuge tubes (e.g. Fisherbrand 0.5 ml polypropylene)   **Reagents**
- analytical HPLC column (Waters Nova Pak 10μm C-18 analytical column)
- HPLC solvent (see below)

- constant temperature Bath (eg. Julabo Exotherm P5)   **Equipment**
- pipeting device (Pipetman P20)
- vortex (Fisher Vortex Genie 2)
- HPLC equipped with an automatic sampling device if many assays are performed.

Procedure

1. The solvents to be used in the HPLC chromatography must be degassed and filtered. A solvent gradient system for separating the substrate from the products must be found for performing the assay. Linear gradients of either acetonitrile/water, methanol/water or isopropanol/water in 0.1% TFA combined with a reverse phase C-18 column is an excellent system for separating peptides.

2. A 95 μl aliquot of buffered fluorescent substrate is placed in a 1.5 ml polypropylene tube in a constant temperature bath. For each substrate solution 5 samples are prepared in this manner. The enzymatic reaction is initiated by adding five μl of enzyme solution to the tube. The solution is gently vortexed and placed back in the temperature bath. The reaction is stopped after specified time intervals (1, 2, etc minutes) by the addition of 5 μl of glacial acetic acid and vortexing a second time. The final concentration of enzyme will be determined by its kinetic parameters but usually will be in the nM to μM range.

3. The samples are placed in an automatic sampler for HPLC analysis. Aliquots of 10 to 50 μl of the reaction are used for the chromatography.

Results

The concentration of product, P_t, at given time t is determined according to equation 1 provided there is no great difference in the absorbance of the products and the substrate at the wavelength chosen.

$$P_t = \frac{P_a + P_b}{P_a + P_b + S_t} \times S_o \qquad (1)$$

P_a, P_b, S_t are, respectively, the integrated areas of product A and B and substrate peaks at time t and S_o is the initial substrate concentration. The enzyme concentration should be chosen to allow no more than 10 % of the

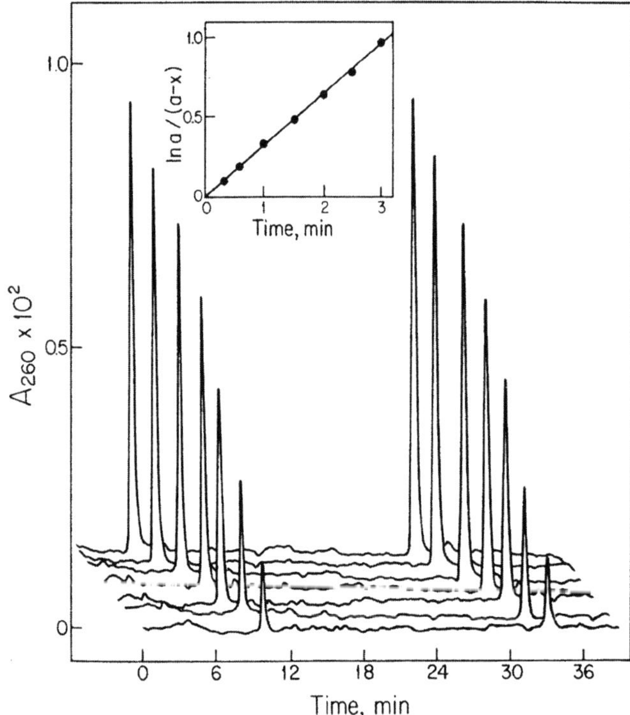

Fig. 2. HPLC assay of the trypsin, 0.05 µM, catalyzed hydrolysis of the peptide, DnsGKYAPWV, 50 µM, in 20 mM Hepes, pH 7.7, at 20 °C. A 15 to 35 % gradient of 75 % acetonitrile/water/0.08 % TFA over 40 min readily resolves the product peaks. The 40 µl reaction solutions are quenched with 5 µl of glacial acetic acid at 20, 35, 60, 90, 120, 150, and 180 sec. The insert shows that the increase in product concentration with time follows a first-order process (Ng and Auld, 1989)

reaction to occur in the time period for the assay if kinetic parameters are being determined. This will give a good initial rate if the product does not bind much stronger than the substrate. If this assay is being used to follow the effect of inhibitors on catalysis then the reaction can be followed at a substrate concentration that is $\geq$ 10 fold below K_m so that first order kinetics are expected. Figure 2 shows the results of such a HPLC assay for trypsin activity. The products of the reaction, Dns-GK (9.9 min) and YAPWV (33.2 min) are readily resolved. The concentration of trypsin in this assay is 5 x 10^{-8} M (2 pM per assay point) and the assay time intervals are 15 to 30 sec. The k_{cat}/K_m for this substrate is 1.3 x 10^5 $M^{-1}s^{-1}$ (Ng and Auld, 1989). If the time interval is increased to 10 min, each assay point would use only 50 fM of enzyme. A one time point assay may be sufficient to follow the reaction for the purpose of an enzyme purification procedure.

Subprotocol 3
Intramolecular Fluorescence Quenching Based Assays: Fluorophore Part of Scissle Bond

Ever since Hartly and Kilby (1954) demonstrated that p-nitrophenyl acetate was a chromophoric substrate for α-chymotrypsin scientists have tried to develop spectrophotometric assays based on the principle of the liberation of a chromophore upon cleavage of the scissle bond. Many of these assays rely on attaching a chromophoric agent as the C-terminal ester or amide. The assays are particularly useful for those endoproteases whose activities are not greatly affected by the P_1 to P_n positions. Two of the most frequently used derivatives incorporate either 7-amino-4-methylcourmarin, AMC, (Zimmeran et al, 1976) or 6, 1 aminonapthalenesulfonamides, ANSN, (Butenas et al, 1992) for examining endoprotease and aminopeptidase catalysis. The free AMC (λ_{ex} 345 nm, λ_{em} 445 nm) and its substrate amide derivative, MCA (λ_{ex} 325 nm, λ_{em} 395 nm) fluoresce to nearly the same degree which has led to the choice of (λ_{ex} 370 nm, λ_{em} 405 nm) for assay conditions. The fluorescent properties of 6, 1-, 6, 2-, 4, 1-, 5, 2-, 5, 1-ANSN derivatives (λ_{ex} 306-346 nm, λ_{em} 424-540 nm) and their use for assaying hydrolytic enzymes has been discussed (Butenas et al, 1995). The product/substrate fluorescent ratio is 2 to 3 orders of magnitude for all but the 5, 1 derivative. The combination of using alkylated sulfonamide groups and the various 4-6 substitutions can lead to changes in catalysis for the serine proteases examined, likely reflecting interactions in the P_1 and P_2 subsites (Butenas et al, 1995).

Fluorophore and Quencher Separated from Scissle Bond

Priciple   Intramolecular and intermolecular quenching of fluorescence can provide a basis for the design of assays based on following the conversion of substrate into product or the formation and breakdown of ES complexes, respectively (Fig. 3). The first radiationless or resonance energy transfer, RET, based protease assays used either the fluorescent tryptophan fluorophore in the substrate or in the enzyme in combination with a dansyl group in the substrate as the basis of either an intra- or intermolecular quenching assay (Latt et al., 1970; 1972, Auld et al, 1972). This donor/acceptor pair was chosen because of the excellent spectral overlap of Trp fluorescence and Dns absorption centered at 340 nm. It had the added advantage of changes in the Dns fluorescence upon the transfer of radiationless energy from the Trp to the Dns group or by direct excitation of the dansyl group at wavelengths where Trp is not excited. In the mid to late 70's the use of the o-aminobenzoyl, Abz, fluorophore and nitroaromatic quenching groups such as p-nitrobenzyl groups, Phe(NO_2) or ONBz lead to effective protease assays that were based on a direct intramolecular encounter of the fluorophore and the quenching agent (Yaron et al, 1979). The heightened interest in protease catalysis in the last two decades has lead to the development of a number of useful pairs of fluorescent donors and quenchers many of which are based on the principle of radiationless energy transfer or as is sometimes now referred to as fluorescence resonance energy trasfer (FRET) (Table 3).

INTRAMOLECULAR QUENCHING ASSAY

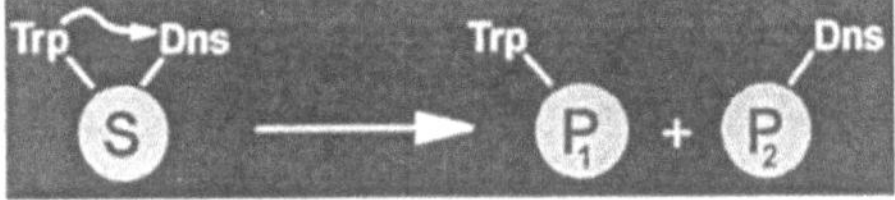

INTERMOLECULAR QUENCHING ASSAY

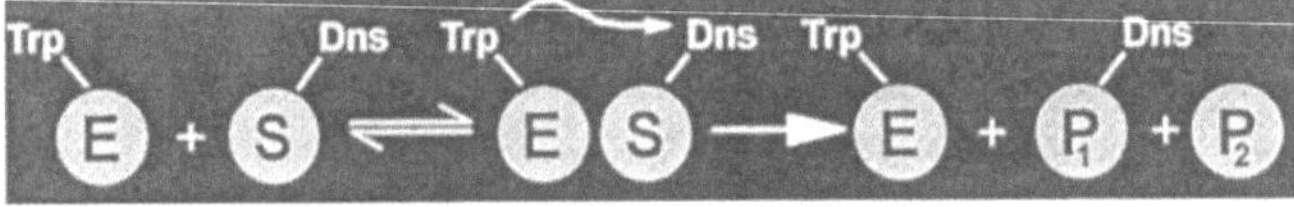

Fig. 3. Schematic of intra- and intermolecular quenching assays

Table 3. Energy Donor-Acceptor Pairs for Intramolecular Quenching Assays

Donor/Acceptor	Enzyme	ΔAa^a	$\lambda_{ex}/\lambda_{em}$	P/S^b	Reference
Abz/ONBzl	Aminopeptidase	1	310/410	18-27	Carmel et al, 1977
Abz/Phe(NO$_2$)	HIV protease	3	337/410	6	Toth and Marshall, 1990
Abz/Tyr(NO$_2$)	Subtilisin, Pepsin	3-15	320/420	33-5	Meldal and Breddam, 1991
Abz/EDDnp	Kallikreins	4	320/420	7-28	Chagas et al, 1991
EDANS/DABCYL	HIV protease	8	340/475	33-40	Matayoshi et al, 1990
Mca/Dnp	MMPs	4	328/393	190	Knight et al, 1992
Trp/Dns	Endoproteases	5	285/360	9-14	Ng and Auld, 1989
Trp/Dnp	MMPs	4	280/360	10	Netzel-Arnett et al, 1991

[a]The value of Δ refers to the number of amino acids between the donor acceptor/acceptor pair. [b]The ratio of the fluorescent yields of product, P and substrate, S.

The choice of an optimal donor/acceptor pair for an intramolecular quenching assay requires consideration of several points. The sensitivity of the assay (ie initial rate determination at the lowest possible enzyme concentration) will be directly proportional to

a) the value of the specificity constant, k_{cat}/K_m, for the substrate,

b) the quantum yield of the fluorophore and

c) the difference in the fluorescence of the substrate and the product.

If the intended assay is to be used to screen the effect of inhibitors without regard to the type of inhibition then a substrate concentration at least 10 fold or lower below K_m is most useful. Under these conditions the K_i for the inhibitor can be obtained directly from the negative x-axis intercept of a plot of the reciprocal of the velocity, v, versus the inhibitor, I, concentration.

$$v = \frac{k_{cat}S}{K_m(1 + I/K_i)} \quad \text{for } S \ll K_m \tag{2}$$

For such an assay any of the donor-acceptor pairs listed in Table 3 are likely to have good enough fluorescence properties. Those that have higher ratios of product to substrate fluorescent and higher quantum yields should give the most sensitive assay. However the most important consideration will be whether or not the added chromophores has any severe adverse effect on the specificity constant for the substrate.

If the assay is to be used to determine the kinetic parameters for the substrate or the type of inhibition then a number of additional considerations become important. Quenching agents with high molar absorptivity values at the maximum of the fluorescence emission will cause trivial quenching (or inner filter) effect of the light emitted as the substrate concentration is increased. Thus the intramolecularly quenched substrate DABCYL-RGVVNA~SSRLA-EDANS (~scissle bond) shows a tenfold enhancement in fluorescence upon bond cleavage by cytomegalovirus protease but assays must be performed below 10 µM due a severe inner filter effect of this donor/acceptor pair (Holskin et al, 1995). This is also true for quenching agents that absorb light strongly at the excitation wavelength of the fluorophore. For example if the acceptor has a molar absorptivity value of 10,000 at the donor excitation wavelength, a 100 µM concentration of the substrate will absorb 90 % of the light within 1 cm. The solubility of the substrate also becomes more important if concentrations above 1 µM are to be used. In general, the larger more hydrophobic chromophores will decrease the solubility of the substrate. The introduction of Arg groups into the peptide substrate and using unblocked C- and N-terminal amino acid groups will increase the solubility of the substrate.

Materials

- 0.45 micron Millex-HV-Filter unit (Millipore Corporation)
- Spectrofluorometer. Low spectral resolution or filter adaptable
- thermostated cell compartment.

Procedure

Intramolecular quenching

1. The spectrofluorometer used for the assays need not be a high resolution type. Band pass or combinations of cut-on/off filters often give the most sensitive assays.

2. The cell holder should have circulating water passing through it that comes from a thermostated source of water. Fluorescence is sensitive to temperature so any drift in the temperature of the reaction solution can lead to time dependent artifactual changes in the fluorescence.

3. Stock solutions of the fluorescent peptide should be stored at 4 °C in dark bottles or bottles wrapped with aluminum foil.

4. The solubility of the substrate and products under the assay conditions should be checked. Light scattering phenomena can occur if either one is not soluble.

5. Since many of the fluorescent labeled substrates are very insoluble in water some amount of organic solvent may be needed for solubility. A few percent of acetonitrile, dimethylsulfoxide or dimethylformamide have often been used for this purpose. However the effect of the solvent on enzymatic activity should be checked. Increasing the organic solvent concentration may increase K_m for the substrate resulting in decreased activity if the assay is being performed at substrate concentrations well below K_m. Time dependent denaturation of the enzyme may also occur.

6. All solutions should be filtered through a 0.45 micron filter to remove any particulate matter since spectrophotometric assays are very sensitive to light scattering phenomena. The presence of any solid material that settles during an assay can cause artifactual time-dependent changes in light levels.

7. The fluorescence of the product as a function of its concentration should be determined. The absorbance of the quenching agent can result in trivial quenching of the product fluorescence and non-linear standard curves.

8. The position of bond cleavage should be verified by separation of the products eg by HPLC and the aminoacid composition of the products determined.

9. The enzyme is added last to the temperature-equilibrated cell. Parafilm is put on the top of the cell and the cell is inverted at least three times, gently shaking each time. All air bubbles should be removed by gentle tapping of the cell before returning it to the thermostated cell holder. This process should be completed in about 30 seconds. Care should be taken not to touch any of the side surfaces of the cell since oils from the hand can quench the fluorescence.

10. The stability of many enzymes requires their storage at temperatures near 0 °C. The assay should therefore be designed to equilibrate the reaction solution (1 to 3 ml) in the cell for a few minutes prior to the addition of a few μl of concentrated enzyme. If the reaction solution minus enzyme has been filtered and temperature equilibrated in this manner the substrate should be in solution and the temperature should remain constant even upon addition of stock enzyme stored on ice.

Figure 4 shows the fluorescence change that occurs in the hydrolysis of 325 μM Dns-VKR~APWV (~ scissle bond) catalyzed by 0.1 μM Astacin under stopped-flow fluorescence conditions. The initial rate is calculated from the linear part of the trace which occurs in the first 5 seconds in the reaction. The double reciprocal plot for triplicate assays performed over the substrate concentration range 40 to 400 μM yields a k_{cat} of 194 s^{-1} and a K_m of 285 μM (Stöcker et al., 1990). The reaction is complete in the next 265 seconds. Progress curve analyzes can be used for the entire reaction if desired.

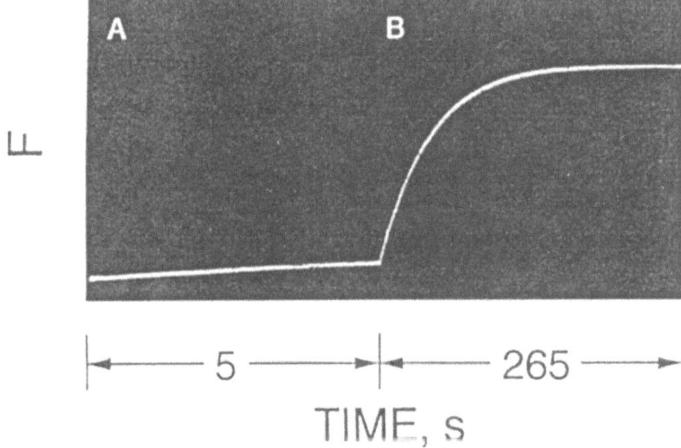

Fig. 4. Intramolecular quenching assay for the astacin (0.1 μM) catalyzed conversion of Dns-VKRAPWV(325 μM) into Dns-VKR + APWV (Stocker et al, 1990). Excitation was at 285 nm and the increase in tryptophan fluorescence was monitored using a 360-band-pass filter.

Subprotocol 4
Intermolecular Fluorescence Quenching Based Assay

Determination of the mechanism of action of an enzyme requires a complete kinetic description of the individual steps involved in substrate binding and catalysis. The direct visualization of the enzyme/substrate, ES, complexes is not an easy task due to the high turnover of an enzyme catalyzed reaction that thermodynamically favors product formation. The combination of low-temperature stopped-flow methodology (Auld, 1994) and the use of radiationless energy transfer as the means to see the ES complex (Auld, 1977, 1987) greatly reduces the difficulties imposed by the high rates of enzymatic reactions. The choice of an energy donor/acceptor pair largely will be dictated by the need for excellent binding characteristics of the substrate to the enzyme. In addition, if one is interested in the mechanism of the reaction the rates of formation of and interconversion of ES complexes will need to be determined at a range of substrate concentrations. This will require the substrate to remain soluble at high concentrations and not greatly quench the fluorescent signal from the enzyme through trivial quenching mechanisms. The spectral overlap between the substrate dansyl group absorbance and the fluorescence of the natural fluorophore Trp is excellent and the dansyl emission spectrum is red shifted far enough not to overlap with its own absorption spectrum, properties that make these an exceptionally good donor/acceptor pair (Auld, 1977). The solubility of the substrate can be increased by introduction of hydrophilic residues such as Arg, Lys or Ser into the peptide substrate and keeping C- and N-terminal amino acid residues unblocked.

This approach retains many of the virtues of conventional initial rate kinetic analyses. Since the ES complex can be observed at enzyme concentrations of 0.01 to 1 μ M where $(E_T)m$, a steady state in ES complex formation and breakdown can be achieved and maintained even though the entire reaction may be complete in a few seconds. Each experiment requires only 1 to 100 pM of enzyme and is recoverable since the technique is non-destructive. Moreover, since in each experiment independent measurements can be made of parameters reflecting the concentration of ES complexes and their rate of breakdown, the kinetic parameters k_{cat} and K_m can be measured in multiple ways (Lobb and Auld, 1980, 1984). The evaluation of k_{cat} and K_m from the progress curve at a single substrate concentration can be of particular value for the screening of substrates and determination of the best conditions for further studies. The speed of the assay performed under stopped-flow conditions combined with computerized data storage and analysis allows both quantitative evaluation within seconds and, if desired, the continuous redesign of experimental strategy. Thus, this approach can allow a facile evaluation of the effects of pH, temperature, ionic strength, solvents, and inhibitors and activators of enzyme activity. These assays should be particularly useful for examining the effect of mutations on enzymatic activity. They can directly determine if loss in activity is due to loss of substrate binding or to decreased catalytic activity, as well as quantitating the changes in activity for any enzyme that is not totally inactivated by the mutation. Figure 5 shows the binding of Dns-Leu-Ala-DED, 10 µM, to *Aeromonas* aminopeptidase 0.1µM, and its subsequent hydrolysis. The rapid decrease in protein Trp fluorescence to reach a minimum within the mixing time of the instrument indicates rapid equilibration of the ES complex(es) at 25 °C (Figure 5 A). This is followed by an increase in fluorescence as substrate is converted to product over the next 90 seconds. The concomitant rapid increase in the dansyl fluorescence upon formation of the ES complex and its subsequent decrease upon its conversion to products is shown in Figure 5 B. Both presteady state and steady state kinetics can be performed in the same experiment (Lobb and Auld, 1979, 1980). Assays at steady state precisely determine k_{cat} and K_m values while analyses at presteady state determine the number of intermediates, the type of reaction scheme, and the individual binding and rate constants.

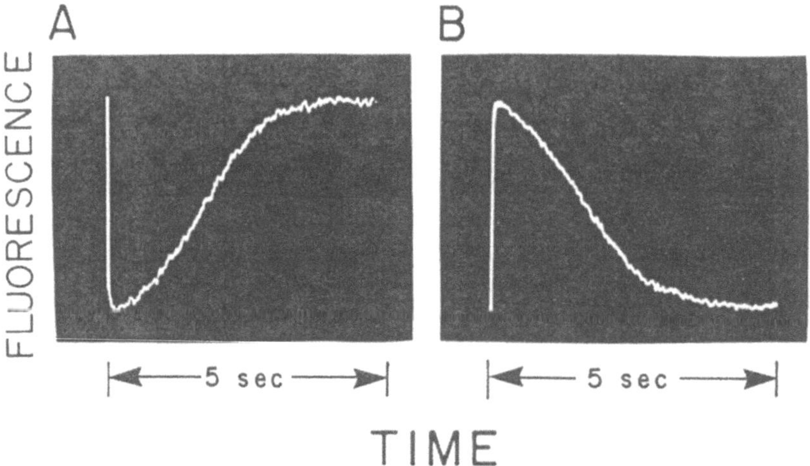

Fig. 5. Intermolecular quenching assay of the formation and breakdown of ES complexes of *Aeromonas* aminopeptidase (0.25 µM) and Leu-Ala-DED (10 µM) at pH 7.5, 20 °C (Auld and Prescott, 1983). Excitation is at 285 nm and either A) quenching of enzyme tryptophan emission (360 nm band pass filter) or B) enhancement of bound dansyl emission above 430 nm is observed.

Materials

- 0.45 micron Millex-HV-Filter unit (Millipore Corporation)
- vacuum chamber (eg Nalgene polycarbonate or polyetherimide jar)
- water vacuum pump aspirator
- Stopped-flow spectrofluorometer. A filter adaptable instrument is preferable.
- thermostated cell compartment.

Equipment

Procedure

The experimental procedure given in subprotocol 3 apply to the stopped-flow studies as well. Probably the biggest artifacts in stopped-flow kinetics are related to the presence of particulates or air bubbles in the solution before or after mixing enzyme with substrate. All solutions that are to be used in the stopped-flow should be degassed by means of vacuum prior to their use. If this is not done air bubbles can evolve during the rapid mixing process. Care must also be taken during the transfer of solution from a reservoir syringe to a drive syringe to not introduce air bubbles into the system. Removal of insoluble material from all solutions is accomplished by filtering solutions through a 0.45 micron filter. Care should also be taken to determine the substrate solubility at the final mixing conditions. The solvent should be only in the substrate solution unless it is known that the solvent has no long term deleterious effect on enzyme activity. However the substrate may be soluble in a 10% solution of some organic solvent but precipitate when the solvent is reduced to 5% upon mixing. The failure to remove particulate matter and/or air bubbles from the system often yield beautiful exponential stopped-flow traces that are nothing more than an air bubble or some solid matter moving in the observation cell. Mixing solutions that have much different indexes of refraction should also be avoided if reactions are to be examined on the ms time scale. The elimination of these potential problems will save time and precious resources.

Experimental considerations

The most critical feature in the use of radiationless energy transfer to observe ES complex formation and breakdown requires the dansylated product to bind much weaker to the enzyme than the substrate. If the dansylated product binds tightly to the enzyme the fluorescent changes will reflect the conversion of the ES complex to an EP complex not free enzyme, thus complicating the steady-state kinetics (Lobb and Auld, 1984). Fortunately one of the products usually is a poor inhibitor allowing the dansyl acceptor to be placed in it. In addition it is generally possible to obtain presteady-state kinetics even when the dansylated product binds. Competitive inhibition by the non-dansylated product does not interfere with the mathematical analyses. In fact the K_i value for this product can be obtained from the change in the apparent k_{cat}/K_m values obtained from the first order decay of dES/dt at the end of the reaction (Lobb and Auld, 1984).

References

Auld DS (1977) Direct observation of transient ES complexes: Implications to enzyme mechanisms. Bioorganic Chem. 1:1-18

Auld DS (1987) Acyl group transfer-metalloproteinases in Page MI, Williams A (eds) Enzyme Mechanisms Royal Society of Chemistry, London pp 240-258

Auld DS, Prescott JM (1983) ES complexes of *Aeromonas* aminopeptidase: Direct observation by stopped-flow fluorescence. Biochem. Biophys. Res. Commun. 111:946-951

Auld DS (1993) Low-temperature stopped-flow rapid-scanning spectroscopy: performance tests and use of aqueous salt croyosolvents Methods in Enzymology 226: 553-565

Auld DS, Latt SA, Vallee BL (1972) An approach to inhibition kinetics. Measurement of enzyme-substrate complexes by electronic energy transfer. Biochemistry 11:4994-4999

Atherton E, Sheppard RC (1989) in Solid Phase Peptide Synthesis a Practical Approach, IRL Press, New York

Barany G, Merrifield RB (1980) in The Peptides. Analysis, Synthesis, Biology (Gross E and Meienhofer, J Eds) Vol. 2A, Academic Press New York, pp 1-284

Bond MD, Auld DS, Lobb RR (1986) A convenient fluorescent assay for vertebrate collagenases. Analytical Biochemistry 155:315-321

Butenas S, Orfeo T, Lawson JH, Mann KG (1992) Aminonapthalenesulfonamides, a new class of modifiable fluorescent detecting groups and their use in substrates for serine protease enzymes. Biochemistry 31:5399-5411

Butenas S, Drungilaite V, Mann KG (1995) Fluorogenic substrates for activated protein C: substrate structure-efficiency correlation. Analytical Biochemistry 225:231-241

Carmel A, Kessler E, Yaron A (1977) Intramolecularly-quenched fluorescent peptides as fluorogenic substrates of leucine aminopeptidase and inhibitors of *Clostridial* aminopeptidase Eur J. Biochem. 73:617-625

Chagas JR, Juliano L, Prado ES (1991) Intramolecularly quenched fluorogenic tetrapeptide substrates for tissue and plasma kallikreins. Analytical Biochemistry 192:419-425

Hartly BS, Kilby BA (1954) The reaction of p-nitrophenyl esters with chymotrypsin and insulin. Biochem. 56:288-297

Holskin BP, Bukhtiyarova M, Dunn BM, Baur, P, de Chastonay J, Pennington MW (1995) A continuous fluorescence-based assay of human cytomegalovirus protease using a peptide substrate. Analytical Biochemistry 226:148-155

Knight CG, Willenbrock F, Murphy G (1992) A novel courmarin-labeled peptide for sensitive continuous assays of the matrix metalloproteinases. FEBS 296:263-266

Latt SA, Auld DS, Vallee BL (1970) Surveyor Substrates: Energy-transfer gauges of active center topography during catalysis. Proc. Natl. Acad. Sci. USA 67:1383-1389.

Latt SA, Auld DS, Vallee BL (1972) Fluorescence determination of carboxypeptidase A activity based on electronic energy transfer. Analytical Biochemistry 50:56-62

Lobb RR, Auld DS (1979) Determination of enzyme mechanisms by radiationless energy transfer kinetics. Proc. Natl. Acad. Sci. USA 76:2684-2688

Lobb RR, Auld DS (1980) Stopped-flow radiationless energy transfer kinetics: Direct observation of enzyme-substrate complexes at steady state. Biochemistry 19:5297-5302

Lobb RR, Auld DS (1984) Direct observation of enzyme substrate complexes by stopped-flow fluorescence: Mathematical analyses. Experientia 40:1197-1206

Maggiora LL, Smith CW, Zhang Z-Y (1992) A general method for the preparation of internally quenched fluorogenic protease substrates using solid-phase synthesis. J. Med. Chem. 35:3727-3730

Matayoshi ED, Wang GT, Krafft GA, Erickson J, (1990) Novel fluorogenic substrates for assaying retroviral proteases by resonance energy transfer. Science 247:954-958

Meldal M, Breddam K, (1991) Anthranilamide and nitrotyrosine as a donor-acceptor pair in internally quenched fluorescent substrates for endopeptidases: Multicolumn peptide synthesis of enzyme substrates for subtilisin Carlsberg and pepsin. Analytical Biochemistry 195:141-147.

Nagase H, Fields CG, Fields GB (1994) Design and characterization of a fluorogenic substrate selectively hydrolysed by stromelysin 1 (matrix metalloproteinase-3). J. Biol. Chem. 269:20952-20957

Netzel-Arnett S, Mallya SK, Nagase H, Birkedal-Hansen H, Van Wart HE (1991) Continuously recording fluorescent assays optimized for five human matrix metalloproteinases. Analytical Biochemistry 195:86-92

Ng M, Auld DS (1989) A fluorescent oligopeptide energy transfer assay with broad applications for neutral proteases. Anal. Biochem. 183:50-56

Stöcker W, Ng M, Auld DS (1990) Fluorescent oligopeptide substrates for kinetic characterization of the specificity of *astacus* protease. Biochemistry 29:10418-10425

Toth MV, Marshall GR (1990) A simple, continuous fluorometric assay for HIV protease. Int. J. Peptide Protein Res. 36:544-550

Welch RA, Holeman CM, Browner MF, Gehring MR, Kan C-C, Van Wart HE (1995) Purification of human matrilysin produced in *Escherichia coli* and characterization using a new optimized fluorogenic peptide substrate. Arch. Biochem. Biophys. 324:59-64

Yaron A, Carmel A, Katchalski-Katzir E (1979) Intramolecularly quenched fluorogenic substrates for hydrolytic enzymes. Analytical Biochemistry 95:228-235

Zimmeran M, Yurewicz E, Patel G (1976) A new fluorogenic substrate for chymotrypsin. Analytical Biochemistry 70:258-262

Abbreviations

Abz	2-aminobenzoyl or anthranoyl
ANSN	aminonapthalenesulfonamide
AMC	7-amino-4-methyl coumarin
γ-Abu	gamma-aminobutyric acid
BOP	benzotriazo-1-yloxytris(dimethylamino)phosphonium hexafluorophosphate
Boc	butyloxycarbonyl
Bzl	benzyl
DABCYL	4-(4-dimethylaminophenylazo)benzoyl
Dns or dansyl-	5-dimethylaminonapthalene-1-sulfonic acid
DED	dansyl ethylenediamine
Dnp	2, 4-dinitrophenyl
Dhbt	3,4-dihydro-4-oxo-1,2,3-benzotriazo-3-yl
Dpa	N-3-(2, 4-dinitrophenyl)-L-2, 3-diaminopropionyl
EDDnp	ethylenediamine 2, 4-dinitrophenyl
EDANS	5-[2-aminoethyl)aminonapthalene-1-sulfonic acid
Fmoc	9-fluorenylmethoxycarbonyl
Mca	(7-methoxycoumarin-4-yl)acetyl
MCA	4-methylcourmaryl-1-amide
MMP	matrix metalloproteinase
ONBzl	p-nitrobenzyloxy
Phe(NO$_2$)	p-nitrophenyl
TFA	trifluoracetic acid
Tyr(NO$_2$)	3-nitro-tyrosine

Chapter 4

Detection of Proteolytic Enzymes Using Protein Substrates

KEVIN K.W. WANG

Introduction

Short peptidic substrates are powerful tools for studying and detecting proteases. However, the natural substrates for the majority of proteases are in fact proteins. The larger size of a protein substrate can allow additional interaction with the protease of interest. In fact, some proteases are much more efficient in hydrolyzing proteins than small peptides. The calcium-activated protease (calpain) present in all mammalian cells is one example (Murachi, 1983; Saido et al., 1994; Wang and Yuen, 1994). Thus, a protein substrate assay can be a sensitive and powerful assay for studying certain proteases. In this chapter, we will describe such an assay, using casein as a protein substrate (Buroker and Wang, 1993). Arguably, for certain protease systems, a protein substrate may better mimic the physiological processes in which the proteases participate. Another area in which a protein substrate is more useful than a peptide is protease zymography (Paech, et al., 1993a; Raser et al., 1995). In this chapter, we will describe two zymography protocols that call for co-polymerizing a protein substrate into an acrylamide gel. Both assays are useful tools in studying purified enzymes, but in analyzing proteolytic activities in tissues or culture cell samples (Buroker et al., 1993; Raser et al., 1995).

Kevin K.W. Wang, Parke-Davis Pharmaceutical Research, Division of Warner-Lambert Company, Laboratory of Neuro-biochemistry, Department of Neuroscience Therapeutics, 2800 Plymouth Road, Ann ArborMichigan, 48105, USA *fax* (734) 622-7178; *e-mail* Kevin.Wang@WL.com)

Outline

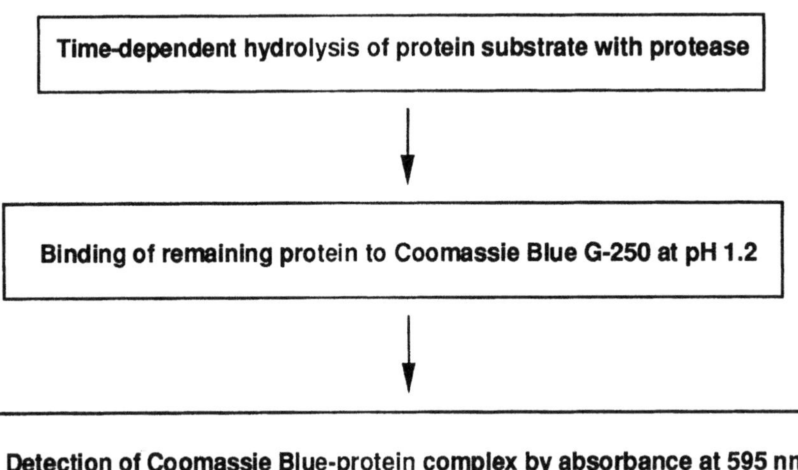

Fig. 1. Schematic flow-sheet demonstrating the proteolytic hydrolysis of protein substrate (e.g., casein) with a protease and the subsequent detection of remaining unhydrolyzed substrate by Coomassie blue-binding

Materials

For Colorimetric Assay

Reagents
- Bio-Rad protein dye reagent concentrate [0.05% (w/v) Coomassie Brilliant Blue G-250, 23.5% (w/v) ethanol, and 42.5% (w/v) phosphoric acid] (Bio-Rad Laboratories, #500-0006)
- Bovine casein (sodium salt) (Sigma, #C8654)
- Porcine m-calpain (Calbiochem, #208715) and μ-calpain (Calbiochem, #208712)
- $CaCl_2$, dithiothreitol (DTT), EGTA (Sigma Chemical Company, reagent grade)
- 96-well microtiter plates

Equipment
- Thermomax microplate reader (Molecular Devices) with printer or computer connection
- 8- or 12-channel multipipettor (Eppendorf)

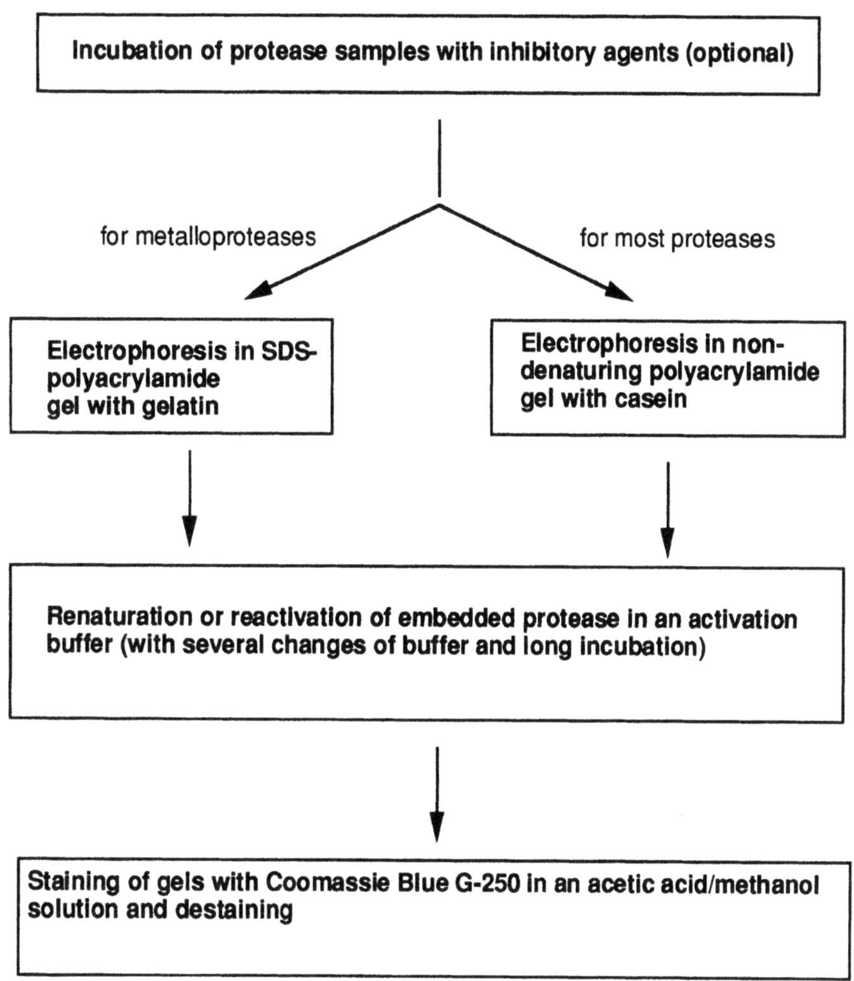

Fig. 2. Schematic flow-chart showing the electrophoresis of protease into protein substrate-containing polyacrylamide gel, the reactivation of the protease and the subsequent detection of proteolytic activity by Coomassie blue staining of the substrate gels (zymograms).

DTCC Soup

Buffers and solutions

33.3 mM DTT

83.3 mM Tris-HCl (pH 7.4 at 25°C)

0.83 mg/ml casein

10 µg/ml µ-calpain (unless otherwise stated) or

14 mµg/ml m-calpain or

14 µg/ml trypsin or

60 ng/ml papain

Coomassie Blue Solution
Bio-Rad protein assay dye reagent : water (1:1)

For Zymography

Reagents
- Bovine casein (sodium salt) (Sigma, #C8654)
- Porcine m-calpain (Calbiochem, #208715) and µ-calpain (Calbiochem, #208712)
- Collagenase (type III; Worthington, #CLS-3)
- Polyacrylamide gel cassettes (1.5 mm thick) (Novex, #NC2015) and 10-well comb (Novex, #NC3510).
- Precasted 0.1% gelatin Tris-glycine 10% polyacrylamide gel (1.0 mm thick, 10 wells) (zymogram gel, Novex, #EC6175)
- N-acetyl-Leu-Leu-Met-H (Calbiochem, #208721)
- E64c, calcium chloride ($CaCl_2$), dithiothreitol (DTT), casein (sodium salt), gelatin, EGTA, Tris-base and glycerol (Sigma Chemical Company, reagent grade)
- Glycine, Coomassie Brilliant Blue G-250, bromophenol blue, 2-mercaptoethanol and sodium dodecyl sulfate (SDS) (all from BioRad, electrophoresis grade)

Equipment
- Xcell II mini-cell electrophoresis unit (Novex, #EI9001)
- Power supply unit (BioRad, model 200/2.0)

Buffers and solutions
- Separating substrate gel solution (25 ml for 4 gels)
  - 12% (w/v) acrylamide
  - 0.32% (w/v) N,N'-methylene-bisacrylamide
  - 375 mM Tris-HCl (pH 8.8)
  - 0.2% (w/v) casein or gelatin
  - 3.5% glycerol
- Stacking gel solution (5 ml for 4 gels)
  - 4% (w/v) acrylamide
  - 0.10% (w/v) N,N'-methylene-bisacrylamide
  - 330 mM Tris-HCl (pH 6.8)
- 4X Sample buffer
  - 20% (v/v) glycerol
  - 2 mM 2-mercaptoethanol (not used for metalloproteases)
  - 0.004% (w/v) bromophenol blue
  - 200 mM Tris-HCl (pH 7.0)

- 4X SDS-sample buffer
  -10% (w/v) SDS
  -20% (v/v) glycerol
  -2 mM mercaptoethanol
  -0.004% (w/v) bromophenol blue
  -200 mM Tris-HCl (pH 6.8)
- SDS-Gel running buffer
  -0.2% (w/v) SDS
  -25 mM Tris-base
  -192 mM glycine (pH 8.3)
- Non-SDS running buffer
  -25 mM Tris-base
  -192 mM glycine
  -1 mM EGTA and 1 mM DTT (pH 8.3)
- Reactivation buffer C (for calpain)
  -20 mM Tris-HCl (pH 7.4 at room temperature)
  -10 mM DTT
  -1-4 mM $CaCl_2$
- Renaturation buffer M1 (for metalloprotease)
  -2.5% (v/v) triton X-100
  -100 mM glycine (pH 8.3)
- Proteolysis buffer M2 (for metalloprotease)
  -100 mM glycine (pH 8.3)
- Fixing/Destaining solution
  -methanol/water/acetic acid (5:4:1)
- Staining solution
  -0.25% (w/v) Coomassie blue R-250 in fixing/destaining solution

Procedure

Colorimetric Assay

1. Pipette 2.5 µl of calpain inhibitor as DMSO or DMF stock into a 96-well **Proteolysis** polyproprolene microtiter plate, to achieve a desirable final concentra- **reaction plates** tion in a total volume of 250 µl (optional).

2. Add 47.5 µl (or 50 µl if no compound is added in step 1 of water.

3. Add 150 µl of freshly made DTCC soup and shake briefly.

4. Start the proteolytic reaction by adding either 50 µl of $CaCl_2$ to positive control wells and wells with inhibitors or 1 mM EDTA to negative control wells.

5. After shaking, the plate is covered with a lid and is incubated at 25°C for 60 min.

Colorimetric development

1. Add 50 µl of water to a separate polystrylene 96-well plate.

2. Transfer two 100 µl-aliquots from each used well of the reaction plate to two separate wells of a new plate (to create duplicates).

3. Add 100 µl of Coomassie Blue solution and shake briefly; burst any bubbles in the wells using a needle.

4. Incubate at room temperature for 10-15 min.

5. Read absorbance at 595 nm in a microplate reader.

Zymogram

PAGE in non-denaturing (native) substrate gels

1. To initiate polymerization, added ammonium persulfate (0.04%, w/v) and TEMED (0.028%, v/v) to the separating casein gel solution.

2. Immediately pour the separating casein gel solution into the empty gel cassettes and allow to polymerize for 30-60 min.

3. After the well-forming combs are inserted into the top of the cassettes, again, ammonium persulfate (0.08%, w/v) and TEMED (0.028%, v/v) are added to the stacking gel solution.

4. Fill the top part of the cassettes with the mixture. Allow the gel to polymerize for 15-30 min. Alternatively, precated casein (0.5h) polyacrylamide gels are now commercially available (Novax).

5. Calpain samples in 50 mM Tris-HCl, 1 mM DTT, 1 mM EGTA is either untreated or incubated in 50 mM Tris-HCl (pH 7.4), 3 mM DTT, 2 mM $CaCl_2$ in the presence of a calpain inhibitor for 5 min on ice (volume 32 µl).

6. Three µl of 100 mM EGTA is added.

7. Five µl of the non-SDS-sample buffer is added (total volume 40 µl) and keep on ice.

8. During step 5, The casein gels were pre-run with the non-SDS running buffer for 15 min in an ice-water bath.

9. Protease samples are then loaded into the wells. Rainbow molecular weight markers (Amersham; RPN756) are also run alongside to monitor the progress of electrophoresis.

10. Electrophoresis is run at 125 V for 3 h in an ice-water bath.

11. The gel is then removed and incubated in reactivation buffer C with two changes within 2 h.

12. The gel is then further incubated overnight (20 to 24 h) at ambient temperature in the same buffer.

PAGE in denaturing substrate gels

1. Commercially precasted gelatin gels are used wherever possible; otherwise, follow the previous section to make your own gelatin gels.

2. Collagenase samples in 50 mM Tris-HCl are either untreated or incubated in 50 mM Tris-HCl (pH 7.4), 2 mM $CaCl_2$, 1 µM ZnCl2 in the presence of an inhibitor for 5 min on ice (volume 32 µl).

3. Three µl of 100 mM EGTA is added.

4. Five µl of the SDS sample buffer is added (total volume 40 µl) and the samples are kept on ice.

Note: For metalloprotease the presence of 2-mercaptoethanol should be avoided.

5. During step 5, the gels are pre-run with the SDS-gel running buffer for 15 min at room temperature.

6. Protease samples are then loaded into the wells. SDS-PAGE molecular weight markers (BioRad; #161-0317) are also run alongside for subsequent molecular weight estimation.

7. Electrophoresis is run at 125 V for 2.2 h at room temperature.

8. The gel is then removed and incubated in reactivation buffer M1 with two changes within 1 h.

9. The gel is then incubated in renaturing buffer M1 for 3 h with three changes.

10. Further incubate the gels overnight in the proteolysis buffer M2 for 16 to 20 h at ambient temperature.

Staining and destaining of zymograms

1. At the end of the proteolysis reaction, incubate the gels in water for 1 h with two changes, followed by 30 min in the fixing solution.

2. Stain the gel with the staining solution for 30 min, followed by the destaining solution with several changes in 2-5 h.

3. Store the gel in 2.5% (v/v) acetic acid.

Results

Colorimetric Assay

This protocol is relatively simple to execute. It involves two major steps: the proteolysis incubation during which protein substrate casein is hydrolyzed by the added protease and the subsequent binding or unhydrolyzed casein to Coomassie blue at very low pH (see Fig. 1). This assay relies on the ability of a protease to cleave the substrate protein at multiple sites. Under these conditions, only large proteins, but not its proteolytic products (e.g. short peptides and amino acids), would bind to Coomassie blue (Bradford, 1976; Buroker et al., 1993). The dye-protein complex is blue while the free dye is brownish yellow. Thus, an absorbance at 595 nm essentially monitors the disappearance of substrate. For example, Absorbance of 0.850 and 1.900 for wells with and without protease activity, respectively. It should be noted that before using the assay for routine experiments, an initial run should be done to first determine the optimal concentration of the protease of interest (Fig. 3). As expected, certain proteases are more readily detected than others. We found that papain, trypsin, chymotrypsin and calpains can be detected with this assay. Proteases that are more restrictive on their cleavage sequence would not be well suited for this assay. This assay can also be used to detect or monitor activity of a protease in tissues, cells (using homogenate) or biological fluids.

We extended the usage of this assay to evaluate the potency of a protease inhibitor. Using μ-calpain and its inhibitor Ac-Leu-Leu-Met-H as an example, the calpain activity was determined at various concentrations of this inhibitor spanning three orders of magnitude (10 nM to 10 μM). Using calpain activities (without inhibitor added) in the absence (-Ca) or the presence of calcium (+Ca) as standards, we calculate calpain activity at x μM of the inhibitor as: A595(X) - A595(+Ca). Thus, the % calpain inhibition at x μM of Ac-Leu-Leu-Met-H is:

$$\frac{A_{595}(X) - A_{595}(+Ca)}{A_{595}(-Ca) - A_{595}(+Ca)} \times 100\%$$

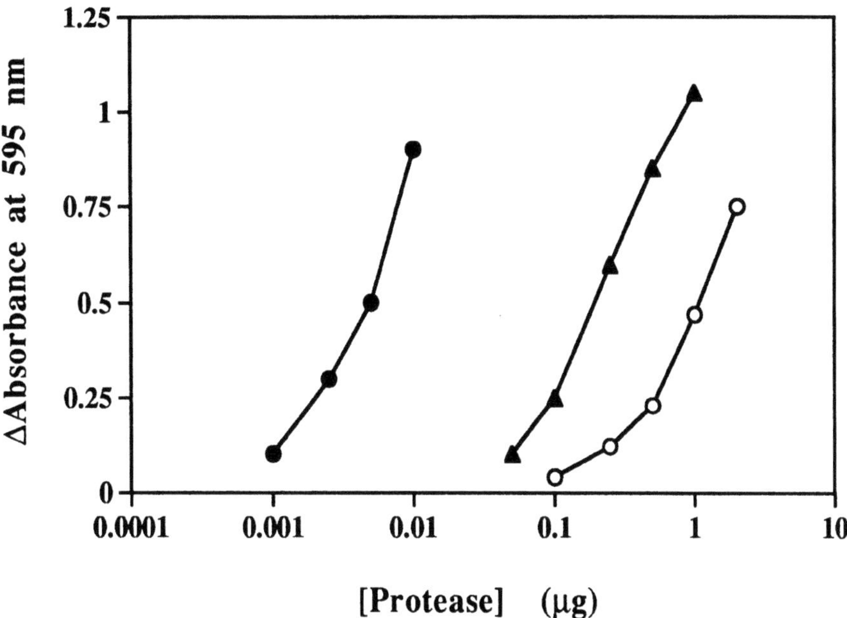

Fig. 3. Protein-based colorimetric protease assay with various protease concentrations. Different amounts of papain (solid circles), μ-calpain (solid triangles) or trypsin (open circles) were used well to determine sensitivity of the assay. Difference in absorbance at 595 nm in the (-protease) well and the (+protease) well reflects protease activity.

From the calculation, a sigmoidal curve of inhibition can be generated and an inhibitor concentration that inhibits 50% calpain activity (IC50) can be estimated by curve fitting (Fig. 4).

Zymography

Two different protocols of protease zymography are described here (see Fig. 2). For metalloproteases, we have the luxury that the proteases can renature functionally after the removal of denaturing detergent SDS (Heussen and Dowdle, 1980; Milton et al., 1992; Paech et al., 1993a,b). This allows for the electrophoresis of protease samples in regular SDS containing Tris-glycine gel system (Laemmli, 1970). The major modification was the co-polymerization of gelatin with the acrylamide/bisacrylamide (see Fig. 2). On the other hand, it is reasonable to assume that members of other protease families are more sensitive to inactivation by SDS. For proteases that contain more than one subunit (e.g., calpains), upon electrophoresis, the different

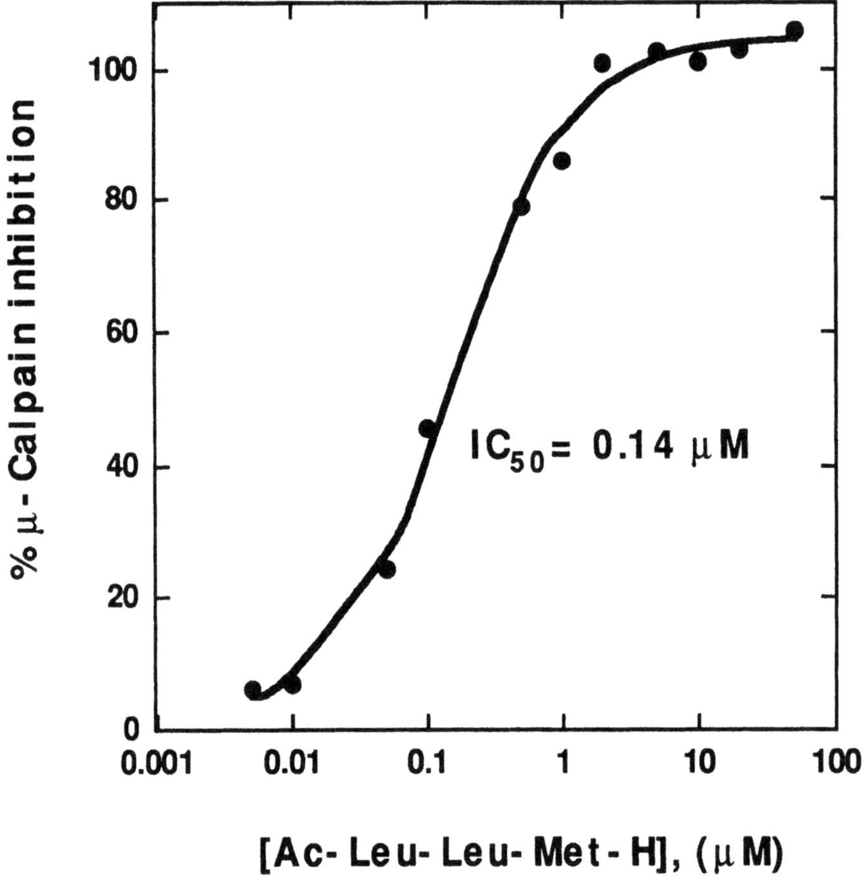

Fig. 4. Determination of potency of a protease inhibitor using a casein-based protease assay. μ-calpain and a calpain inhibitor (Ac-Leu-Leu-Met-H) were used. Using a formula described in the text, the absorbance at 595 nm was converted to % μ-calpain inhibition for each inhibitor concentration. The % μ-calpain inhibitor was plotted against the log [inhibitor]. From that, a sigmoidal plot was generated and an IC50 value was obtained.

subunits are likely to migrate to different distance, thus further diminishing the potential for functional reactivation. Thus, we recommend a protocol that essentially keeps the proteases in a native but inactive state during the course of electrophoresis (e.g., by not using SDS, the addition of calcium chelator EGTA and reducing agent DTT in the electrophoresis buffer and maintaining a low running temperature) (Raser et al., 1995). In the case of calpain, its preferred substrate, casein, was co-polymerized into the acrylamide gel.

At the end of the electrophoresis, the embedded protease would have migrated to a distance from the origin as a single band. The migration distance is a function of apparent molecular weight for SDS-PAGE or total surface charge for native PAGE. At this point, the gels were removed from the casts and placed in renaturing buffer to reactive the embedded protease (see Fig. 2).

Following the reactivation step is the proteolysis incubation (from 4 to 24 h) (see Fig. 2). During this long incubation, the embedded protease would digest the co-localized casein in the gel. The small proteolytic products (short peptides and amino acids) would diffuse out of the gel while the intact protein would remain trapped inside the gel. The casein or gelatin gels are finally stained with Coomassie blue. Upon a uniformly stained background, the protease bands appear as distinct "clearing" bands (see Fig. 5 – 6).

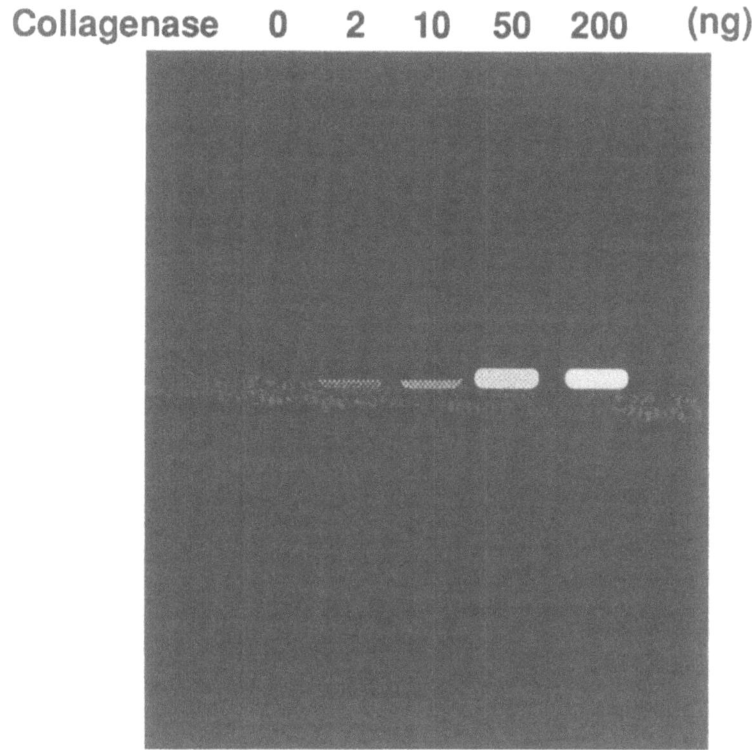

Fig. 5. Gelatin zymography using different amount of collagenase (0, 2, 10, 50 and 200 ng) using SDS-polyacrylamide gel.

Before using this assay for routine purposes, determination of the detection limit for the protease of interest is important. Usually, metalloproteases (such as gelatinase or collagenase) performed well in zymograms, as low as 2-10 ng of protease could be detected (Fig. 5). On the other hand, 20-50 ng μ- and m-calpain may be needed for good detection (results not shown). These assays can be used to detect or monitor protease activity in biological samples (e.g., cell or tissue homogenate or biological fluids).

Lastly, we extended the use of zymography to study protease inhibitors. Pretreatment of μ-calpain samples with an inhibitor was done prior to electrophoresis. Under these conditions, calpain treated with a reversible inhi-

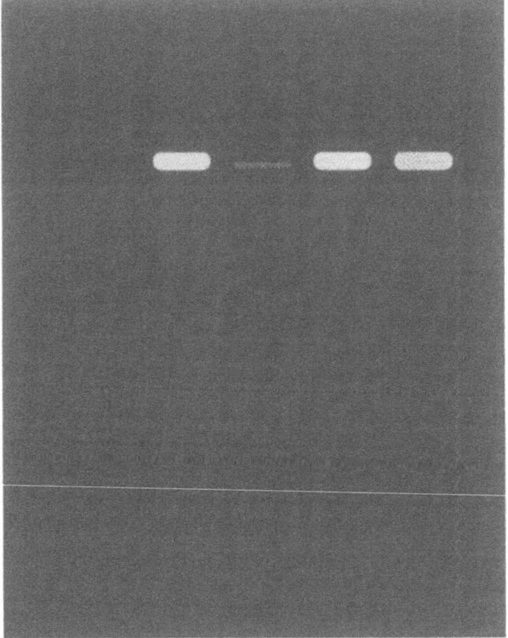

Fig. 6. Casein zymography of untreated and inhibitor-treated μ-calpain in non-denaturing gel. μ-Calpain (1 μg) was either untreated or pretreated with a reversible inhibitor (Ac-Leu-Leu-Met-H; 50 μM), with an irreversible inhibitor (E64c; 16 μM) or pre-exposed to Ac-Leu-Leu-Met-H for 5 min before incubation with E64c.

bitor (e.g., Ac-Leu-Leu-Met-H) which readily dissociates from the embedded calpain, would show no significant reduction of calpain activity in the casein gel (Fig. 6). However, the use of an irreversible active site inactivator (e.g., E64c) resulted in sustained inhibition (see Fig. 5). In addition, pre-exposure of μ-calpain to Ac-Leu-Leu-Met-H prevented the subsequent inactivation of calpain by E64c in the zymogram as both compounds act on the same site (Fig. 6). We argue that this is a useful method to sort out inhibitors that target at different sites of a protease.

Troubleshooting

For Colorimetric Assay

- Variable readings from duplicate wells: may be due to incomplete mixing of the reaction mixture. Try to mix more vigorously with a multipipettor fitted with tips.

- No difference in readings between -Ca and +Ca wells: not enough protease or insufficient amount of co-factors (e.g., Ca_{2+} or reducing agent DTT), check freshness of reagents and protease stock.

- The Bio-Rad dye concentrate is strongly recommended as a source of the Coomassie blue solution, for best results and consistency.

- Cell or tissue homogenate or biological fluids can also be used as source of protease, but pre-optimization of the protease concentration may be needed. If low concentration of the protease is in the original materials or the presence of an excess amount of an endogenous inhibitor is encountered, a sample concentration/fractionation step can be added (e.g., using ion exchange resin DEAE-agarose packed as mini-columns).

For Zymography

- Choosing denaturing or native gel: a general rule of thumb is that metalloproteases can be used in denaturing gel while most other proteases would prefer native gel conditions.

- Pre-casted zymographic gels are recommended for cost-effectiveness and consistency.

- If zymographic gels are to be casted, make acrylamide (30%, w/v) /bis-acrylamide (0.8%, w/v) stock in water, which can be stored for up to 45 days at 4°C; make up enough fresh gel polymerization solution for 4-5 gels at a time which can be stored at 4°C and used within 3 days for time-effectiveness. This also allows you to discard imperfectly set gels.

- Protease "clearing" bands are too faint or too broad: reduce or increase amount of protease loaded; shorten or lengthen proteolysis incubation time.

- If biological samples rather than purified protease are used, pre-optimization may be needed, such as enriching the concentration of the protease sample or removal of an endogenous inhibitor with ion exchange resin (as above).

References

Bradford MM (1976) A rapid and sensitive method for quantitation of microgram quantities of protein utilizing the principle of protein-dye binding. Anal Biochem 72:248-254

Buroker KM, Wang, KKW (1993) A Coomassie brilliant blue G250-based colorimetric assay for measuring activity of calpain and other proteases. Anal Biochem 208:387-392

Heussen C, Dowdle EB (1980) Electrophoretic analysis of plasminogen activators in polyacrylamide gels containing sodium dodecyl sulfate and copolymerized substrates. Anal Biochem 102:196-202

Laemmli UK (1970) Cleavage of structural proteins during the assembly of the head of bacteriophage T4. Nature 227:680-685

Milton DL, Norqvist A, Wolf-Watz H (1992) Cloning of a metalloprotease gene involved in the virulence mechanism of Vibrio anguillarum. J Bacteriol 174:7235-7244

Murachi T (1983) Calpain and calpastatin. Trends Biochem Sci 8:167-169

Paech C, Christianson T, Maurer K-H (1993a) Zymogram of proteases made with developed film from nondenaturing polyacrylamide gels after electrophoresis. Anal Biochem 208:249-254

Paech C, Christianson T, Blasig S (1993b) Enhanced zymograms of proteases. Anal Biochem 213:440 441

Raser KJ, Posner A, Wang KKW (1995) Casein zymography: a method to study μ-calpain, m-calpain and their inhibitory agents. Arch Biochem Biophys 319:211-216

Saido TC, Sorimachi H, Suzuki K (1994) Calpain: new perspectives in molecular diversity and physiological-pathological involvement. FASEB J 8:814-822

Wang KKW, Yuen P-W (1994) Calpain inhibition: an overview of its therapeutic potential. Trends Pharmacol Sci 15:412-419

Zymography, Casein Zymography, and Reverse Zymography: Activity Assays for Proteases and their Inhibitors

GARY W. OLIVER, WILLIAM G. STETLER-STEVENSON AND
DAVID E. KLEINER

Introduction

Proteases and their inhibitors play an important role in the extracellular matrix (ECM) remodeling that occurs in a number of physiological and pathological processes. These include such diverse processes as trophoblast implantation, mammary gland involution, rheumatoid arthritis and tumor cell invasion (Liotta et al. 1991a; Liotta et al. 1991b). Matrix metalloproteinases involved in ECM degradation (Birkedal-Hansen et al., 1993; Stetler-Stevenson et al., 1993) have been found to be subject to regulatory controls at multiple levels including transcription, mRNA stability, translation, secretion, activation of proenzymes, degradation and inhibition by specific endogenous inhibitors known as Tissue Inhibitors of Metalloproteinases or TIMPS (Matrisian, L.M., 1992; Stetler-Stevenson et al., 1993b). The need to rapidly and easily screen tissue extracts, biological fluids or the conditioned media of a number of cell types has resulted in the adoption of the zymogram technique in many laboratories. The reverse zymogram is a related technique which detects metalloproteinase inhibitors. These techniques utilize electrophoresis for the separation of the various protease and inhibitory species coupled with the detection of the activity within the polyacrylamide gel. These techniques are simple to perform, highly sensitive in detecting secreted protease and protease inhibitory activity, and flexible enough to be utilized for a variety of proteases and their inhibitors.

Activity stains have been used for many years to detect enzyme species in polyacrylamide gels. In general, the enzymes are electrophoresed and then

Gary W. Oliver
William G. Stetler-Stevenson
Correspondence to: David E. Kleiner, Extracellular Matrix Pathology Section, Laboratory of Pathology, National Institutes of Health/National Cancer Institute, 9000 Rockville Pike, Building 10, Rm 2N212, BethesdaMaryland, 20892, USA (*phone* 301-594-2942; *fax* 301-480-9488; *e-mail* kleiner@helix.nih.gov)

incubated in a solution containing buffer and substrates. The detection of protease activity using macromolecular substrates in polyacrylamide gels was first introduced by Granelli-Piperno and Reich (1978). These investigators detected plasminogen activator activity in situ following gel electrophoresis by overlaying the original polyacrylamide gel with a second agar gel that had been copolymerized with two sequential substrates, plasminogen and fibrin. After incubating the gels together, bands of plasminogen activator activity were visible under background illumination as localized dark zones of fibrinolysis against an opaque white background of undegraded fibrin.

This technique has been expanded in recent years to include incorporation of macromolecular protease substrates directly in the acrylamide gel during electrophoresis, as well as the use of a wider variety of protease substrates. The macromolecular protease substrate, such as gelatin, casein or fibrinogen, is copolymerized within the polyacrylamide gel. As such it is fixed within the gel and does not migrate during the electrophoretic separation of the protease sample. Following electrophoresis the gel is washed to remove detergent which may interfere with the protease activity and then it is incubated in buffer which favors proteolytic activity for the enzymes of interest. In this procedure, popularized by Heussen and Dowdle (1980), protease activity is revealed by areas of clearing following staining of the gel for protein. The method is simple and requires little more than the standard equipment and solutions used for sodium dodecylsulfate polyacrylamide gel electrophoresis (SDS-PAGE). This assay is most useful for detecting the gelatinase activity of members of the matrix metalloproteinase (Matrixin) family. The gelatin zymogram has proven to be a rapid, highly sensitive screening method for these enzymes in both tissue samples and culture media from cell lines. Modifications of zymography for other members of the matrix metalloproteinase family (stromelysins) will also be described.

Reverse zymography is, by contrast, designed to detect proteinase inhibitor activity rather than protease activity. The technique has similar conceptual roots to a fibrinogen-agarose plate assay for protease inhibitory activity in which a digesting agent was added to grooves after electrophoresis and allowed to diffuse into the gel (Heimburger 1962). The original method had very crude sensitivity and resolving power but was improved with a complete submersion technique developed later (Kueppers 1970). Fibrinogen-agarose electrophoresis was later described for the detection of protease inhibitors using crude horse serum as a source of inhibitors (Pellegrini 1984). The method was later modified and improved by directly incorporating a proteinase source into the acrylamide-substrate gels (Heussen 1980,

Herron 1986). The method was modified for the detection of metalloproteinase inhibitor activity using the conditioned media of LA24-infected chicken embryo fibroblasts and referred to as protease/substrate gel electrophoresis (Staskus 1991). Modern methods have focused on the detection of matrix metalloproteinase inhibitory activity using conditioned media with high concentration of gelatinases. We provide a method using HT1080 fibrosarcoma cell conditioned media. The drawback is that conditioned media contains other biologically active substances and is therefore subject to variation from lab to lab and lot to lot. To minimize the number of variables and make reverse zymography more consistent we have characterized a method using recombinant gelatinase A or B. For the gelatinase A gels there is linear response to inhibition over at least two orders of magnitude which begins at the detection limit. Differential sensitivities in picogram ranges are observed for TIMP-1 and TIMP-2.

Outline

Subprotocol 1 **Zymography**

Materials

- Tris   Reagents
- Glycine
- Sodium dodecylsulfate (SDS)
- Acrylamide/bis-acrylamide (National Diagnostics)
- Gelatin (Porcine, EIA grade) (Bio Rad)
- Ammonium persulfate 10% (w/v), prepared fresh (Bio Rad)
- TEMED (N,N,N',N'-tetramethlyethylenediamine) (Bio Rad)
- NaCl
- $CaCl_2$
- Brij-35 (Flow)
- Triton X-100
- Coomassie brilliant blue G-250 (Bio Rad)
- Methanol Acetic acid
- Glycerol
- Bromophenol blue

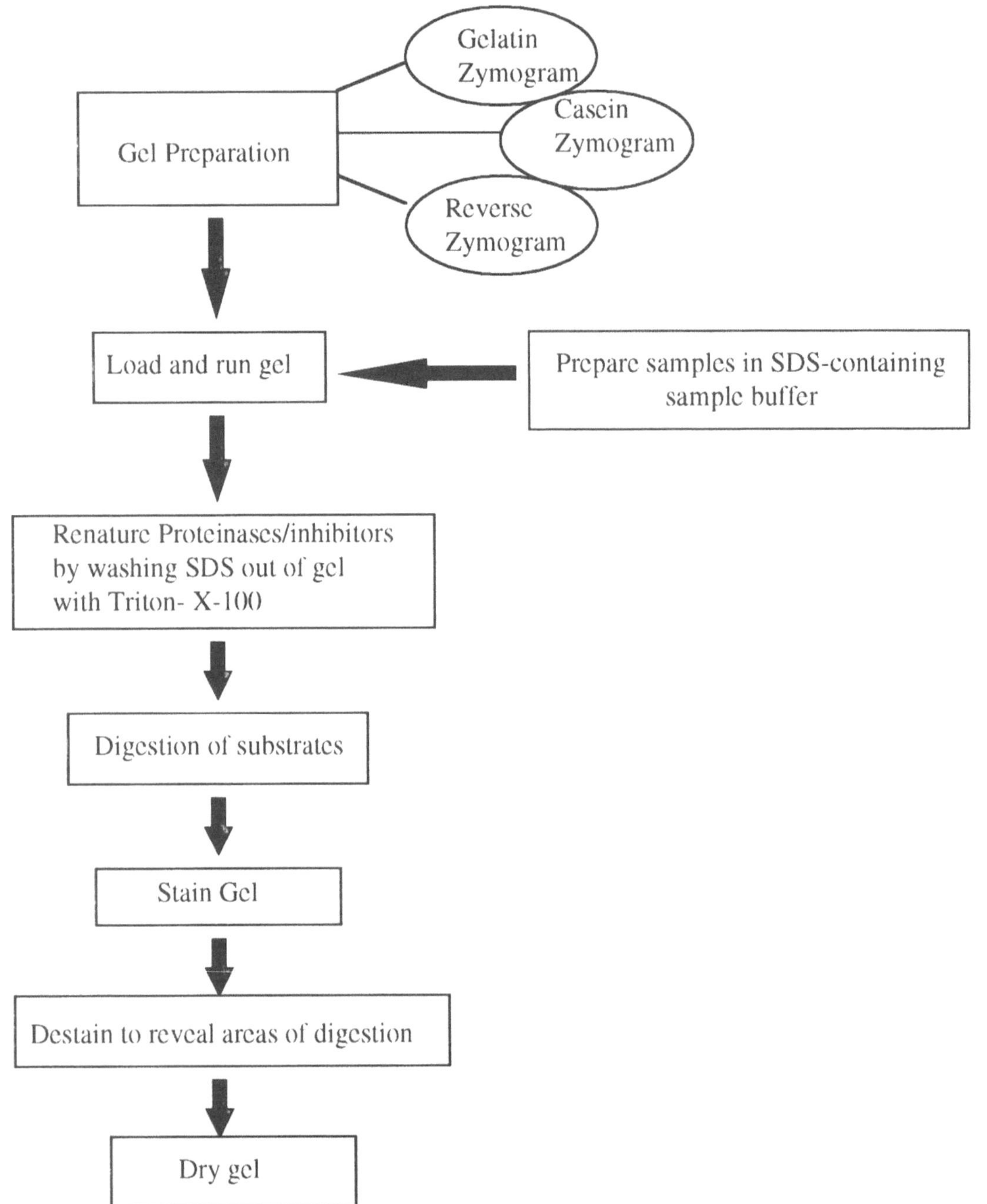

Fig. 1. Schematic flow sheet demonstrating the method of zymography and reverse zymography

- Electrophoresis gel apparatus with plates, spacers and combs or precast
  plastic cassettes and combs
- Gel incubation boxes
- Electrophoresis power supply
- Incubator
- Ph-meter

Equipment

- Separation gel buffer: 1.5 M Tris-HCL, pH 8.8
- Stacking gel buffer: 0.5 M Tris-HCl, pH 6.8
- Gelatin stock: 10 mg/ml (porcine) in separation gel buffer
  Prepare suspension and heat at 55 °C to dissolve. Store refrigerated at
  minus 20 °C. Stock solution will freeze and should be remelted at 55
  °C before preparation of the final running gel polymerization solution.
- Sample buffer (5x): 0.4 M Tris, pH 6.8, 5% SDS, 20% glycerol, and 0.03%
  bromophenol blue
- Running buffer (10x): 29g Tris, 144g glycine, 10g SDS, dissolved in a final
  volume of 1 liter of distilled water.
- Wash solution: 2.5% (v/v) Triton X-100 in destilled water
- Incubation buffer: 50 mM Tris, pH 7.5, 200mM NaCl, 5mM CaCl2, 0.02%
  (w/v) Brij-35
- Gel staining solution: 30% (v/v) methanol, 10% (v/v) acetic acid, 0.5%
  Coomassie brilliant blue G-250. Filter before use.
- Gel destaining solution: 30% (v/v) methanol, 10% (v/v) acetic acid
- Gel pre-drying solution: 30% (v/v) methanol, 5% (v/v) glycerol

Solutions and buffers

Procedure

1. Assemble electrophoresis plates and seal the bottom end of the gel cassette with tape.

Gel preparation

2. Prepare separation gel mix without adding ammonium persulfate or
   TEMED as indicated below.
   Zymogram separation gel mix: for 10 ml total volume add: 3.3ml 30%
   acrylamide 0.8% bis-acrylamide, 0.1ml 10% SDS, 2.5 ml 1.5 M Tris,
   pH 8.8, 100ul 10% gelatin, 2.8ml distilled water, 100 ul 10% ammonium
   persulfate, 0.01 ml TEMED.

3. Add the ammonium persulfate and the TEMED to the solution and mix.
   Quickly pour the gel solution into the cassette or between the plates leaving 1.5-2.0 cm for the stacking gel and comb.

Note: Do not allow the solution to bubble while mixing as this may result in uneven polymerization.

4. Layer 500 ul of butanol gently to the top of the separation gel to give a flat top to the gel. Allow the gel to polymerize for at least 30 minutes.

5. Wash the butanol off the surface of the lower layer gently with distilled water or separation gel buffer.

6. Prepare the stacking gel mix like the separation gel but with 4% acrylamide, 0.1% bis-acrylamide and stacking gel buffer. The ammonium persulfate and TEMED are added just before pouring onto the bottom layer and adding the comb. Allow the stacking gel to polymerize for 20-30 minutes.

Sample preparation and running gel

1. Prepare the samples in 5x sample buffer and allow them to incubate at room temperature for a minimum of 10 minutes.

Note: Do not heat the samples or add reducing agents such as 2-mercaptoethanol. These will destroy enzyme activity.

2. Mount the gel cassette in the electrophoresis apparatus and add 1x electrophoresis buffer to the outer and inner chambers.

3. Remove the comb and load the samples into the wells. Run the electrophoresis at current and voltage levels appropriate to the size of the gel. For the Novex mini-gel system, 20 mA constant current or 120 V constant voltage is appropriate. The gel should be run until the dye front approaches the bottom portion of the gel.

4. Remove the gel and place it in a container with sufficient Triton-X-100 to completely submerge the gel. For a 1mm mini-gel, 100 ml of Triton solution may be used. Incubate with gentle shaking for one hour.

5. Transfer the gel to the incubation buffer, using sufficient buffer to cover the gel. Incubate at 37 °C for a length of time appropriate for the amount of enzyme in the gel. An overnight incubation (15-18 hours) at 37 °C is sufficient to detect picogram quantities of the human gelatinases.

6. Stain the gel using the 0.5% Coomassie blue solution on a shaking platform for at least 3 hours. The concentration of Coomassie blue is higher than that which is usually used for protein detection in order to increase the contrast between the clear areas of digestion and the blue background.

7. Destain the gel with several changes of 30% methanol, 10% acetic acid until there is a sharp contrast between the clear bands and the background.

At this point the gel may be photographed or dried between sheets of cellophane. If the gel will be dried using cellophane place the gel into 30% methanol, 5% glycerol solution and shake for 30 minutes to help prevent cracking. Cellophane drying allows for easy storage and convenient photography or digital scanning of stored gels.

Subprotocol 2
Casein Zymogram Alternate Procedure

Materials

- 1% non-fat dry milk (filtered with a 22 μm filter)   Reagents
- Enzyme buffer: 50 mM Tris, pH 7.6, and 10 mM $CaCl_2$
- Staining solution- 30% isopropanol, 0.5% Coomassie blue R-250
- Destaining solution- 30% isopropanol

Procedure

1. Prepare the separating gel as for the zymogram replacing the 1% gelatin with 1% non-fat dry milk.

2. Prepare and run gel as for the zymogram until the Triton-X 100 step.

3. Wash the gels in 2.5% Triton X-100 for 3 h, changing the wash solution each hour.

4. Incubate the gel for at least 36 hours at 37 °C in 50 mM Tris, pH 7.6, 10 mM $CaCl_2$.

5. Stain for at least 8 h in the 0.5% Coomassie R-250 staining solution.

6. Destain the gel in 30% isopropanol.

7. Proceed as outlined in the standard method, (step 7)

Subprotocol 3
Reverse Zymogram (Alternate Procedure for Detection of Protease Inhibitors)

Materials

In addition to the materials for the zymogram the following are needed.

HT1080 cells
- Gelatinase source: 150 ng/ml gelatinase A or gelatinase B, 200 ng/ml Gel A-Timp-2 complex, or 2.5 ml HT1080 conditioned media per 10 ml of separating gel volume (lower gel).
- Serum free conditioned media is made by growing the cells to 40% confluence in 100x20 mm culture dishes in DMEM with 10% fetal calf serum, washing 4x in PBS to remove the serum, and then adding 15 ml of serum free media and allowing it to be conditioned for 48 hours. The media is then removed and stored at -20 or -80 °C. For consistency, it is best to store the media in large pooled batches as the digesting ability of each batch and other variables may vary slightly. HT1080 cells can be obtained from ATCC (CLL 121).
- Reverse zymogram separating gel mix: prepare with final concentrations: 2.25 mg/ml porcine gelatin, 0.25M Tris-Hcl pH 8.8, 0.125% SDS, 1ul/ml N,N,N,N'-Tetramethylethylenediamine; 0.4mg/ml ammonium persulfate; 15% (w/v) acrylamide and 0.4% bisacrylamide, 0.125% SDS, gelatinase source to the final concentration given above.

Procedure

1. Prepare the gel as in the standard zymogram method substituting the reverse zymogram separating gel mix.

2. Run gel as for the zymogram until the Triton X-100 step.

3. Wash the gels in 2.5% Triton X-100 for 3 hr, changing the wash solution each hour.

4. Digestion time is 15 hours at 37 °C in incubation buffer.

5. Stain a minimum of 4-6 hours in gel staining buffer (30% (v/v) methanol, 10% (v/v) acetic acid).

6. Destain until the background is clear and inhibitors are clearly visualized. Areas of inhibition are visualized as positively staining regions of the gel.

7. Proceed as outlined in the standard method.

Results

Zymogram

Figure 2 illustrates the band pattern of purified human gelatinase A. The gelatinases have degraded the gelatin and the bands are areas of decreased staining due to the loss of gelatin. Since exposure to SDS and refolding activates latent gelatinase, both progelatinase A (72 kD) and active gelatinase A (62 kD) are visualized. The method is highly sensitive with picogram detection after an overnight incubation at 37 °C. The reduction in absorbance of the band (amount of digestion) relative to undigested backgorund is proportional to the amount of enzyme present. The method can be used for quantitative purposes (Kleiner DE, 1994) by plotting an unknown against a standard curve.

Human cells secrete gelatinase A and gelatinase B, members of the matrix metalloproteinase family, generally in a latent form which requires cleavage of an N-terminal segment in order to become catalytically active (Stetler-

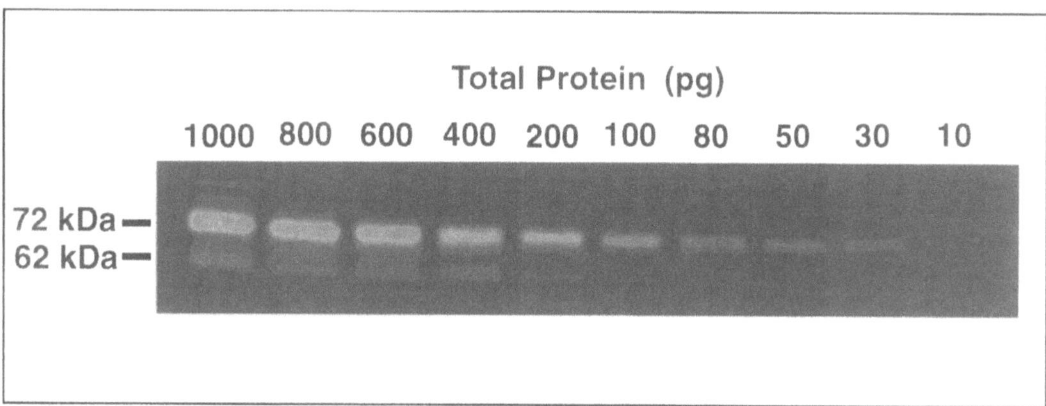

Fig. 2. Typical digestion pattern on a zymogram. This figure shows an 18 hour digestion with partially activated gelatinase A. Known amounts of purified gelatinase are loaded to make a standard curve from 10 to 1000 pg

Stevenson, et al., 1989). The latent 72 kDa progelatinase A is converted to a 65 kD intermediate and a 62 kD active form. The 92kD progelatinase B is converted to an 82 kD intermediate and then to a 72 kD active form. Thus the zymogram may show several bands corresponding to the latent and active forms of the enzymes. Further autolytic cleavage may result in other lower molecular weight bands. Other proteases with the capability to digest gelatin to small fragments may also be detected with the zymogram.

Reverse Zymograms

Figure 3 illustrates an HT1080 conditioned media reverse zymogram (optimized for sensitivity) and a typical gel A reverse zymogram. Both figures show prominent dark bands in the areas of inhibition. TIMP-1 runs at 29 kD, and TIMP-2 runs at 21.5 kD. For the gelatinase A gels there is linear response to inhibition over at least two orders of magnitude which begins at the detection limit (Oliver 1997). Differential sensitivities in picogram ranges are observed for TIMP-1 and TIMP-2.

The reverse zymogram gels have different background staining patterns even when they are allowed to incubate without samples. This is because the gels prepared with conditioned media contain gelatinases, inhibitors and non-gelatin proteins which migrate differentially during electrophoresis. This results in gels with band-like variation in background staining. The inhibitors are small and migrate towards the bottom of the gel. Thus, the background at the bottom of the conditioned media gel is darker. The labels to the left of the conditioned media gel in Figure 3 illustrate the inhibitors and gelatinases present in the background bands. The reverse zymogram prepared with purified gelatinases do not show background bands due to TIMPs. However, on both gels there is one additional band of darker blue at the top of the gel. This corresponds to an area from which the incorporated gelatinases have migrated and are now absent or decreased. In this area the gelatin remains intact. In this region of the reverse zymogram, gelatinases in the sample may be seen as they are in a gelatin zymogram, although the separation of the higher molecular weight enzymes is not optimal in the high percentage acrylamide gels used for reverse zymography.

Troubleshooting

- No band visualized in samples
  The most common error may be treating the samples with a reducing
  agent prior to loading them or heating them.
  A Triton X-100 step that is too short or incomplete will result in reduced
  gelatinolytic activity as the SDS has not been completely removed.

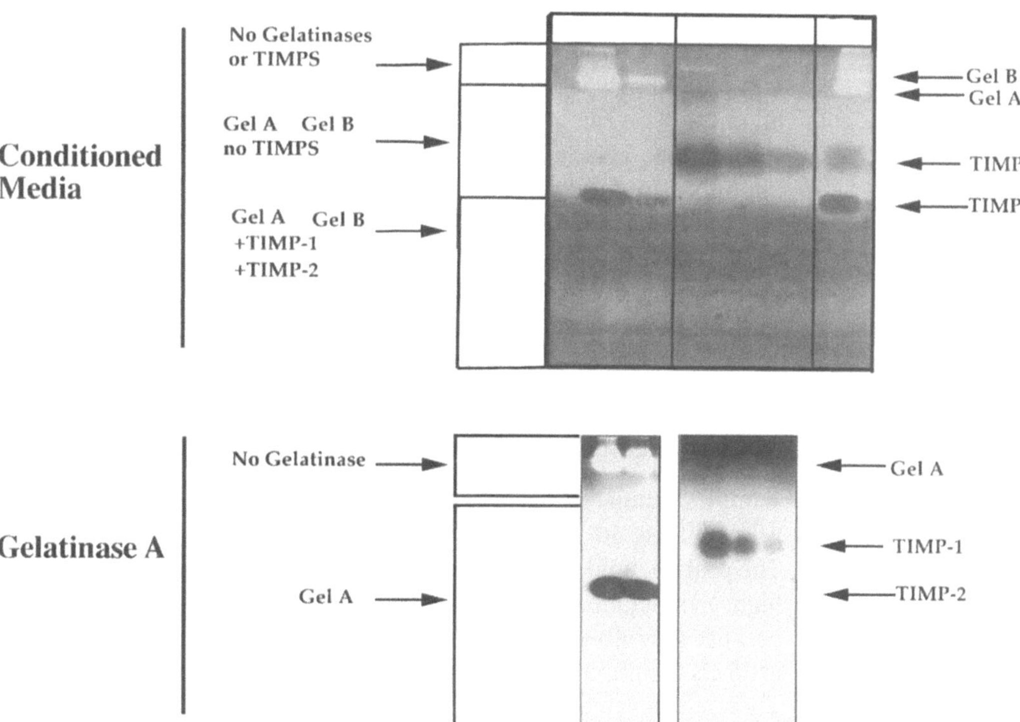

Fig. 3. Reverse zymograms. Illustrates an HT1080 conditioned media reverse zymogram
(upper gel) and gelatinase A reverse zymograms (lower gels). Both types of gel detect multiple
inhibitors simultaneously. Also the gels detect gelatinases in the top portion of the separating
gel. The incorporated gelatinases do not remain completely immobilized as the samples are
electrophoresed. Because of this, the gels have different background patterns which corre-
spond to the nature of their incorporated digesting agent. The conditioned media gel shows
background shades where the major gelatinases and inhibitors incorporated in the gel have
migrated. The gelatinase A gel, by contrast does not have the inhibitor-related background
bands in the lower portion of the gel. From 200+ to approximately 55kd there is a darker blue
background corresponding to an area from which the incorporated gelatinases have migrated. In this area gelatinases in the samples can be observed.

If the gel is not shaken after adding the incubation buffer a similar problem may result; alternatively for the reverse zymogram, uneven digestion may take place, resulting in a variable background.

- Bands too light
  Any problem which may cause the bands not to be visualized may also result in light or pale bands. If those problems are ruled out, the sample may need to be concentrated. In addition, a longer digestion time may increase the sensitivity of gelatin zymograms. For reverse zymography concentration of the samples may be beneficial, but if the background is clear and the bands are light, the digestion time should be reduced, not increased. Increasing the staining time may also be beneficial. The staining solution may become depleted of stain, resulting in light bands. 500 ml of the stain can be used to stain 30-40 gels before the stain becomes sub-optimal. The use of a standard containing gelatinases or proteinase inhibitors (for the reverse zymogram) is recommended in order to monitor the adequacy of the experiments. Serum-free conditioned media such as HT1080 (fibrosarcoma) or A2058 (melanoma cell) make good gelatinase controls. HT1080 conditioned media also makes a good control for TIMP 1 and 2.

- Samples do not leave the well
  This is most commonly seen with samples that contain high salt. Dialysis or a desalting spin column may be used to remove excess salts.

- Samples "tunnel" or seem to run in a straight line down the middle of the gel or diffuse out in a cone shape.
  The protein, lipid, or detergent concentration in the sample may be too high. Zymograms seem to be more sensitive to this problem, perhaps because they already contain protein in the acrylamide. Electrophoretic artifacts common to all electrophoresis methods may be encountered. These are covered in standard texts.

References

Birkedal-Hansen H, Moore W G I, Bodden M K, Windsor LJ, Birkedal-Hansen B, DeCarlo A, and Engler J A. (1993) Matrix Metalloproteinases: A Review. Crit. Rev. Oral Biol. Med. 4, 197-250.

Brown P D, Bloxidge R E, Anderson E, and Howell, A (1993) Expression of Activated Gelatinase in Human Invasive Breast Carcinoma Clin Exp Metastasis 11, 183-189

Collier IE, Wilhelm SM. Eisen AZ, Kronberger A. He C, Bauer EA and Goldberg GI (1988) H-ras oncogene transformed human bronchial epithelial cells (TBE-1) secrete a single metalloprotease capable of degrading basement membrane collagen. Journal of Biological Chemistry 263:6579-6587

Davies B, Miles DW, Happerfield L C, Naylor M S, Bobrow LG, Rubens RD, and Balkwill FR (1993) Activity of Type-IV Collagenases in Benign and Malignant Breast Disease Br J Cancer 67, 1126-1131

Fridman R, Bird RE, Hoyhtya M, Oelkuct M, Komarek D, Liang CM, Berman ML, Liotta LA, Stetler-Stevenson WG, Fuerst TR (1993) Expression of human recombinant 72 kDa gelatinase and tissue inhibitor of metalloproteinase-2 (TIMP-2): characterization ofcomplex and free enzyme. Biochem J, J an 15;289 (Pt 2):411-6

Granelli-Piperno A and Reich E (1978) A study of proteases and protease-inhibitor complexes in biological fluids. Journal of Experimental Medicine 148:223-234

Herron GS, Banda MJ, Clark EJ, Gavrilovic J and Werb Z (1986) Secretion of metalloproteinases by stimulated capillary endothelial cells. II. Expression of collagenase and stromelysin activities is regulated by endogenous inhibitors. Journal of Biological Chemistry 261: 2814-2818

Heussen C and Dowdle EB (1980) Electrophoretic analysis of plasminogen activators in polyacrylamide gels containing sodium-dodecyl sulfate and copolymerized substrates. Analytical Biochemistry 102:196-202

Kato Y, Nakayama Y, Umeda M, and Miyazaki, K (1992) Induction of 103-kDa Gelatinase/Type-IV Collagenase by Acidic Culture Conditions in Mouse Metastatic Melanoma Cell Lines J Biol Chem 267, 11424-11430

Kleiner DE, Stetler-Stevenson WG, (1994) Quantitative Zymography: Detection of Picogram Quantities of Gelatinases Analytical Biochemistry 218 325-329

Kleiner DE, Unsworth EJ, Krutzsch HC and Stetler-Stevenson WG (1992) Higher-order complex formation between the 72-kilodalton type IV collagenase and tissue inhibitor of metalloproteinases-2. Biochemistry 31:1665-1672

Liotta LA and Stetler-Stevenson WG (1991b) Tumor invasion and metastasis: an imbalance of positive and negative regulation. Cancer research 51 (supplement): 5054a-5059a

Liotta LA, Steeg PS and Stetler-Stevenson WG (1991a) Cancer metastasis and angiogenesis: an imbalance of positive and negative regulation. Cell 64: 327-336

Odekon LE, Sato Y and Rifkin DB (1992) Urokinase-type plasminogen activator mediates basic fibroblast growth factor-induced bovine endothelial cell migration independent of its proteolytic activity. Journal of Cellular Physiology 150: 258-263

Oliver GW, Leferson ID, Steter-Stevenson WG and Kleiner DE (1997) Quantitative reverse zymography: Analysis of picogram amounts of metalloproteinase inhibitors using gelatinase A and B reverse zymograms. Anal Biochem 244:161.166

Pellegrini A, Hageli G, Gretz D and Von Fellenberg R (1984) Natural protease inhibitors: qualitative and quantitative assay by fibrinogen-agarose electrophoresis. Analytical Biochemistry 138: 335-339

Staskus PW, Masiarz FR, Pallanck LJ, Hawkes SP (1991) The 21-kDa protein is a transformation-sensitive metalloproteinase inhibitor of chicken fibroblasts J. Biol. Chem. 266, 449-454

Stetler-Stevenson WG, Krutzsck HC, and Liotta LA (1989) The Activation of Human Type IV Collagenase Proenzyme. Sequence identification of the major conversion product following organomercurial activation. J. Biol. Chem. 264, 17374-17378

Stetler-Stevenson WG, Liotta LA and Kleiner DE, Jr (1993) Extracellular matrix 6: role of matrix metalloproteinases in tumor invasion and metastasis FASEB J., 7, 1434-1441

Wilhelm SM, Collier IE, Marmer BL, Eisen AZ, Grant GA and Goldberg GI (1989) SV40-transformed human lung fibroblasts secrete a 92-kDa type IV collagenase which is identical to that secreted by normal human macrophages. Journal of biological Chemistry 264: 17213-17221

Suppliers

30% (w/v) acrylamide and 0.8% (w/v) bisacrylamide stock solution, protein and sequencing electrophoresis grade, gas stabilized
(National Diagnostics 1013-1017 Kennedy Blvd. Manville, NJ 08853)

N,N,N,N'-Tetramethylethylenediamine
(BIO-RAD LABORATORIES 2000 ALFRED NOBEL DR., HERCULES, CA 94547)

Gel cassettes and pre-poured gelatin zymogram gels
(NOVEX 11040 Rosell St., San Diego, CA 92121)

HT1080 Cells
(ATCC 12301 Parklawn Dr. Rockville, MD 20852-1776)

Chapter 6

Genetic – Based Assays of Viral Proteases

C. RIZZO, Y-S. E. CHENG AND B. KORANT

Introduction

Many viruses code for specialized proteolytic enzymes which are essential for virus assembly (Korant, 1994). The viral enzymes appear to belong to one of the recognized protease classes, including serine, cysteine and aspartic active sites, but some novel combinations have also been identified, including a cysteine active site residue participating in a structure reminiscent of a serine protease fold (Bazan, 1989). In general, the viral proteases are able to discriminate and hydrolyze viral substrates with primary recognition involving five or more amino acids. This leads to rather limited proteolysis, even of viral precursors, and only rarely do the viral proteases preferentially attack a non-viral substrate.

Some of the viral proteases are interesting therapeutic targets, with the HIV protease attracting the most effort. Cell-based assays for putative inhibitors of viral proteases have therefore become relevant. However, not all laboratories are equipped to do experiments safely using virus-infected cells, in particular, with hazardous human pathogens such as HIV. Furthermore, some viruses, for example the diverse hepatitis group, grow poorly if at all in cultured cells. Therefore direct assays for the viral proteases in genetically-modified cells can provide a useful surrogate system. Assays for the HIV protease are now available in bacteria, yeast, animal cells and transgenic mice (Flexner 1988, Baum 1990, Murray 1993, Rizzo 1994, Tumminia 1996) and we will provide some examples of our own design which offer a variety of options to permit the assay of viral protease genes in *E. coli* or uninfected animal cells.

C. Rizzo
Y-S. E. Cheng
Correspondence to: B. Korant, DuPont Life Sciences, Dept of Molecular Biology, Experimental Station Bldg 336, rm 22, Wilmington DE, 19880-0336, USA (*phone* 302 695 9493; *fax* 302 695 9420; *e-mail* bruce.d.korant@dupontpharma.com)

Subprotocol 1
In Situ Assay for HIV-1 Protease in *E. coli* Based on Cell Viability

HIV-1 protease (HIVPR) is cytotoxic when expressed in *E. coli* cells (Hostomskey et.al., 1989). The following protocol describes a simple, low-cost, and semi-quantitative assay for testing the cytoprotective activity of protease inhibitors. Protease expression in this system is regulated by the T7 promoter and occurs only in host cells producing the T7 RNA polymerase (Studier 1986).

1. Transform bacteria with protease expression plasmid

2. Spread transformed cells (50-500 colony forming cells/dish) on agar plates containing cytoprotective chemical.

3. Count colonies and photograph

Materials

- HIV PR expression plasmids in pET vector (Cheng 1990a, Cheng 1990b).
- T7 RNA polymerase producing *E. coli* strains BL21(DE-3) or HMS174(DE-3) and control strain JM109 [NOVAGEN #69450-10 or 69453-1 and CLONTECH #C1005-1].1
- Luria Broth (LB) + ampicillin (150 µg/ml) plates and LB + ampicillin plates supplemented with HIV PR inhibitors.
- S.O.C. buffer (Gibco/BRL #15544-018)

Procedure

Bacterial transformation

1. BL21 (DE-3), HMS174(DE-3), and JM109 cells are treated with 0.05M $CaCl_2$ and transformed with 10 - 100 ng of plasmid DNA as described (Hanahan 1983).

2. After transformation, cells are incubated for 45 min. in 0.8 - 0.9 ml of S.O.C. and spread immediately using a sterile glass rod onto LB ampicillin plates or LB ampicillin plates supplemented with inhibitors.

1. Inhibitor plates are prepared in one of two ways:
    1.1. 0.2 ml of a solution containing a single concentration of inhibitor (e.g. 10 µg/ml) is spread evenly using a sterile glass rod on the surface of a LB ampicillin plate and allowed to soak in completely prior to plating cells or
    1.2. after the transformed cells are spread evenly over the plate 1, 5, and 10 µl aliquots of a 10 µg/ml inhibitor solution (e.g. in 100% DMSO) are spotted on the plate directly or on a 1cm filter paper disk which is then placed on the hardened agar.

2. The plates are incubated at 37°C overnight to allow colonies to form.

3. Colonies are counted and photographed.

Preparation of
inhibitor plates

Results

As seen in figure 1a, HIV PR expressing cells were able to form colonies only when grown on media containing an HIVPR inhibitor, e.g. DMP323 (Lam 1994). Relative potencies of inhibitors could be grossly examined as shown in figure 1b. Dose dependent protection of cells from HIVPR activity is seen with inhibitor III but not inhibitor II at these levels.

Troubleshooting

Solubility characteristics and cytotoxicity of inhibitor compounds should be first tested on control cells (without viral protease).

A simple but useful variation on this assay can detect mutations in HIV-PR leading to resistance to inhibitors. In the presence of an inhibitor, drug-sensitive versions of the enzyme yield a "large, healthy" colony, similar to control bacteria with no active viral protease present. In comparision, drug-resistant colonies may appear small and/or translucent, even in high concentrations of drug, to which the parental enzyme was sensitive (in press).

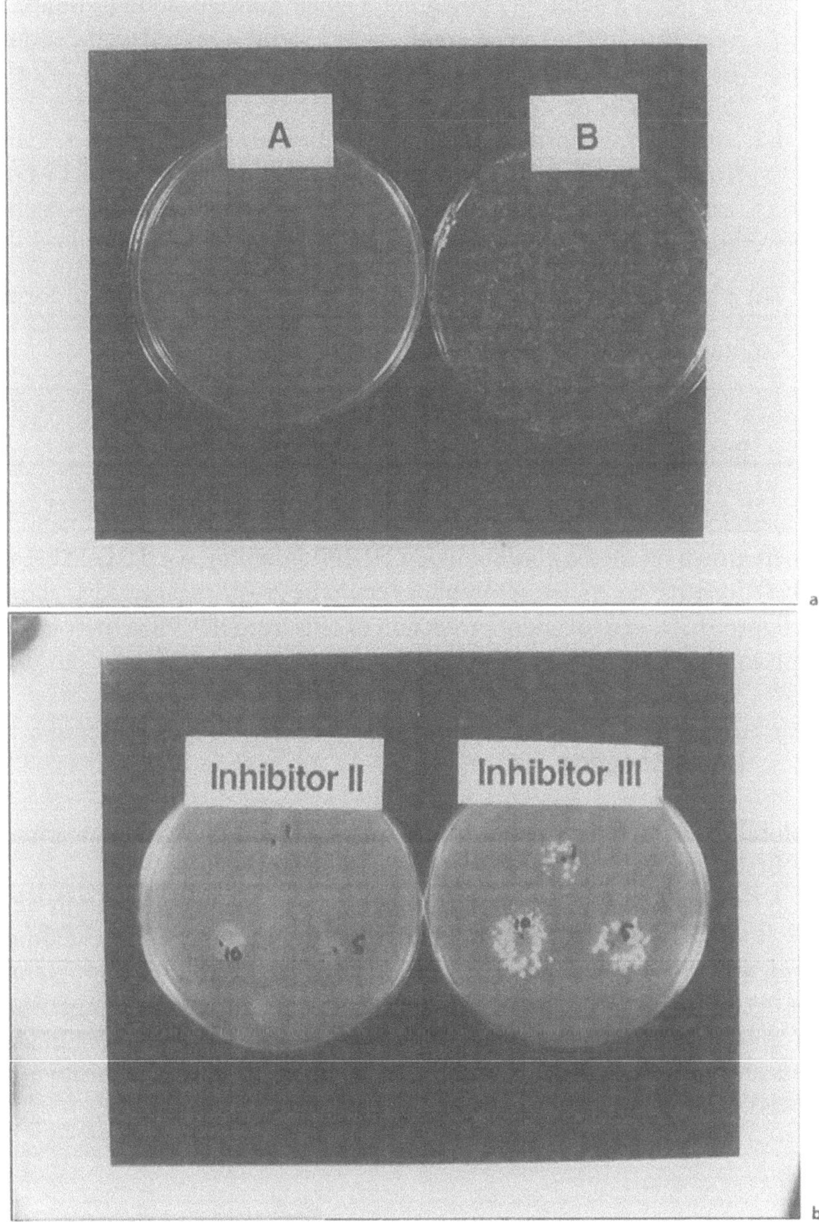

Fig. 1. Bacterial growth after 20 hours of incubation at 37°C. In figure **a** protease transformed cells were plated on either LB ampicillin (dish A) or LB ampicillin supplemented with 10 µg/ml of a protease inhibitor (dish B). In figure **b** a dose range of 1-10 µg of two inhibitors was spotted on the LB ampicillin plates after transformed cells were spread.

Subprotocol 2
Use of a β-Galactosidase Assay to Indirectly Measure Activity and Inhibition of HIV-1 Protease

E. coli transformed with a plasmid containing the HIV-1 protease gene expresses active protease when treated with isopropylthio-β-D-galactoside (IPTG). Under the same conditions, synthesis of endogenous β-galactosidase (β-gal) is also induced in most strains of *E. coli*. However, expression of active HIV PR interrupts normal protein synthesis and β-gal synthesis is reduced. When cultures are grown in media containing HIV protease inhibitors, protein synthesis in the cell is protected as determined by a measurable increase in β-gal activity. We have adapted a microtiter plate assay (Eustice 1991) to indirectly measure HIV protease activity and its inhibition.

1. Transform *E. coli* with protease expression plasmid

2. Grow and induce cultures +/- inhibitors.

3. Add substrate and measure β-gal activity on a plate reader

Materials

- Protease expression plasmid pET11PR (pET11vector available from Novagen).
- *E. coli* strain BL21 (DE-3) [Novagen #69450-1]
- LB broth and plates
- Isopropylthio-β-D-galactoside (IPTG) [e.g. SIGMA #15502]
- Lysis buffer-100mM $NaPO_4$ pH7.2, 10mM KCI, 1mM $MgSO_4$, 50mM β-mercaptoethanol, 0.175% NP40, 2.5mM EDTA.
- SDS - PAGE and blotting apparatus
- Protein binding membranes, such as Immobilon (Millipore)
- Lysing solution (NLS)-8M Urea, 1% NP-40, 1% β-mercapto-ethanol, 1% Ampholytes 3-10 (LKB -Pharmacia #80-1125-87).
- Custom antisera
- Chlorophenol red D-galactopyranoside (CPRG), 4 mg/ml in lysis buffer- (chromogenic substrate for β-gal).
- 96 well plate reader with 575 nm filter

Procedure

An *E. coli* expression plasmid (pET11PR) was constructed by cloning synthetic ds DNA coding for the 99 amino acid HIV-1 protease and a 54 amino acid N terminal peptide (p6) into the T7 translation vector pET11C (Novagen #69438-1).

Growth of bacterial cultures

1. Grow pET11PR transformed BL21(DE-3) cells to mid-log (0.4 OD600) in LB + ampicillin, 50 µg/ml. Grow untransformed control cells to mid-log in LB.

2. Aliquot 50 µl samples to wells of a 96 well microtiter dish

3. The cells are then either left untreated or treated as follows:
   3.1. incubated for 45' at 37°C with 1mM IPTG; or
   3.2. incubated for 3' at 24°C with HIV PR inhibitor followed by 45' incubation at 37°C with 1mM IPTG, with HIVPR inhibitor present throughout.

4. Place the microtiter plate on ice for 10' to arrest growth.

Spectrophotometric assay of β-gal

1. Add 40 µl of lysis buffer to each well and incubate plate at 37°C for 10'.

2. Using a multichannel pipettor, add 10 µl of 4 µg/ml CPRG

3. Measure OD575 on plate reader. Measurements may be taken as end point reading at a single point (e.g. 30') or at intervals from 0-30 minutes.

Optional: Western blot analysis

Duplicate samples may be prepared containing total protein extracts for Western blot analysis: Cells are centrifuged briefly at full-speed in a microfuge; all liquid is removed, the cells lysed in 10 µl of NLS, and boiled for 2-3 minutes. Proteins are separated by SDS-PAGE and blotted to membranes by standard methods.

Results

Treatment of transformed bacterial cultures with specific HIV-1 protease inhibitors prior to expression of the HIV-PR improved the production of endogenous β-gal as measured by hydrolysis of CPRG (Fig 2). β-gal levels of induced, untransformed *E. coli* cultures were unchanged by addition of these compounds. Uninduced cultures and induced cultures expressing PR with no inhibitor present produce almost no β-gal activity.

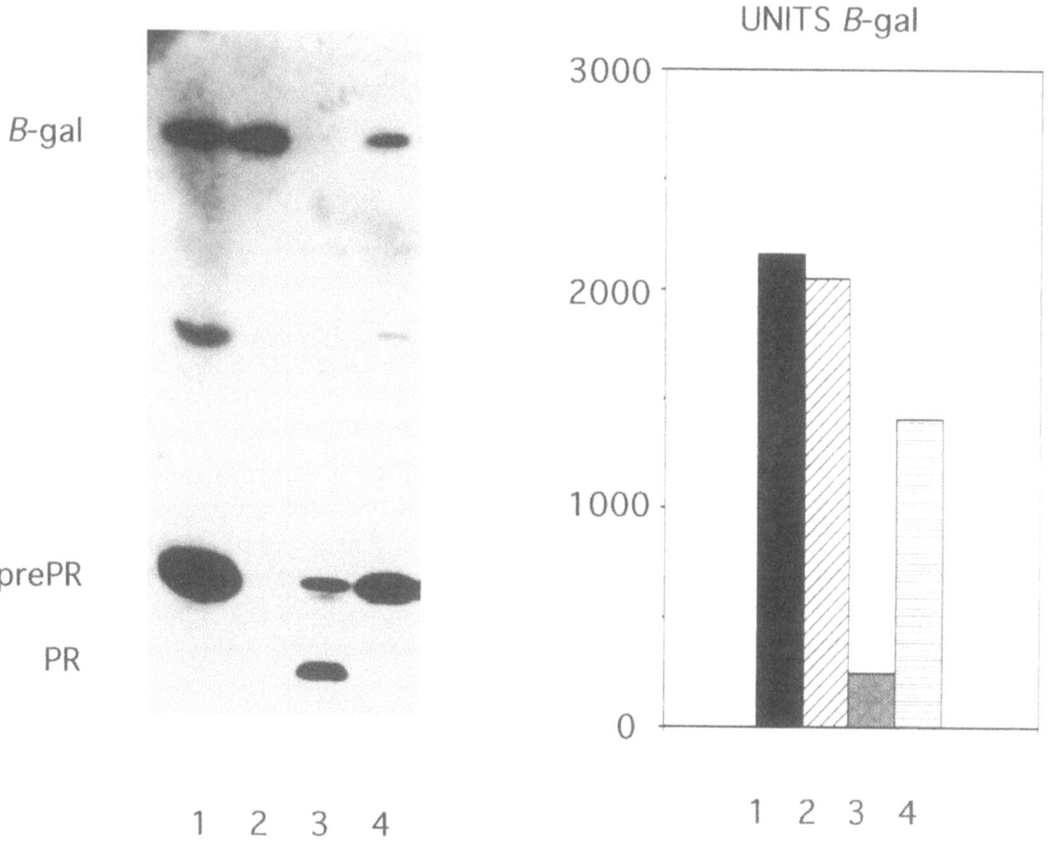

Fig. 2. Western blot and β-gal assay results showing effects of HIV-PR on β-galactosidase production in *E. coli* with or without HIV-PR inhibitors present. Left panel shows Western blotted proteins immunostained with antibodies to both β-galactosidase and HIV protease and detected using enhanced chemiluminescence. Right panel is measurement of units of β-galactosidase based on change in OD575. Lane 1 – BL21(DE-3) transformed with expression plasmid containing the inactive protease mutant asp25 -› gly. Lane 2 – BL21(DE-3) untransformed and treated with 2 µg/ml Ro-31-8959 (Roberts 1990). Lane 3 – BL21(DE-3) transformed with wild-type protease expression plasmid. Lane 4 – BL21(DE-3) transformed with wild-type protease and treated with 2 µg/ml Ro-31-8959.

Troubleshooting

Triplicate wells of each treatment should be done to evaluate reproducibility and reliability. Western blots may be useful in monitoring protease expression level and auto-processing. Freezing aliquots of culture of a single clone as inoculum for daily cultures may reduce variability in the induction and activity levels of the protease.

Subprotocol 3
Assays of Mammalian Cells Transiently Expressing HIV PR and Either Alkaline Phosphatase (AP) or Firefly Luciferase (luc)

HIV infects human lymphocytes, therefore we devised a method for measuring the activity levels of HIV protease in mammalian fibroblasts and lymphocytes. This strategy relies on co-transfection of cells with vectors that will express HIV PR and a reporter gene. We chose vectors which contain the HCMV IE promoter for unregulated transcription of the cloned gene. The vector CMVSEAP (TROPIX, Bedford, MA) features a secreted form of alkaline phosphatase, which allows *in situ* assays to be performed (Cullen 1992, Bronstein 1994).

1. Transfect tissue cultures with appropriate plasmid vectors by electroporation, lipofection or other means.

2. Culture transfected cells +/- HIVPR inhibitor.

3. Collect medium and cells separately; lyse cells

4. Assay for AP or luc activity

Materials

- Protease and reporter expression plasmids
- Cultured cell lines (e.g. COS-7 or MT2)
- Electroporator (e.g. Bio-Rad Gene Pulser) and sterile electroporation chambers (e.g. Bio-Rad #165-2088).
- Reduced serum medium (Opti-MEM I, GIBCO/BRL #31985-021)
- Assay reagents for luciferase (Promega #E1500) and alkaline phosphatase (Tropix #BP100)
- Luminometer and 96 well microtiter plates suitable for luminescence assays (Dynatech Labs)

Procedure

The following protocol has been optimized for transfection of COS 7 (MK) and 293 (HEK) cell lines. Conditions for other cell lines must be determined empirically.

1. Grow in T 150 flasks until cells are approximately 60-80% confluent

2. 4×10^6 cells should be used per transfection

3. Prepare cells for electroporation by washing briefly 2X with 10 mls of PBS, detach cells with 2.5 mls. of trypsin/EDTA for 5 minutes. Resuspend the cells in 10 mls of growth media, and centrifuge cells at 1500 RPM for 3' at room temperature in sterile centrifuge tubes.

4. Wash cell pellet 2X with PBS by resuspending and centrifuge as above.

5. Resuspend cells in 10 ml of Opti-MEMI (reduced serum medium OM-1). Gently pass through a sterile syringe with 18g needle to disrupt clumps if necessary.

6. Count and determine number of cells/ml and divide into 4×10^6 cell aliquots for electroporation.

7. Centrifuge cells as above, discard supermatant

8. Resuspend cells in 60 µl of OM-1 (total volume with pellet is approximately 80 µl).

Preparation of cells

1. Transfer to sterile electroporation chamber and add plasmid DNA: 1µg of reporter + 5 µg of protease for co-transfection, 1 µg of reporter + 5 µg of parent vector as control. NOTE: Only high purity DNA preps should be used (e.g. CsCI or Qiagen purified).

2. Adjust volume to 100 µl with OM-1

3. Set parameters for the electric charge to be delivered for gene pulser: capacitance = 500 µF, voltage = 150 volts.

4. Place chamber in electroporator and deliver charge

5. Incubate the cells 10' at room temperature

6. Dilute cells in 10ml of growth media and transfer to culture dishes as follows: 100l/well for 96 well dishes (approx. 4×10^4 cells/well); 2 mls./well for 6 well dishes (approx. 8×10^5/well). In some experiments HIV

Electroporation of cells

protease inhibitors may be included in growth media if desired. Assay cells for enzymatic activity at various times after transfection (for example, 6, 24, and 48 hours).

Luciferase assay

1. Remove all medium

2. Add 50 µl of reporter lysis buffer per well and incubate 10' at room temperature.

3. Pipet several times to dislodge all cells and transfer 20µl to microtiter 96 well luminometer plates. Alternatively, lysate may be stored at -80°C for several weeks.

4. Add 80 µl of luciferase assay reagent which contains luciferin, the substrate for luciferase.

5. Load samples into luminometer for measurement of light produced.

Alkaline phosphatase assay

1. Remove 50 µl aliquots of medium from transfected cells and place in 1.5 ml microfuge tubes. (triplicate samples are suggested).

2. Add 150 µl of dilution buffer and incubate at 65°C for 30'. (Eppendorf repeater pipettors are convenient for large numbers of samples).

3. Cool to room temperature

4. Add 50 µl of assay buffer and incubate 5 minutes at room temperature.

5. Add 50 µl of reaction buffer containing CSPD substrate and incubate 20 minutes at room temperature.

6. Load samples into luminometer and measure light produced for a 5-second interval.

Note: Alternative procedures for measuring alkaline phosphatase are commercially available. For example, SIGMA procedure #DG1245 uses the chromogenic substrate p-nitrophenyl phosphate. Alkaline phosphatase activity is accurately measured by absorbance change at 405nm in a conventional spectrophotometer (no luminometer required).

Results

As seen in figure 3, co-expression of active HIV-1 protease together with reporters in cells reduced the enzymatic activity of luciferase and alkaline phosphatase compared to cells expressing only the reporter. Firefly lucifer-

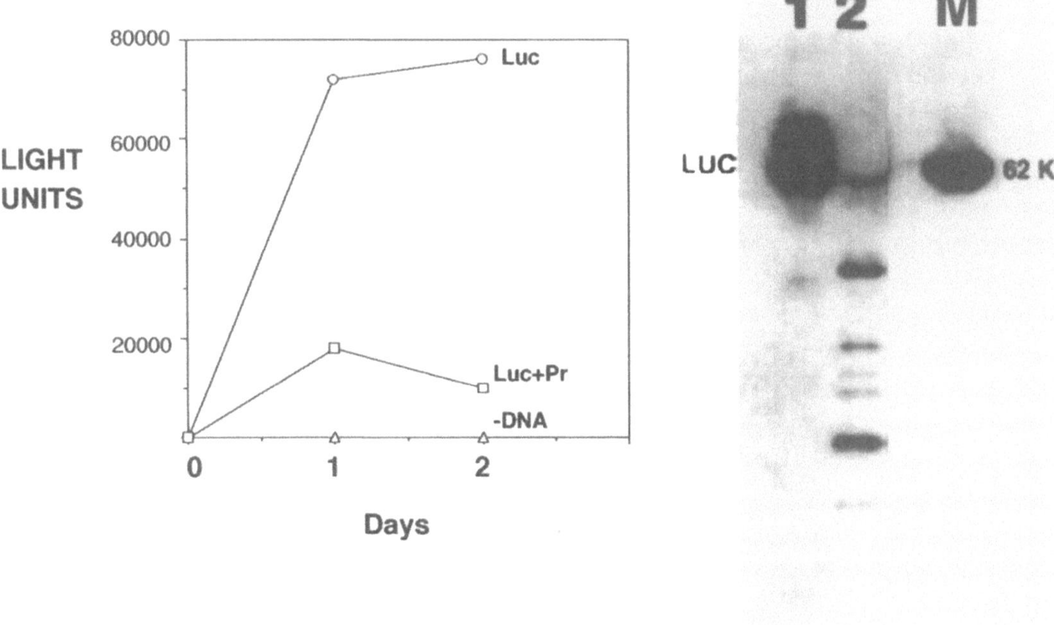

Fig. 3. Luciferase assay and Western blot results from a luciferase/protease co-transfection experiment. Expression of the active HIVPR gene (5 μg of plasmid DNA) reduced luciferase activity by nearly 90% at 2 days after transfection. Western blot data shows that luciferase expressed in COS-7 cells (lane 1) is a single 62 KDa band when immunostained with a rabbit polyclonal antibody (East Acres Biologicals #RaLuc-gG). By day 2 in COS-7 co-transfected with HIVPR (lane 2) luciferase had been hydrolyzed into several smaller products.

ase was cleaved by the HIV-PR into specific fragments. Culturing co-transfected cells in the presence of an HIV-1 protease inhibitor restored some of the reporter activity. Substitution of a plasmid expressing an inactive HIV-PR (asp25 -> gly) resulted in reporter levels similar to the reporter alone.

Troubleshooting

Using cultures that have reached confluency reduces transfection efficiency. Repeated exposure to light and air causes a pH change and concurrent color change in OM-1 which can reduce transfection efficiency. Successful transfection of the human T-lymphocyte cell line MT2 required altered conditions. Washes and incubations were done at 4°C and voltage delivered was increased to 200 volts. Lower voltages resulted in reduced transfection efficiency and reporter signal. Above 200 volts there were unacceptably low levels of cell viability.

References

Baum EZ, Bebernitz GA and Gluzman Y (1990) β-Galactosidase containing a human immunodeficiency virus protease cleavage site is cleaved and inactivated by human immunodeficiency virus protease. Proc. Natl. Acad. Sci. USA 87:10023-10027.

Bazan, JF and Fletterick, RJ (1988) Viral cysteine proteases are homologous to the trypsin-like family of serine proteases: structural and functional implications. Proc. Natl. Acad. Sci. USA 85:7872-7876.

Bronstein I, Fortin J, Stanley PT, Stewart GSAB, Kricka LJ (1994) Chemiluminescent and bioluminescent reporter gene assays. Analyt. Biochem 219:169-181.

Cheng Y-SE, McGowan MH, Kettner CA, Schloss JV, Erickson-Viitanen S, Yin FH (1990a). High-level synthesis of recombinant HIV-1 protease and the recovery of active enzyme from inclusion bodies. Gene, 87:243-248.

Cheng Y-SE, Yin FH, Foundling S, Blomstrom D, Kettner CA (1990b) Stability and activity of human immunodeficiency virus protease: comparison of the natural dimer with a homologous, single-chain tethered dimer. Proc Natl Acad Sci USA 87:9660-9664.

Cullen B, Malim M (1992) Secreted placental alkaline phosphatase as a eukaryotic reporter gene in: Methods Enzymol, 216: 362-368.

Eustice D, Feldman P, Colberg-Poley A, Buckery R, Neubauer R (1991) A sensitive method for detecting β-galactosidase in transfected mammalian cells. Biotechniques 11:739-742.

Flexner C, Broyles SS, Earl P, Chakrabarti S and Moss B (1988) Characterization of human immunodeficiency virus gag/pol gene products expressed by recombinant vaccinia viruses. Virology 166:339-349.

Hanahan D (1983) Studies on transformation of E. coli with plasmids. J. Mol Biol 166:557-580.

Hostomosky Z, Appelt K and Ogden RC (1989) High-level expression of self-processed HIV-1 protease in Escherichia coli using a synthetic gene. Biochemical and Biophysical Research Communications 161:1056-1063.

Korant BD (1994) in Biological Functions of Proteases and Inhibitors (Katumuma N, Suzuki K, Travis J, and Fritz H, eds.) 149-160, Karger, Basel.

Lam P, Jadhav PK, Eyerman CJ, Hodge CN, Ru Y, Bachelor LT, Meek JL, Otto MJ, Rayner MM, Wong NY, Chang C-H, Weber PC, Jackson DA, Sharpe TR, Erickson-Viitanen S (1994) Cyclic, non-peptide HIV protease inhibitors: "de novo" design and characterization of potent, orally bioavailable cyclic ureas. Science 263:380-384.

Murray MG, Hung W, Sadowski I, Das Mahapatra B (1993) Inactivation of a yeast transactivator by the fused HIV-1 proteinase: a simple assay for inhibitors of the viral enzyme activity. Gene 134:123-128.

Rizzo, CJ, Korant BD (1994) Genetic approaches designed to minimize cytotoxicity of a retroviral protease. Methods Enzymol 241:16-29

Roberts NA, Martin JA, Kinchington D, Broadhurst aV, Crain JC, Duncan IB, Galpin SA, Handa BK, Kay J, Kröhn A, Lambert RW, Merrett JH, Mills JS, Parkes KEB, Redshaw S, Ritchie AJ, Taylor DL, Thomas GJ, Machin PJ (1990) Rational design of peptide-based HIV proteinase inhibitors. Science 248:358-361.
Studier RW, Moffatt B (1986) Use of bacteriophage T7 RNA polymerase to direct selective high-level expression of cloned genes. J. Mol Biol 189:113-130.
Tumminia SJ, Jonak GJ, Focht RJ, Cheng, Y-S, Russell P. Cataractogenesis in transgenic mice containing the HIV-1 protease linked to the lens α-crystallin promoter. (1996) J. Biol. Chem. 271:425-433.
Wood KV (1991) in: Stanley P and Kricka L (ed.) Bioluminescence and Chemiluminescence: Current Status. 11-14 and 543-546 John Wiley and Sons, Chichester.

Suppliers

BIO-RAD LABORATORIES, 2000 Alfred Nobel Drive, Hercules CA, 94547, USA

NOVAGEN, INC., 597 Science Drive, Madison WI, 53711, USA

BOEHRINGER MANNHEIM CORPORATION, 9115 Hague Road, P.O. Box 50414, Indianopolis IN, 46250-0414, USA

PHARMACIA BIOTECH, 800 Centennial Avenue, P.O. Box 1327, Piscataway NJ, 1327, USA

CLONTECH LABORATORIES, INC., 1020 East Meadow Circle, Palo Alto CA, 94303-4230, USA

PROMEGA CORP., 2800 Woods Hillow Road, Madison WI, 53711-5399, USA

DYNATECH LABORATORIES, INC., 14340 Sullyfield Circle, Chantilly VA, 22021, USA

SIGMA CHEMICAL CO., P.O. Box 14508, St. Louis MO, 63178, USA

GIBCO/BRL (LIFE TECHNOLOGIES), Gaithersburg MD, 20884-9980, USA

TROPIX, INC., 47 Wiggins Avenue, Bedford MA, 01730, USA

MILLIPORE PRODUCTS, 80 Ashby Road, Bedford MA, 01730, USA

Chapter 7

Strategies for Inhibiting Proteases of Unknown Mechanism

RUSSELL L. WOLZ

Introduction

Hydrolysis of proteins was one of the first catalytic functions attributed to enzymes. Since the discovery of trypsin by W. Kühne more than a century ago (Kühne, 1886), the study of proteases has continued to be at the forefront of research in biochemistry. Proteolytic activities are now known to be involved in nearly every type of physiological process including simple digestion, enzyme activation, blood coagulation, blood pressure regulation, bone growth, embryonic development, snake venom toxicity, and neuroparalytic diseases, to mention just a few. To fully understand the molecular and/or physiological functions of a protease, it is important to know how to inhibit its activity. This chapter will provide some strategies and methodology for characterizing a protease through its inhibition. At present, there are four known catalytic mechanisms by which an enzyme can cleave a peptide bond. These are named after the chemical group most directly involved in bond cleavage and comprise

- the serine proteases,
- the cysteine proteases,
- the aspartic proteases, and
- the metalloproteases.

Class specific inhibitors are known for each of these mechanistic classes, thus when characterizing a protease of unknown mechanism, the obvious strategy is to test the effect of inhibitors from each class. There are, however, cases when this simple strategy appears to fail. This chapter will describe the use of inhibitors for each of the mechanistic classes, including suggestions for overcoming or avoiding such possible failures.

Russell L. Wolz, Commonwealth Biotechnologies, Inc. 601 Biotech Drive Richmond VA, 23235, USA 8007359224 or 8046483820 8046482641 rwolz@cbi-biotech.com)

For the purpose of this chapter, it will be assumed that the researcher already has a method for measuring the activity of the protease in question. There are, of course, many cases where the protein in question is a putative protease i.e. that its amino acid sequence is homologous to a protein with demonstrated proteolytic and/or peptidolytic activity, but that such activity has not actually been observed in vitro. The identification of a suitable substrate presents another problem which will not be addressed in detail here, although inhibition by certain reagents can also be useful in determining substrate specificity.

Two general experimental protocols will be given. Sections will then be presented for each mechanistic class of protease. Each section will begin with a general statement, then specific information about various inhibitors will be summarized in the following way (Table 1).

There are two general approaches to inhibition of an enzyme and subsequent assay for residual activity: **General protocols**

a. Preincubate a concentrated enzyme solution with inhibitor, remove an aliquot containing an amount of enzyme appropriate for assay, and begin the assay reaction by adding the inhibited enzyme aliquot to a buffered solution of substrate

b. Preincubate the enzyme (at an appropriate assay concentration) with inhibitor, then begin the assay reaction by adding substrate.

Both approaches have advantages.

1. Preincubation of a concentrated enzyme solution has the advantage of providing multiple aliquots of exactly the same sample which can be
   • assayed at various times to follow the time-course of inactivation,
   • assayed in conditions which are different from the preincubation conditions (e.g. different pH or lower concentration of organic solvent), or
   • tested for reversal of inhibition.

2. Preincubation of enzyme at assay concentration has the advantages of
   • allowing variation of the preincubation conditions while conserving enzyme
   • insuring identical conditions in the preincubation and assay.

With both protocols, the final concentration of enzyme in the assay should be high enough to give a strong activity in the uninhibited control. The wider the difference between the uninhibited control and the inhibited sample, the more significant the result. Also, as in all scientific experiments,

Table 1. Characteristics of inhibitors

Compound abbreviation (full name and synonyms)	
Mechanism of inactivation	The mechanism of inactivation is given if known.
Specificity	A statement about the specificity of the given inhibitor within its class, as well as effects on other types of proteases or other enzymes. The most general inhibitors for each mechanistic class are given first, then others are listed in order of increasing selectivity for proteases within that class.
Comments	Other infomation about the reagent or its use
Solutions	
FW	Formula Weight
Solubility	Solubility in given solvents. The limit of solubility is given if known.
Stock solution	Suggested stock solution concentration given in molarity and mg/ml
Stability	Stability of the stock solution under given conditions
Conditions for use	
Preincubation concentration	Inhibitor concentration (usually given as a range) to be used during pre-incubation of the protease with the reagent, and during activity assay as noted.
Preincubation conditions	Special conditions are given if needed for preincubation and/or assay. No entry indicates that the same conditions may be used for preincubation and assay.
Preincubation time	Time required for pre-incubation of the protease with the reagent to give essentially complete inactivation. If partial inhibition is observed and full inhibition is desired, the pre-incubation time should be increased.
Availability	Availability is given from Bachem, Boehringer, Enzyme Systems Products, or Sigma, but reagents are also available from other sources.
References	For most inhibitors, the original reference, or earliest readily available reference is given. For some more general reagents, such as EDTA, a review article appropriate for inhibition of proteases is cited.

proper controls must be performed including control preincubations and control assays. Notably, many inhibitors need to be dissolved in organic solvents such as alcohol, DMF, DMSO, or acetonitrile. The presence of organic solvent in the assay often significantly inhibits protease activity. It is therefore necessary to include control preincubations and/or assays in the absence of inhibitor, and in the absence and presence of organic solvent.

Preincubation of Concentrated Protease Solution

The preincubation enzyme concentration should be 10 - 100 times higher than the final assay concentration. Assay concentrations for proteases usually range from 0.1 to 10 g/ml, thus preincubation concentrations can range from 1 g/ml to 1 mg/ml.

Enzyme concentration

The preincubation buffer should be the same as the assay buffer if possible, however for some inhibitor/protease combinations, the preincubation and assay conditions must be different. Such cases are noted in the Preincubation conditions subheading for each inhibitor when needed. Many inhibitors are soluble only in the presence of organic solvent. The concentration of organic solvent should be kept to a minimum. In the preincubation, the concentration of organic solvent should be less than 10% to avoid denaturation of the protein. If the inhibitor is an irreversible inhibitor, no additional organic solvent needs to be included in the buffer for the ensuing assay. Thus, when the concentrated preincubation aliquot is added to the substrate solution for assay, the organic solvent is diluted to less than 1%.

Preincubation conditions

Prepare the assay reaction mixture to include buffer, substrate, and any other required components (except enzyme). If the inhibitor is an irreversible inhibitor, no additional inhibitor or organic solvent needs to be included in the buffer for the assay. If the inhibitor is a reversible inhibitor, the assay mixture must also contain inhibitor at the same final concentration in the assay as was used in the preincubation. This prevents reactivation due to dissociation of the inhibitor from the enzyme.

Assay conditions

Remove an aliquot from the preincubation mixture which contains the appropriate amount of enzyme for one assay, and add it to the assay mixture to begin the reaction.

Assay

Preincubation of Protease at Assay Concentration

Enzyme concentration
Preincubation buffer

The final concentration of enzyme in the assay should be high enough to give a strong activity in the uninhibited control. Assay concentrations for proteases usually range from 0.1 to 10 g/ml. The preincubation buffer is the same as the assay buffer. If it was necessary to use organic solvent to dissolve the inhibitor, the final assay concentration of the solvent should be kept to less than 1% if possible.

Assay

Add concentrated substrate to the preincubated enzyme to begin the assay reaction. The volume of concentrated substrate added should be less than 5% of the final assay volume (less than 50L added per ml of assay reaction). If the added volume is greater than 5%, and the inhibitor is a reversible inhibitor, then it is necessary to include inhibitor in the stock substrate solution so that the final concentration of inhibitor in the assay is the same as was used in the preincubation.

Serine Protease Inhibitors

Covalent irreversible inhibitors for serine proteases and other hydrolases with a serine-based mechanism (e.g. acetylcholine esterase) have long been known. Although much is also known about reversible inhibitors for this class of enzyme, this section will include only the covalent type. Irreversible inhibition provides a more conclusive result by limiting the chances for reactivation. Furthermore, covalently modified active site residues can be isolated and identified to more thouroghly characterize the enzyme.

Table 2. DFP (diisopropyl fluorophosphate)

Mechanism of inactivation	Irreversible, covalent modification of the active site serine.
Specificity	General inhibitor of all serine-based hydrolases, including acetylcholine esterase. May also inhibit some cysteine peptidases, but in that case, inhibition can be reversed by the addition of thiol reagents such as 2-mercaptoethanol.
Comments	Potent neurotoxin due to inhibition of acetylcholine esterase. (Wilson and Walker, 1974). Use with caution!
Solutions	
FW	184.1
Solubility	Readily soluble in propanol

Table 2. Continuous

Stock solution	200 - 500 mM (36.8 - 92.1 mg/ml) in propanol
Stability	Stable in propanol at -70C
Conditions for use	
Preincubation concentration	100 M (18.3 g/ml)
Preincubation time	30 min
Availability	Sigma (not-mailable)
Reference	Cohen et al., 1967

Table 3. PMSF (phenylmethylsulfonylfluoride)

Mechanism of inactivation	Irreversible, covalent modification of the active site serine.
Specificity	General inhibitor of all serine proteases. Does not inhibit serine protease zymogens. May also inhibit some cysteine proteases, but in that case, inhibition can be reversed by the addition of thiol reagents such as 2-mercaptoethanol.
Comments	Safer alternative to DFP.
Solutions	
FW	174.2
Solubility	Soluble to > 10 mg/ml (57.4mM) in alcohols (ethanol, methanol, propanol); not appreciably soluble in H_2O
Stock solution	10 mM (1.7 mg/ml)
Stability	Unstable in aqueous solution, but stable in alcohol for at least 9 months at 25C
Conditions for use	
Preincubation concentration	100 - 1000 M (17 - 174 g/ml)
Preincubation time	30 min
Availability	Bachem, Boehringer, Sigma
Reference	Fahrney and Gold, 1963

Table 4. DCI (3,4-dichloroisocoumarin)

Mechanism of inactivation	Acylates the active site serine to form a reactive acyl-chloride, which then esterifies a second site residue in or near the active site (probably histidine).
Specificity	Does not inhibit cysteine proteases, metalloproteases, acetylcholine esterase, or b-lactamase
Comments	Rapidly reversible in the presence of hydroxylamine, or in 24-100 hours in the absence of hydroxylamine.
Solutions	
FW	215.0
Solubility	Soluble in DMSO or DMF. Sparingly soluble in H_2O
Stock solution	10 mM (2.2 mg/ml)
Stability	Stable in DMF at -20C. Unstable in H2O and aqueous buffers.
Conditions for use	
Preincubation concentration	5 - 200 M (1.1 - 44 g/ml)
Preincubation time	30 min
Availability	Boehringer, Sigma
Reference	Harper et al., 1985

Table 5. TPCK (N-tosylamidophenylalanine chloromethyl ketone), also named L-1-chloro-3-tosylamido-4-phenyl-2-butanone

Mechanism of inactivation	Irreversible, covalent modification of a His57 (of the catalytic triad) in chymotrypsin.
Specificity	Specific for chymotrypsin and serine peptidases with chymotrypsin-like specificity. Does not inhibit trypsin or serine proteases with trypsin-like specificity.
Comments	
Solutions	
FW	351.8
Solubility	Soluble in ethanol to 20 mg/ml
Stock solution	30 mM (10.5 mg/ml)
Stability	Stable in aquous solution at pH 6

Table 5. Continuous

Conditions for use	
Preincubation concentration	300 M (105 g/ml)
Preincubation time	60 min.
Availability	Bachem, Boehringer, Sigma
Reference	Schoellmann and Shaw, 1963

Table 6. TLCK (N-tosyl-lysine chloromethyl ketone)

Mechanism of inactivation	Irreversible, covalent modification of His57 (of the catalytic triad).
Specificity	Specific for trypsin and other serine proteases with a primary specificity for basic groups (Arg or Lys) in P1. Does not inhibit chymotrypsin. Also can inhibit some cysteine proteases including papain and ficin. (Stein and Liener, 1967)
Comments	
Solutions	
FW	333.8 (free acid)
Solubility	Salts soluble in H_2O to 20 mg/ml (60 mM)
Stock solution	15 mM (5.0 mg/ml)
Stability	Stable in aqueous solution at pH 6
Conditions for use	
Preincubation concentration	150 M (50 g/ml)
Preincubation time	60 min.
Availability	Bachem, Boehringer, Sigma
Reference	Shaw et al., 1965

Cysteine Protease Inhibitors

One general requirement for the activity of cysteine proteases is that the active site cysteine must be in its reduced form. This can be accomplished by pre-activation of the enzyme with thiol reducing agents such as 2-mercaptoethanol, dithiothreitol, or free cysteine, or by inclusion of these low molecular weight reducing agents in the assay mixture. If the peptidase is found to be active only under reducing conditions, it could be an indication that the peptidase is of the cysteine class. The researcher must also be aware, however, that some metallo- and serine peptidases also show a form of dependence on thiol reducing agents. This can cause misinterpretation of inhibition/activation data unless other observations are taken into account. For example, metalloproteases, whether thiol dependent or not, are inhibited by high concentrations (2-20mM) of thiol-containing compounds. A few metallopeptidases, however, are activated by low concentrations (10-100 M) of the same reagents. Thus, for example, if a low concentration of dithiothreitol (DTT) activates a protease, but a high concentration of DTT is inhibitory, then the protease might be a thiol-dependent metalloprotease (such as thimet) rather than a cysteine protease.

The reagents listed in this section are covalent inhibitors which can react with the reduced active site cysteine residue, but which (as indicated) may also react non-specifically with other reduced cysteine residues in the protein, or with free cysteine or other low molecular weight thiol compounds. Thus, to avoid side-reactions which would destroy the non-specific reagents, the putative cysteine peptidase must first be pre-incubated with the thiol activator, then this enzyme activator must be removed (by rapid ultrafiltration or size exclusion chromatography) or significantly diluted before the attempted enzyme modification.

Table 7. IAA (iodoacetic acid) and IAM (iodoacetamide)

Mechanism of inactivation	Covalent alkylation of the active site cysteine residue.
Specificity	Reacts with reduced cysteine residues in any protein.
Comments	Also reacts with low molecular thiol-containing compounds such as 2-mercaptoethanol, dithiothreitol, or free cysteine.
Solutions	
FW	IAA 185.9 IAM 185.0 IAM Na salt 207.9
Solubility	Soluble in H_2O or alcohol
Stock solution	10 mM (1.9 mg/ml)
Stability	Prepare fresh
Conditions for use	
Preincubation concentration	100 M

Table 7. Continuous

Preincubation conditions	Thiol activators must be removed or diluted before incubation with the inhibitor.
Preincubation time	60 min.
Availability	Sigma
Reference	Gurd 1967

Table 8. E64 (L-trans-epoxysuccinyl-leucylamido-(4-guanidino)butane), also named L-3-carboxy-trans-2,3-epoxypropyl-leucylamido-(4-guanidino)butane, also named trans-3-carboxyoxiran-2-carbonyl-leucylagmatine

Mechanism of inactivation	Covalent modification of the active site cysteine residue
Specificity	Reacts rapidly with cysteine peptidases of the papain or calpain families. Reacts poorly with cysteine peptidases of the clostripain and streptopain families. Does not inhibit (ID50 > 250 g/ml) serine proteases or aspartic proteases. Not reactive with low molecular weight thiol reagents such as 2-mercaptoethanol.
Comments	Originally isolated as a natural product of the soil mold Aspergillus japonicus TPR-64.
Solutions	
FW	357.4
Solubility	Soluble in DMSO to 100 mM
Stock solution	Dissolve in pure DMSO to 100mM (35.7 mg/ml), then dilute with water to 1 mM (final 3.57 mg/ml)
Stability	Stock solution stable for several days at 4°C
Conditions for use	
Preincubation concentration	10^{-7} to 10^{-5} M or at least 5-times the enzyme concentration
Preincubation time	15 min.
Availability	Boehringer, Sigma
Reference	Hanada et al., 1978; Barrett et al., 1982

Table 9. Amino acid and peptide diazomethylketones:
Z-Phe-CHN$_2$ (benzyloxycarbonyl-phenylalanine diazomethylketone) Z-Lys-CHN$_2$ (benzyloxycarbonyl-lysine diazomethylketone) Z-Phe-Ala-CHN$_2$ (benzyloxycarbonyl-phenylalanyl-alanine diazomethylketone) Z-Phe-Phe-CHN$_2$ (benzyloxycarbonyl-phenylalanyl-phenylalanine diazomethylketone) Z-Leu-Val-Gly-CHN$_2$ (benzyloxycarbonyl-Leu-Val-Gly-diazomethylketone)

Mechanism of inactivation	Time and concentration dependent (see incubation concentrations) covalent modification of the active site cysteine.
Specificity	The reactivity depends on matching the inhibitor side chain(s) to the specificity of the putative cysteine protease. Examples are given below with the incubation concentrations. Metalloproteases are not inhibited. Inhibition of serine peptidases is possible, but rare. Aspartic proteases may be inhibited only if Cu(II) is present in the incubation (Delpierre and Fruton, 1966)
Comments	Optimally reactive at pH 6.0. Diazomethylketones are stable in the presence of thiol reducing compounds such as DTT and 2-mercaptoethanol.
Solutions	
FW	Z-Phe -CHN$_2$ 323.4 Z-Lys-CHN$_2$ 305.4 Z-Phe-Ala-CHN$_2$ 394.4 Z-Phe-Phe-CHN$_2$ 470.4 Z-Leu-Val-Gly-CHN$_2$ 445.5
Solubility	Soluble in DMSO or acetonitrile
Stock solution	10 mM
Stability	Crystaline solids are stable for years at room temperature, but solutions are much less stable. Stock solution in DMSO is stable 1 week at -20°C; Use freshly diluted.
Conditions for use	
Preincubation concentration	5×10^{-7} to 2.5×10^{-4} M (0.5 - 250 M), depending on protease specificity.
Preincubation time	Depends on protease specificity. The following are examples of inhibitor concentration and time required for 50% inactivation of two different cysteine proteases by the same peptide diazomethylketone, and the same protease by two different diazomethylketones (Green and Shaw, 1981): cathepsin B + 5×10^{-7} M Z-Phe-Ala-CHN2 $t_{1/2}$ = 18.5 min clostripain + 2.5×10^{-4}M Z-Phe-Ala -CHN2 $t_{1/2}$ > 1000 min. clostripain + 1.8×10^{-6}M Z-Lys-CHN2 $t_{1/2}$ = 14.1 min
Availability	Bachem, Enzyme Systems Products
Reference	Leary et al., 1977; Green and Shaw, 1981; Shaw and Green, 1981; Shaw, 1995

Aspartic Protease Inhibitors

The first indication that a protease might be of the aspartic type is if it is optimally active at low pH. This evidence alone is by no means conclusive and must be considered along with inhibition data. Both reversible and irreversible inhibitors of aspartic proteases will be presented here. The inhibitor considered to be most clearly diagnostic for an aspartic protease is pepstatin. Pepstatin is a tight binding reversible inhibitor which does not inhibit non-aspartic proteases, but which also does not inhibit some proteases with a low pH optimum (Barrett, 1995). Thus, inhibition by pepstatin is indicative of an aspartic mechanism, but lack of inhibition by pepstatin does not necessarily exclude an aspartic mechanism.

Table 10. Pepstatin

Mechanism of inactivation	Tight, but reversible binding. It is a transition state analog.
Specificity	Specific for most known aspartic proteases.
Comments	
Solutions	
FW	685.9
Solubility	Soluble in methanol to 1 mg/ml. Slowly (12 hr at room temperature) soluble in ethanol to 1mg/ml. Soluble in 6 M acetic acid to 0.3 mg/ml.
Stock solution	1 mg/ml in methanol or ethanol
Stability	Stable at least one week at 4°C
Conditions for use	
Preincubation concentration	1 M (0.7 g/ml) in preincubation and in assay.
Preincubation time	30 min.
Availability	Bachem, Boehringer, Sigma
Reference	Umezawa, 1976

Table 11. Diazomethylketones
DPTB (1-diazo-4-phenyl-3-tosylamidobutanone) DPB (1-diazo-4-phenylbutanone)
Z-Phe-CHN$_2$ (benzyloxycarbonyl-phenylalanine diazomethylketone)

Mechanism of inactivation	In the presence of Cu(II) ion, a carbene is generated from the diazomethyl ketone or ester. The carbene reacts with the carboxyl group of one of the active site Asp side chains (Asp215 in pepsin) to form a covalent adduct.
Specificity	Can inhibit cysteine proteases and aspartic proteases, but inhibits aspartic proteases only in the presence of Cu(II). Pepsinogen is not inhibited. Inhibition of serine peptidases is possible, but rare. Metalloproteases are not inhibited.
Comments	The reagents are reactive between pH 5 to pH 6. Thus, the inactivation must be performed in this pH range, but the activity assay should be done at the pH optimum for the uninhibited enzyme (e.g. pH 1.8 for optimum activity of pepsin)
Solutions	
FW	DPTB 343.4 DPB 174.2 Z-Phe -CHN$_2$ 323.4
Solubility	Soluble in ethanol, methanol, DMSO, or acetonitrile
Stock solution	1.5 mM in ethanol
Stability	Crystaline solids stable for years at room temperature, but solutions are much less stable. Stock solution in DMSO stable 1 week at -20°C. Use freshly diluted.
Conditions for use	
Preincubation concentration	150 M diazomethylketone with 1 mM CuCl$_2$
Preincubation time	60 min.
Availability	Bachem, Enzyme Systems Products
Reference	Delpierre and Fruton, 1966; Ong and Perlmann, 1967

Table 12. EPNP (1,2-epoxy-3-(p-nitrophenoxy)-propane)

Mechanism of inactivation	Two molecules of EPNP react with pepsin to modify two carboxyl groups on the enzyme. One carboxyl group is in the active site, but the second is apparently not required for enzyme activity. In HIV-protease, only one active site residue (Asp25) is modified.
Specificity	Specific for aspartic proteases if the preincubation is done at pH 5 to 6. At higher pH, epoxides may also react with cysteine proteases (see E-64) and other enzymes with catalytic cysteine residues. EPNP is a substrate for some glutathione reductases.

Table 12. Continuous

Comments	The EPNP content of the modified protein can be determined spectrophotometrically by absorbance at 315 nm. The bound EPNP groups can be released from the protein at alkaline pH.
Solutions	
FW	195
Solubility	Limited solubility in H_2O; may be used in suspension Soluble in 10% DMSO
Stock solution	2 mg/ml (10mM) in suspension
Stability	Stable as a dry solid
Conditions for use	
Preincubation concentration	0.2 mg/ml (1 mM) in suspension
Preincubation conditions	Preincubation should be between pH 5 and pH 6 to avoid reaction with cysteine residues.
Preincubation time	A minimum of 2 hours is probabaly required. It may require 24 hours or more for complete inactivation.
Availability	Sigma
Reference	Tang, 1971; Salto et al., 1994

Metalloproteases

Although irreversible inhibitors are known for some metalloproteases, there is so far no covalent reagent with broad enough reactivity to inhibit all or even most of the enzymes in this class. Thus, this section will present reversible inhibitors which, by various mechanisms, inactivate metalloenzymes by interacting with the catalytically essential metal (usually zinc). These reagents can, with varying affinity, also chelate other transition metals, and may lead to false determination of a protease mechanism. For example, some serine- and cysteine proteases (e.g. subtilisin and calpain, respectively) require Ca^{2+} for full activity, and thus are inhibited by EDTA [Beynon]. 1,10-phenanthroline, on the other hand, has approximately a 10^6-fold weaker affinity for calcium than for zinc, and therefore inhibits zinc-enzymes, but does not affect Ca^{2+} dependent proteases of other classes.

Table 13. ortho -Phenanthroline (1,10-phenanthroline)

Mechanism of inactivation	Can simply chelate free metal in solution, or may actively extract metal from the active site by formation of transient ternary enzyme-metal-phenanthroline complex, followed by dissociation of chelated metal from the complex to leave inactive apo-enzyme.
Specificity	Can inhibit all metalloproteases and other metalloenzymes. Binds Zn_2+ more tightly than Ca_2+, therefore will not inhibit non-metalloproteases which may require Ca_2+ for stability and/or full activity. Affinities for other metals are summarized in Auld (1988).
Comments	
Solutions	
FW	monohydrate 198.2
Solubility	Slowly soluble in H_20 to 10mM. Directly soluble in alcohol, acetone, or DMSO to $>$ 100 mM.
Stock solution	10 mM in aqueous buffer or 100 mM in organic solvent.
Stability	Aqueous solution stable at least one week at room temperature. Stable for months at $-20°C$ in organic solvent
Conditions for use	
Preincubation concentration	1–10 mM in preincubation and in assay
Preincubation conditions	30 min.
Preincubation time	Sigma
Availability	Auld, 1988; Auld, 1995

Table 14. EDTA (ethylenediaminetetraacetic acid, also named edetic acid)

Mechanism of inactivation	Chelates free metal in solution (usually)
Specificity	Can inhibit all metalloproteases and other metalloenzymes. Forms complexes with Ca_2+, Cu_2+, Fe_2+, Pb_2+, Mg_2+, Mn_2+, Ni_2+, and Zn_2+. Affinities for metals are summarized in Auld (1988). Can inhibit Ca_2+ dependent serine- and cysteine proteases.

Table 14. Continuous

Comments	
Solutions	
FW	free acid 292.2 Na$_2$EDTA(2H$_2$O) 372.2
Solubility	Soluble in H$_2$O to 500 mM at pH 8-9. The disodium salt will not dissolve unless the pH of the solution is adjusted to near pH 8 with NaOH.
Stock solution	500 mM. Dissolve 18.6 g Na$_2$EDTA(2H$_2$O) in 70 ml H$_2$O, adjust pH to 8.0 with 10 N NaOH (approx. 5 ml), then add H$_2$O to a final volume of 100ml.
Stability	Stable in solution at 4C for at least 6 months.
Conditions for use	
Preincubation concentration	5 -10mM (1.9 - 3.7 mg/ml) in preincubation and in assay.
Preincubation time	30 min. for most metalloenzymes, but sometimes may require days. (Stöcker et al., 1988).
Availability	Boehringer, Sigma
References	Auld, 1988; Auld, 1995

Table 15. Amino acid hydroxamates

Mechanism of inactivation	The hydroxamate group displaces H$_2$O from the active site, and complexes with the metal to form a stable ternary enzyme-metal-hydroxamate complex. Readily reversible by dilution.
Specificity	Inhibition is improved by matching the hydroxamate side chain with the substrate specificity of the protease.
Comments	Useful aid for determining substrate specificity
Solutions	
FW	FW of amino acid plus 15
Solubility	Most amino acid hydroxamates are soluble to 10mM in H$_2$O. Tyr-NHOH and Ile-NHOH are soluble to 1mM. Leu-NHOH has limited solubility (100 M)
Stock solution	10 mM in aqueous buffer
Stability	Stock solutions stable at -20°C.

Table 15. Continuous

Conditions for use	
Preincubation concentration	1 to 5 mM in assay
Preincubation conditions	Optimal inhibition from pH 7 to pH 9
Preincubation time	No preincubation required.
Availability	Sigma
References	Nishino and Powers, 1978

References

Auld, D.S., Methods Enzymol. 158, 110-114 (1988)

Auld, D.S., Methods Enzymol. 248, 228-242 (1995)

Barret, A.J., Kembhavi, A.A., Brown, M.A., Kirschke, H., Knight, C.G., Tamai, M., and Hanada, K., Biochem. J. 201, 189 (1982)

Cohen, J.A., et al., Methods Enzymol. 11: 686 (1967)

Delpierre, G.R., and Fruton, J.S., Proc. Natl. Acad. Sci. U. S. A. 56, 1817-1822 (1966)

Fahrney, D.E., and Gold, A.M., J. Am. Chem. Soc. 85, 997-1000 (1963)

Green, G.D.J., and Shaw, S., J. Biol. Chem. 256, 1923-1928 (1981)

Gurd, F.R., Methods Enzymol. 11, 532-541 (1967)

Hanada, K., Tamai, M., Yamagishi, M., Ohmura, S., Sawada, J., and Tanaka, I., Agric. Biol. Chem. 42, 523-528 (1978)

Harper, J.W., Hemmi, K., and Powers, J.C., Biochemistry 24:1831-1841 (1985)

Kühne, W., Verhandlung des Naturhist.-Med. Vereins zu Heidelberg. N.F. III. Bd. 4. Heft, 463-466 (1886)

Leary, R., Larsen, D., Watanabe, H., Shaw., E., Biochemistry 16, 5857-5861 (1977)

Nishino, N., and Powers, J.C., Biochemistry 17, 2846-2850 (1978)

Ong, E.B., and Perlmann, G.E., Nature 215, 1492-1494 (1967)

Salto, R., Babe, L.M., Li, J., Rose, J.R., Yu, Z., Burlingame, A., De Voss, J.J., Sui, Z., de Montellano. P.O., and Craik, C.S., J. Biol. Chem. 269, 10691-98 (1994)

Schoellmann, G., and Shaw, E., Biochemistry 2:, 252 (1963)

Shaw, E. and Green, G., Methods Enzymol. 80, 820 (1981)

Shaw, E., Meth. Enzymol. 244, 649-656 (1995)

Shaw, E., Mares-Guia, M., and Cohen, W., Biochemistry 4:, 2219-2224 (1965)

Stein, M.J. and Liener, I.E, BBRC 26, 376-382 (1967)

Stöcker, W., Wolz, R.L., Zwilling, R., Strydom, D.J., and Auld, D.S. (1988) Biochemistry 27, 5026-5032

Tang, J., JBC 246, 4510-4517 (1971)

Umezawa, H., Methods Enzymol. 45, 678 (1976)

Wilson, B.W. and Walker, C.R., Proc. Natl. Acad. Sci. U. S. A. 71, 3194-8 (1974)

Part II

Purification, Structure, Kinetics and Expression of Proteolytic Enzymes

Chapter 8

Purification of Proteases

NIGEL M. HOOPER

Introduction

There are no set rules for the purification of a protease. The particular chromatographic steps employed will depend on the properties of the protease in question. Conventional chromatographic procedures will exploit one or other of the basic physicochemical properties of the protein; its charge, its size, or its solubility. More specific chromatographic procedures will exploit the affinity of the protease for another molecule, usually either an antibody or a small molecule inhibitor. In general, the more one knows about a particular protease the likelier the possibility of being able to exploit an affinity chromatographic step. If the affinity of the protease for the ligand being used is both selective and strong enough, it may be possible to purify the enzyme in essentially a single step. This was the case for the purification of angiotensin converting enzyme (EC 3.4.15.1) from porcine brain striatum using the competitive inhibitor lisinopril as the affinity ligand, where the enzyme was purifed 53 500-fold with an overall yield of 13% (Hooper and Turner 1987). Unfortunately though, not many proteases are inhibited by such selective and potent compounds that are suitable as affinity chromatography ligands. Thus, in the first instance, conventional chromatographic procedures often have to be used and refined to purify the protease of interest.

In this chapter I cannot outline a blueprint for the purification of each and every protease. What I will do, however, is detail protocols which we have used successfully for the purification of a number of proteases, and which should provide the reader with a starting point from which to go on and modify the application for their particular protease. The techniques

Nigel M. Hooper, The University of Leeds, School of Biochemistry and Molecular Biology, Leeds, LS2 9JT, U.K. (*phone* +44 113 233 3163; *fax* +44 113 233 3167; *e-mail* n.m.hooper@leeds.ac.uk)

that I will address include ion exchange chromatography, hydrophobic interaction chromatography and immunoaffinity chromatography which are generally applicable to most proteases, assuming, in the latter case, that an antibody to the protease of interest is available. I will also describe an inhibitor affinity chromatography procedure which, although because of the ligand involved is only directly applicable to one enzyme, can readily be adapted for other ligands.

Examples of proteases purified by the methods described herein, with the relevant primary reference, are provided in Table 1.

Table 1. Summary of the proteases used as examples and their modes of solubilization and purification.

Protease	Mode of membrane solubilization	Mode of purification	Reference
X-Pro aminopeptidase (EC 3.4.11.9)	PI-PLC	conventional immunoaffinity	Hooper et al 1990 Lloyd et al 1996
Membrane dipeptidase (EC 3.4.13.19)	PI-PLC Detergent (n-octyl-β-D-glucopyranoside)	inhibitor affinity inhibitor affinity	Littlewood et al 1989 Hooper and Turner 1989
Angiotensin converting enzyme (EC 3.4.15.1)	Trypsin Detergent (Triton X-100)	inhibitor affinity inhibitor affinity	Hooper et al 1987 Hooper et al 1987
Neprilysin (EC 3.4.24.11)	Detergent (Triton X-100)	immunoaffinity	Relton et al 1983
Glutamyl aminopeptidase (EC 3.4.11.7)	Trypsin	conventional	Hesp and Hooper 1997

Outline

A summary of the procedures which I will be describing for the purification of both soluble and membrane-bound proteases from a suitable starting tissue is presented in Figure 1.

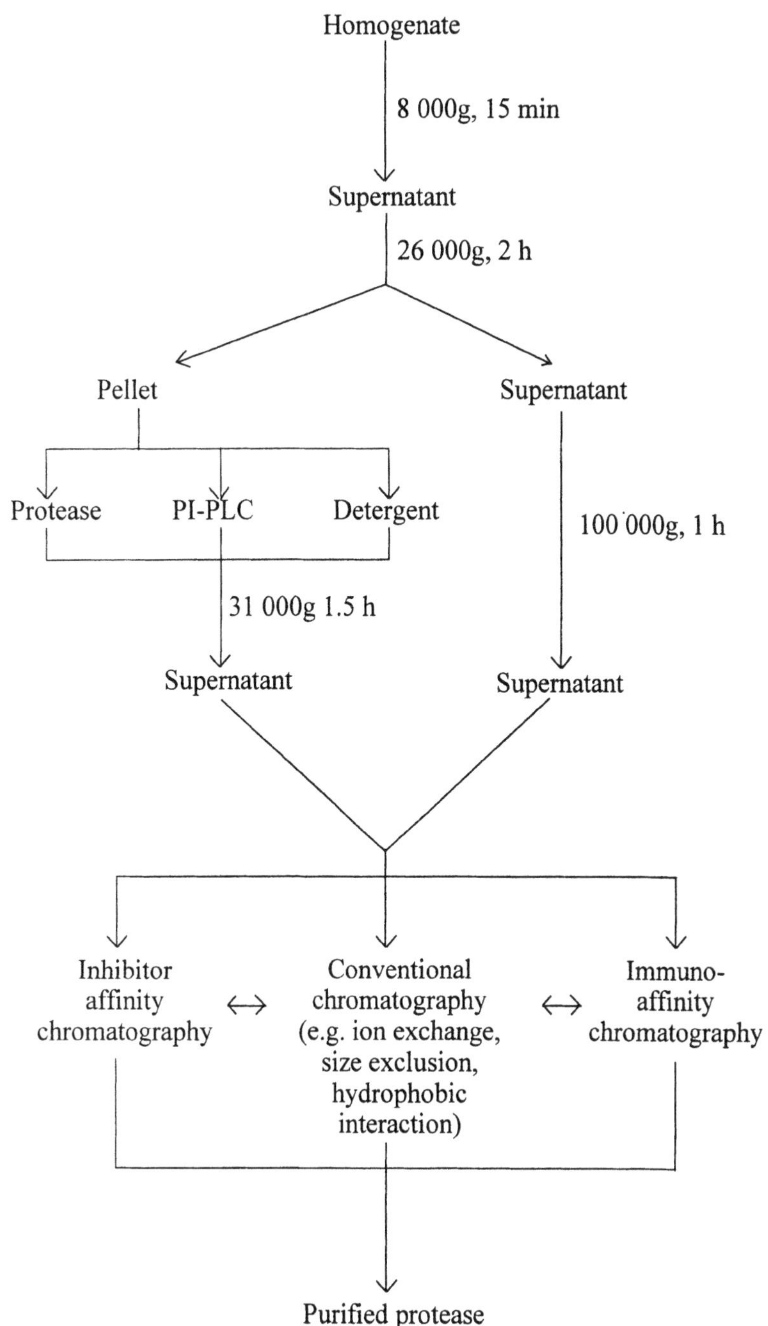

Fig. 1. Scheme showing the various stages in the purification of membrane-bound and soluble proteases.

▨ Materials

For subcellular fractionation	– Waring blender or other homogenizer.
	– Preparative ultracentrifuge and rotors (for example Sorvall RC-5B refrigerated centrifuge; GSA and SS-34 rotors).
	– Homogenization buffer
	  50 mM Hepes/NaOH, pH 7.4, 0.33 M sucrose
	– Resuspension buffer
	  10 mM Hepes/NaOH, pH 7.4

For membrane solubilization

– Trypsin
– Phosphatidylinositol-specific phospholipase C (PI-PLC). Recombinant forms of *Bacillus thuringiensis* or *Bacillus cereus* PI-PLC are available from a number of suppliers including Boehringer Mannheim, Calbiochem, and Peninsula Laboratories. A crude phospholipase C preparation which contains varying amounts of PI-PLC as a contaminant is available from Sigma (catalogue no. P6135) and Fluka (catalogue no. 79484) which is considerably more cost effective for large scale membrane solubilization (for more information see Hooper 1992).
– Triton X-100 as a 20% (weight/vol) solution in water. N.B. This takes several hours to stir into a homogenous solution.
– n-octyl-β-D-glucopyranoside

For large scale anion exchange chromatography

– DEAE-cellulose (DE-52; Whatman)
– Gradient maker (250 ml total volume)
– Column buffer A
  10 mM Tris/HCl, pH 7.6

For hydrophobic interaction chromatography

– HR5/5 alkyl-Superose f.p.l.c. column (Pharmacia)
– Column buffer B
  20 mM sodium phosphate, pH 7.5

For anion exchange chromatography

– HR5/5 MonoQ f.p.l.c. column (Pharmacia)
– Column buffer C
  10 mM Tris/HCl, pH 8.0

For inhibitor affinity chromatography

– CNBr-activated Sepharose 4B
– Sepharose CL-4B
– Inhibitor – cilastatin
– Coupling buffer A
  0.1 M NaHCO$_3$, pH 8.3

- Column buffer D
  50 mM Tris/HCl, pH 7.5.
- 0.5 M NaCl
- Column buffer E
  5 mM Tris/HCl, pH 8.0

1. Add 1 g of CNBr-activated Sepharose 4B to 20 ml of 1 mM HCl and allow Sepharose to swell for 15 min.

2. Wash the Sepharose on a scintered glass (G2) funnel under suction with 4 x 20 ml of 1 mM HCl.

3. Dissolve 5 mg of inhibitor in 10 ml of coupling buffer A, and mix with the Sepharose on a flask shaker for 6 h at room temperature.

4. Wash the Sepharose extensively on a scintered glass funnel under suction with coupling buffer A.

5. Store at 4°C in the presence of 0.1 % sodium azide.

For coupling of ligand to Sepharose

- CNBr-activated Sepharose 4B
- Antibody
- Coupling buffer B
  0.1 M $NaHCO_3$, pH 8.7
  0.5 M NaCl
- Blocking buffer
  0.2 M glycine/NaOH, pH 8.0
- Washing buffer
  0.1 M sodium acetate, pH 4.0
  0.5 M NaCl
- Column buffer F
  50 mM Tris/HCl, pH 7.5
  0.5 M NaCl
  0.1% Triton X-100

For immunoaffinity chromatography

1. Extensively dialyse the antibody against coupling buffer B.

2. Add 1 g of CNBr-activated Sepharose 4B to 20 ml of 1 mM HCl and allow Sepharose to swell for 15 min.

3. Wash the Sepharose on a scintered glass (G2) funnel under suction with 4 x 20 ml of 1 mM HCl, and then with 5-10 ml of coupling buffer B.

For coupling of antibody to Sepharose

4. Mix the dialysed antibody with Sepharose (1 g CNBr-activated Sepharose will bind 20-30 mg of protein) in a total volume of 8 ml made up with coupling buffer B and place on a flask shaker for 2 h at room temperature.

5. Filter the Sepharose on a scintered glass funnel under suction to remove unbound material, and then block remaining reactive sites by incubation with blocking buffer for 2 h at room temperature on a flask shaker.

6. Wash the Sepharose on a scintered glass funnel alternatively with 20 ml aliquots of coupling buffer B and washing buffer.

7. Store at 4°C in the presence of 0.1 % sodium azide.

Procedure

Subcellular Fractionation

The tissue or cell source employed will obviously depend on the protease to be purified. All the procedures detailed here are routinely used in my laboratory to purify proteases from porcine kidney cortex, but are equally applicable to other mammalian tissues. If cell lines are being used which express either a naturally occuring or a recombinant form of the protease then identical procedures can be used, but with the major limitation being that the amount of starting material will be substantially less.

All procedures should be carried out at 4°C with pre-chilled buffers to minimise unwanted proteolysis. Specific protease inhibitors can also be included in the homogenisation and resuspension buffers to further reduce unwanted proteolysis as long as the protease of interest is not itself inhibited. Useful lists of protease inhibitors and their properties can be found in (Beynon and Bond 1989; Hooper 1997) and a leaflet titled 'The easy way to customize your protease inhibitor cocktail' produced by Calbiochem.

1. Homogenise 100g of porcine kidney cortex in 10 volumes of homogenization buffer using a Waring blender.

2. Centrifuge at 8 000g for 15 min.

3. Carefully decant off the supernatant and centrifuge this at 26 000g for 2 h.

4. Resuspend the resulting microsomal pellet in resuspension buffer to a final concentration of approx. 10 mg of protein per ml. This resuspended microsomal pellet can then be used as a source of membrane-bound proteases, while the supernatant can either be used directly or centrifuged further at 100 000g for 1 h and used as a source of soluble proteases.

Membrane Solubilization

If the protease of interest is membrane-bound it is necessary to solubilize it from the membrane prior to its purification. The method of solubilization will depend on how the protease is associated with the lipid bilayer (Fig. 2). For those proteins anchored by a covalently attached glycosyl-phosphatidylinositol moiety, the entire protein minus the hydrophobic fatty acid

Phosphoinositol-phospholipase C

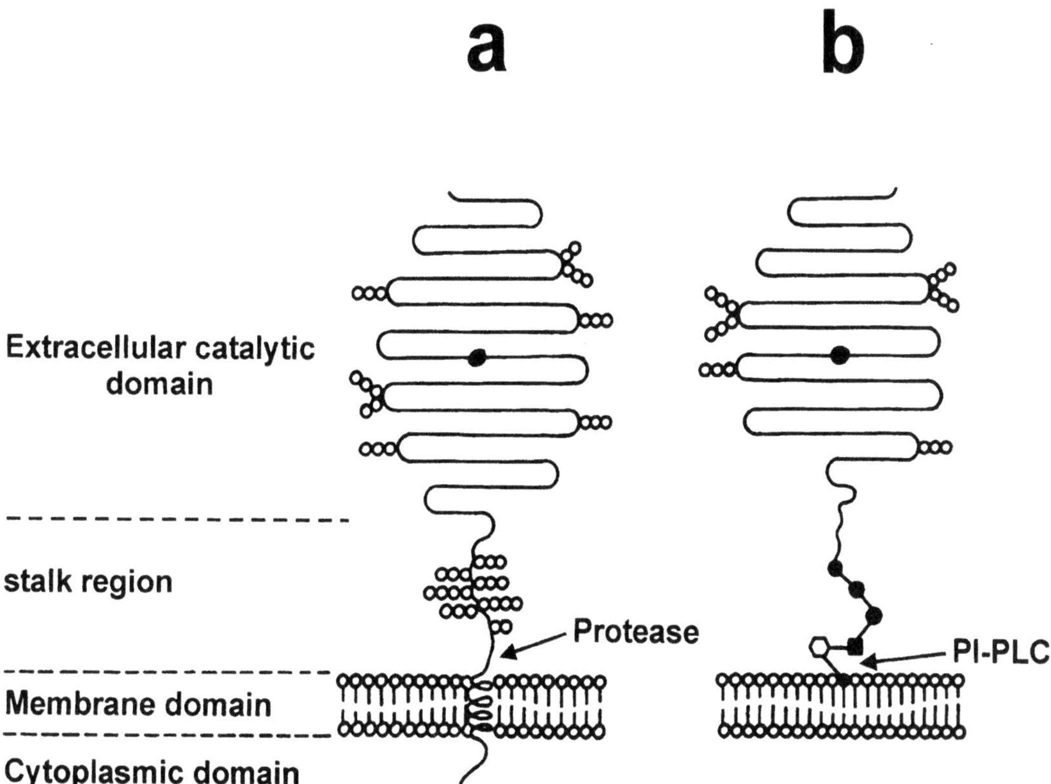

Fig. 2. Types of anchorage of membrane-bound proteases. **a** membrane spanning polypeptide anchor at either the N- or the C-terminus, eg membrane alanyl aminopeptidase (EC 3.4.11.2) or angiotensin converting enzyme (EC 3.4.15.1), showing the site of cleavage in the stalk region by proteases; **b** C-terminal glycosyl-phosphatidylinositol anchor, eg X-Pro aminopeptidase (EC 3.4.11.9) or membrane dipeptidase (EC 3.4.13.19), showing the site of cleavage by bacterial PI-PLC. The protease or PI-PLC treatment results in the bulk of the protein, including the active site, being released in a soluble, hydrophilic form. Both types of membrane anchored protein can also be solubilized with detergents in which case the solubilized protein retains the hydrophobic anchoring domain and is amphipathic in nature.

chains of the anchor can often be readily cleaved from the bilayer with bacterial PI-PLC (Hooper 1992). The resulting released form of the protein is hydrophilic in nature and so purification can proceed without the requirement for detergent in the buffers. Those proteins with a single membrane spanning domain, where the bulk of the protein including the active site is located on one or other side of the bilayer, are often susceptible to cleavage in the stalk region by specific proteases such as trypsin or papain (Fig. 2). Again the released form of the protein is then hydrophilic in nature. Both PI-PLC cleavage and protease cleavage generally solubilize maximally 10-20% of the total membrane protein therefore providing a simple but effective initial purification step.

Detergends Alternatively, membrane-bound proteases can be solubilized using detergents, of which numerous types are now available (both Calbiochem and Boehringer Mannheim produce leaflets listing a range of detergents and their properties). For large scale purifications we generally use Triton X-100, although those proteins anchored by a glycosyl-phosphatidylinositol moiety are not effectively solubilized by this detergent and an alternative detergent such as n-octyl-β-D-glucopyranoside is required (Hooper and Turner 1988). The other major limitation with Triton X-100 is its high u.v. absorbance making the monitoring of total protein elution from a column by measuring absorbance at 280nm difficult. If a detergent is used to solubilize the membrane-bound proteins it is essential to include detergent in all the buffers throughout the subsequent stages of the purification. Failure to do so will almost certainly result in the loss of the protein, due to aggregation of the hydrophobic moieties and their non-specific binding to the surfaces of columns or tubes.

All procedures should be carried out at 4°C unless otherwise stated.

1. Incubate the resuspended microsomal membrane pellet with one of the following:
   A. trypsin at a ratio of 1:10 trypsin:protein (weight:weight) for 1 h at 37°C;
   B. PI-PLC (0.1 unit per mg of protein) for 2h at 37°C;
   C. Triton X-100 at a ratio of 7:1 Triton X-100 20% (w/v) solution:protein (vol:weight, e.g. 7 ml Triton X-100 to 1g protein) for 1 h at 4°C;
   D. n-octyl-β-D-glucopyranoside at a final concentration of 60 mM for 1 h at 4°C.

2. Centrifuge the sample at 31 000g for 90 min, decant off the resulting supernatant and keep it as the source of solubilised membrane-bound proteases.

Large Scale Anion Exchange Chromatography

Following tissue disruption and, where applicable, membrane solubilization the sample is often in a relatively large (100-300 ml) volume. In order to reduce the sample volume and to provide a reasonable purification step we routinely employ a large scale anion exchange chromatographic step at this stage. Alternative methods of concentrating samples include ammonium sulphate precipitation or the use of an ultrafiltration unit such as the Amicon Diaflo.

All procedures should be carried out at 4°C.

1. Extensively dialyse the sample to be chromatographed against column buffer A.

2. Centrifuge the sample at 31 000g for 90 min to remove precipitated material.

3. Apply the resulting supernatant to a DEAE-cellulose column (30 ml bed volume) equilibrated in column buffer A at a flow rate of 20-30 ml/h.

4. Once all the sample has been applied, wash the column with 1-2 column volumes of column buffer A.

5. Elute bound protein with a 200 ml linear gradient of 0-0.5 M NaCl in column buffer A and collect 5 ml fractions.

6. Measure the absorbance at 280 nm of each fraction and assay each for the protease activity of interest.

Hydrophobic Interaction Chromatography

The following two chromatographic steps are just examples of the types of chromatography that can be applied to the purification of the protease of interest. Both of the procedures described here involve the use of a Pharmacia fast protein liquid chromatography (f.p.l.c.) system, although suitable column material is available to allow such purification steps to be performed conventionally. The procedures are carried out at room temperature using the f.p.l.c. system, but should otherwise be performed at 4°C.

1. Extensively dialyse the sample against column buffer B.

2. Mix the dialysed sample with an equal volume of 4 M $(NH_4)_2SO_4$ in column buffer B and apply to an HR5/5 alkyl-Superose column equilibrated

in 2 M $(NH_4)_2SO_4$ in column buffer B. N.B. Do not filter or centrifuge the sample once it has been mixed with the $(NH_4)_2SO_4$ as the precipitated protein will be lost.

3. Elute bound protein with a gradient of 2.0-0.0 M $(NH_4)_2SO_4$ in column buffer B and collect 0.5-1.0 ml fractions.

4. Measure the absorbance at 280 nm of each fraction and assay each for the protease activity of interest.

Anion Exchange Chromatography

1. Extensively dialyse the sample against column buffer C.

2. Apply the sample to an HR5/5 MonoQ f.p.l.c. column equilibrated in column buffer C.

3. Elute bound protein with a gradient of 0.0-0.5 M NaCl in column buffer C and collect 0.5-1.0 ml fractions.

4. Measure the absorbance at 280 nm of each fraction and assay each for the protease activity of interest.

Inhibitor Affinity Chromatography

The method detailed here has been developed for the purification of membrane dipeptidase (EC 3.4.13.19) using the inhibitor cilastatin immobilised on CNBr-Sepharose (Campbell et al 1984), but is applicable to any ligand which can be coupled through a primary amine group. An alternative coupling procedure, again through a primary amine group on the ligand, but also incorporating a long spacer between the ligand and the matrix has been extensively used for the purification of angiotensin converting enzyme (Hooper and Turner 1987; Hooper et al 1987). In the example given here for the purification of membrane dipeptidase, the inclusion of NaCl in the column buffer is to minimise non-specific ionic interactions between the sample and the column material; it is not required for the enzyme to bind to the inhibitor. It is important to extensively wash the inhibitor affinity column with the column buffer prior to elution in order to remove non-specifically bound protein. The bound protein is eluted by including free inhibitor in the buffer which displaces the enzyme from the column. The inhibitor then has to be removed from the enzyme in order to regain functional activity. This can either be done by extensive dialysis or through the use of sample concentrators such as Amicon Centricons.

All procedures are performed at 4°C.

1. Extensively dialyse the sample against column buffer D.

2. Apply the sample to a cilastatin-Sepharose affinity column (5-10 ml bed volume) with a pre-column of unmodified Sepharose CL-4B (20 ml bed volume) equilibrated in column buffer D at a flow rate of 10-20 ml/h. (The pre-column helps to reduce non-specific binding to, and prolong the life of, the affinity column.)

3. Once all the sample has been applied remove the precolumn and wash the affinity column with 300-400 ml of column buffer D.

4. Elute bound enzyme with 10 ml of column buffer D containing 10 mg of cilastatin and collect 1-2 ml fractions.

5. Pool those fractions absorbing at 280 nm and dialyse extensively against column buffer E to remove the cilastatin.

Immunoaffinity Chromatography

The availability of monoclonal or specific polyclonal antibodies may allow one to purify the protease of interest by immunoaffinity chromatography. Two major hurdles have to be overcome with this procedure. First, the antibody must be capable of recognising the protease in its native state (this may not be the case for anti-peptide antibodies or antibodies generated against denatured protein), and secondly, it must be possible to disrupt the antibody-antigen interaction once the protein has bound to the column without irreversibly denaturing either the protease or the antibody. A number of alternative elution buffers are listed; the choice of which to use will depend on the protease-antibody interaction. Again, the immunoaffinity column is extensively washed with column buffer prior to elution of the bound protein in order to minimise non-specific interactions. The procedure described here has been used for the purification of neprilysin (EC 3.4.24.11) from porcine tissues following detergent solubilization from the membrane and employs a monoclonal antibody (Relton et al 1983).

All procedures are performed at 4°C.

1. Extensively dialyse the sample against column buffer F.

2. Apply the sample to the antibody-Sepharose column (1-5 ml bed volume) equilibrated in column buffer F at a flow rate of 2-4 ml/h. (A

pre-column of unmodified Sepharose can be used to minimise non-specific interactions (see section Inhibitor Affinity Chromatography)

3. Once all the sample has been applied wash the column with 200 ml of column buffer F.

4. Elute bound enzyme with one of the following:
   a. 2 mM Tris/HCl, pH 7.4;
   b. 0.1 M NaHCO$_3$, 0.5 M NaCl, pH 10.6 and immediately adjust to pH 7.5 with 0.2 M Tris/HCl, pH 7.0;
   c. 0.2 M glycine, pH 2.3 and immediately adjust pH to 7.5 with 1 M Tris;
   d. 0.1 M ethanolamine, pH 10.5 and immediately adjust pH to 7.5 with 1 M Tris;
   e. 2 M NaI in buffer F.

5. Dialyse eluted enzyme against column buffer E containing 0.1% Triton X-100.

Concentration of Samples

It is often necessary to concentrate a protease containing sample either during or at the end of a purification procedure. This is particularly so following elution of a protease from either an inhibitor affinity column or an immu-

Fig. 3. Purification of X-Pro aminopeptidase from porcine kidney cortex Following solubilization of the microsomal membranes with bacterial PI-PLC (see Procedure) the sample was chromatographed on cilastatin-Sepharose to remove contaminating membrane dipeptidase (see Procedure). The run through fraction from this column was then applied to a DEAE cellulose column (**a**) as detailed in the Procedure. Bound protein was eluted with a linear gradient of 0-0.5 M NaCl, 5.0 ml fractions collected and assayed for X-Pro aminopeptidase activity and protein (A_{280}). Fractions 10-16 were pooled and dialysed against column buffer A before being applied to an HR5/5 MonoQ column equilibrated in column buffer A (**b**). Bound protein was eluted with a non-linear gradient of 0-0.5 M NaCl, 1.0 ml fractions were collected and assayed for X-Pro aminopeptidase activity and protein (A_{280}). Fractions 29-35 were pooled and dialysed against column buffer B before being applied to an HR5/5 alkyl-Superose column (**c**). Bound protein was eluted with a non-linear gradient of 2.0-0 M (NH$_4$)$_2$SO$_4$, 1.0 ml fractions were collected and assayed for X-Pro aminopeptidase activity and protein (A_{280}). Fractions 24-26 were pooled and then chromatographed on a mixed affinity column of cilastatin- and lisinopril-Sepharose. The run through from this column was finally dialysed against column buffer C and applied to an HR5/5 MonoQ column equilibrated in column buffer C (**d**). Bound protein was eluted with a non-linear gradient of 0-0.5 M NaCl, 1.0 ml fractions were collected and assayed for X-Pro aminopeptidase activity and protein (A_{280}). Fractions 32 and 33 were pooled and used as purified X-Pro aminopeptidase. ∘-∘, X-Pro aminopeptidase activity; •-•, protein. Reproduced with permission from (Hooper et al 1990).

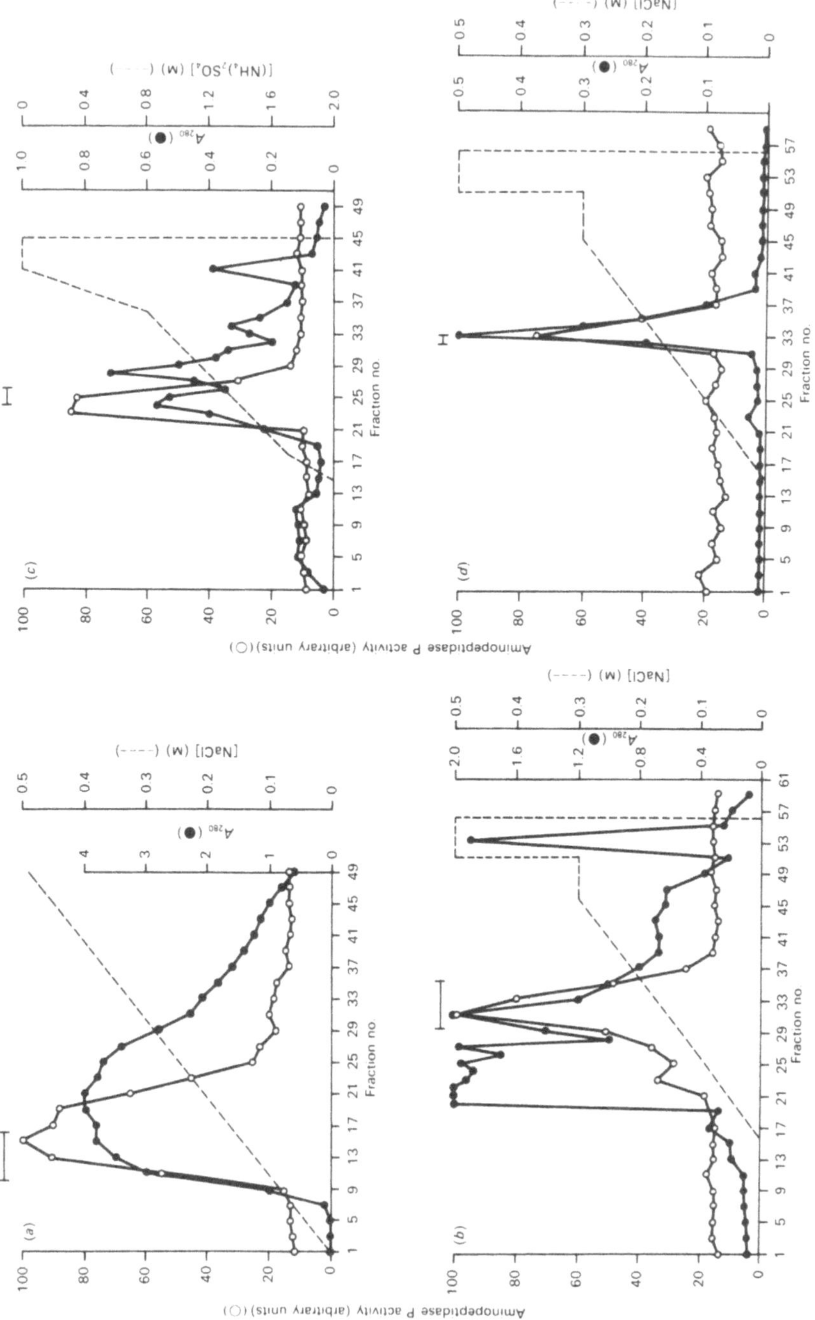
[(NH₄)₂SO₄] (M) (----)
A₂₈₀ (●)
Fraction no.
Aminopeptidase P activity (arbitrary units) (○)
(c)
[NaCl] (M) (----)
A₂₈₀ (●)
Fraction no.
Aminopeptidase P activity (arbitrary units) (○)
(d)
[NaCl] (M) (----)
A₂₈₀ (●)
Fraction no.
(a)
[NaCl] (M) (----)
A₂₈₀ (●)
Fraction no.
Aminopeptidase P activity (arbitrary units) (○)
(b)

noaffinity column when the purified enzyme is in a relatively large (10-30 ml) volume. A very simple, cheap and effective means of concentrating most proteins is to adsorb the sample on to a small anion exchange column and then elute with a high salt containing buffer into a small (0.1-0.5 ml) volume. Alternatively, sample concentrators can be used, such as the Amicon Centricons. We have observed that those proteins solubilized with detergent, and therefore possessing a hydrophobic membrane anchoring domain, can bind irreversibly to the membrane in such sample concentrators. In contrast, good recovery of material is obtained using a small anion exchange column with 0.1 % Triton X-100 included in the buffers.

All procedures are performed at 4°C.

1. Extensively dialyse the sample against column buffer E.

2. Apply the sample to a DEAE cellulose column (0.2-1.0 ml bed volume) equilibrated in column buffer E.

3. Once all the sample has been applied briefly wash the column with column buffer E and then elute bound protein with 0.5 ml aliquots of 0.7 M NaCl in column buffer E directly into 1.5 ml microcentrifuge tubes.

Results

Figure 3 shows the results obtained for the purification of X-Pro aminopeptidase (EC 3.4.11.9) from porcine kidney cortex following its solubilization from the membrane with bacterial PI-PLC (Hooper et al 1990). This combination of anion exchange and hydrophobic interaction chromatographies resulted in a virtually homogenous sample of the enzyme as assessed by sodium dodecyl sulphate polyacrylamide gel electrophoresis and by assaying for other related activities. Contaminating membrane dipeptidase and angiotensin converting enzyme were removed during the purification by adsorption to cilastatin-Sepharose and lisinopril-Sepharose, respectively. Fractions were pooled so as to maximise the purity of the enzyme at each stage rather than to maximise the recovery.

References

Beynon RJ, Bond JS (1989) Proteolytic Enzymes: A Practical Approach. IRL Press, Oxford
Campbell BJ, Forrester LJ, Zahler WL, Burks M (1984) β-lactamase activity of purified and partially characterised human renal dipeptidase. J Biol Chem 259:14586-14590
Hesp JR, Hooper NM (1997) Proteolytic Fragmentation reveals the oligomeric and domain structure of porcine aminopeptidase A. Biochemistry 36: 3000-3007

Hooper NM (1992) Identification of a glycosyl-phosphatidylinositol anchor on membrane proteins. In: Hooper NM, Turner AJ (eds) Lipid Modification of Proteins: A Practical Approach. IRL Press, Oxford, pp 89-115

Hooper NM (1997) Characterization of neuropeptidases using inhibitors. In: Williams CH, Irvine GB (eds) Neuropeptide Protocols. Humana Press, New Jersey, pp 369-381

Hooper NM, Hryszko J, Turner A J (1990) Purification and characterization of pig kidney aminopeptidase P. A glycosyl-phosphatidylinositol-anchored ectoenzyme. Biochem J 267:509-515

Hooper NM, Keen J, Pappin DJC, Turner AJ (1987) Pig kidney angiotensin converting enzyme. Purification and characterization of amphipathic and hydrophilic forms of the enzyme establishes C-terminal anchorage to the plasma membrane. Biochem J 247:85-93

Hooper NM, Turner AJ (1987) Isolation of two differentially glycosylated forms of peptidyl-dipeptidase A (angiotensin converting enzyme) from pig brain: a re-evaluation of their role in neuropeptide metabolism. Biochem J 241:625-633

Hooper NM, Turner AJ (1988) Ectoenzymes of the kidney microvillar membrane. Differential solubilization by detergents can predict a glycosyl-phosphatidylinositol membrane anchor. Biochem J 250:865-869

Hooper NM, Turner AJ (1989) Ectoenzymes of the kidney microvillar membrane. Isolation and characterization of the amphipathic form of renal dipeptidase and hydrolysis of its glycosyl-phosphatidylinositol anchor by an activity in plasma. Biochem J 261:811-818

Littlewood GM, Hooper NM, Turner AJ (1989) Ectoenzymes of the kidney microvillar membrane. Affinity purification, characterization and localization of the phospholipase C-solubilized form of renal dipeptidase. Biochem J 257:361-367

Lloyd GS, Hryszko J, Hooper NM, Turner AJ (1996) Inhibition and metal-ion activation of pig kidney aminopeptidase P: dependence on nature of substrate. Biochem Pharmacol 52: 229-236

Relton JM, Gee NS, Matsas R, Turner AJ, Kenny AJ (1983) Purification of endopeptidase-24.11 ('enkephalinase') from pig brain by immunoadsorbent chromatography. Biochem J 215:519-523

Abbreviations

f.p.l.c.	fast protein liquid chromatography
PI-PLC	phosphatidylinositol-specific phospholipase C

Chapter 9

Crystallization of Proteinases

MARGIT M.T. BAUER AND MILTON T. STUBBS

Introduction

The three dimensional structure of an enzyme is a rich source of information for the modern biochemist. It can give valuable insights into the function and specificity of your proteinase, providing a framework on which to focus previous biochemical data and a sound basis for the design of new experiments. Structural data on your proteinase can have a profound influence on the search and design of novel specific inhibitors ("rational drug design"). The role played by proteinases in a wide range of pathological conditions has provided a spur to this latter theme.

Structures are now available for representatives of most classes of endo- and exo-peptidases (see Appendix of the manual), allowing the use of homology modelling as a first step to understanding structural aspects of your proteinase. Experience shows, however, that the subtleties in *e.g.* specificity can only be explained on the basis of an experimental structure. Recent advances in many aspects of X-ray crystallography have made the goal of determining a specific target proteinase a realistic possibility. Revolutionary developments have taken place in X-ray detectors, radiation sources and computing hardware and software for the solution of X-ray crystal structures (Pflugrath 1992, Helliwell 1992, Finzel 1993, Kottke and Stalke 1994). Probably the single most important advance has come from molecular biology, as it allows the production of large quantities

Correspondence to: Margit M.T. Bauer, Boehringer Ingelheim Pharma KG, Abteilung Chemische Forschung, Gruppe Strukturforschung, Birkendorfer Str. 65, Biberach/ Riß, 88397, Germany (*phone* 49-(0)7351-544795; *fax* 49-(0)7351- 54 5137; *e-mail* Margit.Bauer@bc.boehringer-ingelheim.com)
Milton T. Stubbs, Institut für Pharmazeutische Chemie der Philipps-Universität Marburg, Leiter, Röntgenstrukturanalyse, Marbacher Weg 6, Marburg, 35032, Germany (*phone* 49-(0)6421-285999; *fax* 49-(0)6421-288994; *e-mail* stubbs@mailer.uni-marburg.de)

of protein from a single source, devoid of heterogeneities such as e.g. gly-cosylation, and the design of fragments or mutants more amenable to crys-tallization or derivatization (Price and Nagai 1995, Skelly and Madden 1996). The chances of solving a structure given suitably diffracting crystals have therefore never been so good. It is now the production of such crystals that remains the bottleneck in the successful structure determination of a required enzyme. The number of proteins crystallized has increased expo-nentially in the last 30 years (Roussell et al. 1990), and many articles and reviews have appeared on the subject of protein crystallization (Blundell and Johnson 1976, McPherson 1982, Wiegand 1990, Ducruix and Giegé 1992, McRee 1993, McPherson et al. 1995). Protein crystallization requires little or no specialized equipment, and can be carried out in almost any biochemical laboratory. It is the aim of this article to enable the practicing proteinase biochemist to make the first steps towards achieving this goal.

A well-founded knowledge of the physico-chemical characteristics of your proteinase will certainly be of use in determining successful crystallization conditions. Take for example the digestive enzyme pepsin: the protein was dissolved at 280 mg/ml in 0.5M H_2SO_4 pH 3.6, warmed to 35°C, and then allowed to cool slowly in a Dewar, resulting in diffraction quality crystals (Bernal and Crowfoot 1934). This treatment is not recommended for all proteinases!

Initial considerations

Complexation with a **tight binding inhibitor** is of particular importance in the crystallization of proteinases. The long-term or irreversible binding of an inhibitor (natural or synthetic) prevents self degradation or aging of the proteinase, and very often stabilizes the substrate binding area during the long time necessary for crystal nucleation and growth. Bear in mind that crystallization experiments may stand for several months. Frequently used small synthetic inhibitors are 3,4-dichloroisocoumarin (3,4-DCI) and (for endopeptidases) diisopropylfluorophosphate (DFP) for serine proteinases (careful handling, DFP is very toxic), 1,10-phenanthroline for metallo-proteinases, L-3-carboxy-*trans*-2,3-epoxypropyl-leucylamido(4-guanidi-no) butane (E64) for cysteine proteinases and pepstatin for aspartic protei-nases (Barrett 1994).

It may be that the structure of your target proteinase is already solved, and you are interested in the **interaction of a specific inhibitor** (for example, for development of a thrombin or HIV proteinase inhibitor). In this case, it might be possible to use the published crystallization conditions. Note, however, that such conditions can change radically according to the protein source (let alone species) and inhibitor (see appendix). If the published structure contains a non-covalently bound inhibitor, it may be possible

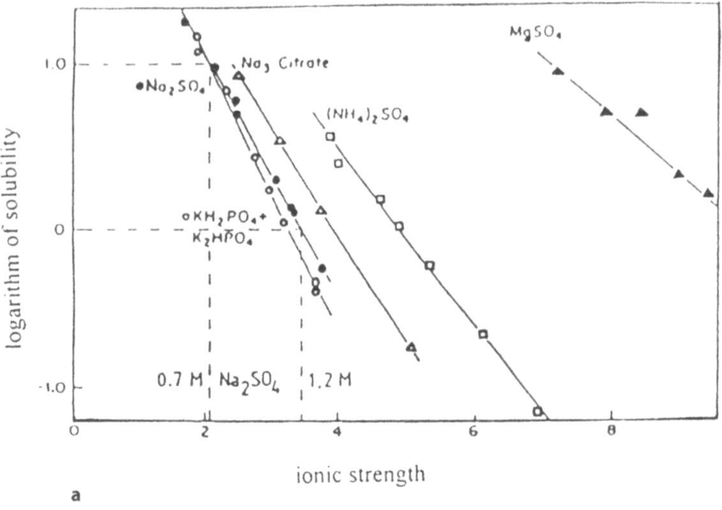

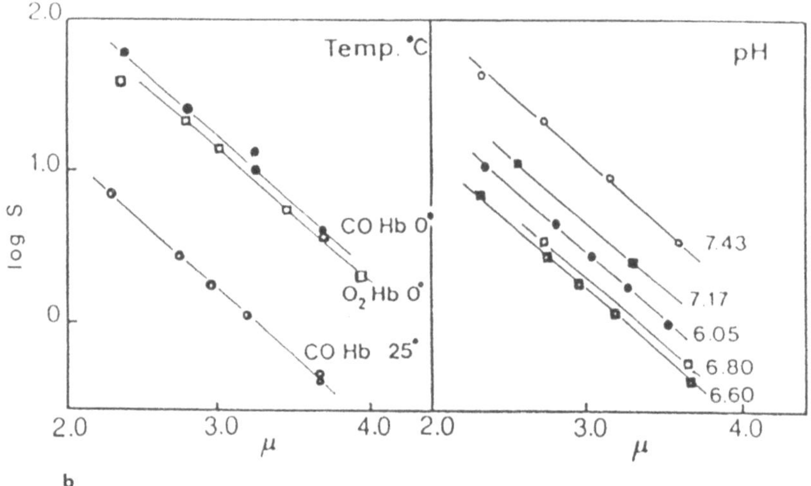

Fig. 1. (after Wiegand, 1990). **a** Solubility of carboxyhemoglobin at 25°C and pH 6.6 in concentrated solution of various electrolytes; **b** Solubility of hemoglobin in concentrated phosphate buffer, at varying temperature and pH values

to replace it by diffusing your inhibitor into the same crystal (**"soaking"**; see case study below).

Protein crystals grow from the equilibration of supersaturated solutions. The thermally unstable state equilibrates either to an amorphous precipitate, or (more rarely) to a crystalline state. Protein solubility in an aqueous solution is a complex function of many variables that are dependent on the physical and chemical constitution of the protein and its environment. Compared to small molecules, proteins exhibit enhanced flexibility and multiple conformations. Due to the exposed charged and polar residues on the surface of the protein, they can be regarded as large polyvalent ions (Scopes 1982).

Principles of protein crystallization

In general, proteins are more soluble in low ionic strength solutions than in pure water, but can be precipitated by very high salt (e.g. ammonium sulphate) concentrations. Thus, at low ion concentrations, the solubility of the protein can be decreased either by further decreasing ("salting in") or increasing ("salting out") the precipitant concentration (Fig. 1). This decrease in solubility can also be accomplished through variation of the wide range of parameters listed in Table 1.

The **precipitants** used most frequently for protein crystallization are **salts** containing polyvalent anions (sulphate, citrate, tartrate or phosphate) with suitable cationic counterions. The choice of ions follows the Hofmeister series (Hofmeister 1888, von Hippel and Schleich 1969), according to their ability to precipitate hen egg white proteins; for anions, sulphate > phosphate > acetate > citrate > tartrate > bicarbonate > chromate > chloride > nitrate > chlorate > thiocyanate (the latter are chaotropic, *i.e.* they destabilize the native conformation), and for cations lithium > sodium > potassium > ammonium > magnesium. Small ions with high charge have been found to be more effective than large ions with low charge (such as KCl). The organic polymer **polyethylene glycol (PEG)** $(HO\text{-}(CH_2\text{-}CH_2\text{-}O\text{-})_nH)$, which comes in various lengths (mean molecular weight 400 – 20 000 kDa) is also a very effective precipitant of proteins.

Salts

Protein solubility is also affected by **organic solvents** such as ethanol, acetone, MPD, isopropanol, acetonitril and tertiary butanol. These lower the dielectric constant, increasing the Coulombic attraction between opposite charges on the protein surfaces and thereby decreasing the solubility. Furthermore, the organic molecule can displace water molecules from the protein surface, facilitating contacts between polar or nonpolar side chains of the protein molecules. As organic solvents tend to denature proteins, they should generally be used in the cold.

Solvents

Table 1. Parameters affecting crystallization (after Ducruix and Giegé, 1992)

Physico-chemical parameters:

- concentration of protein and precipitants

- temperature, pH, pressure, electric and magnetic fields, surface exposed to the air, interface effects

- time to reach supersaturation

- ionic strength and purity of the chemicals used

- density and viscosity effects, speed of diffusion and convection

Biochemical and biophysical parameters:

- sensitivity of the protein against physical parameters like pH, temperature etc.

- binding of ligands like inhibitors, cofactors, metal ions etc.

- specific additives like reducing agents, detergents etc.

- properties of the protein like oxidation, hydrophobicity, hydrophilicity etc.

- age of the protein samples (degradation, denaturation, redox-effects)

Biological parameters:

- amount of protein available

- different biological sources of proteins

- contaminations (funghi or bacteria)

Purity of protein:

- macromolecular contaminants

- microheterogeneities by glycosylation, degradation or conformation (flexible domains, oligomerisation, aggregation, conformer equilibrium)

- batch differences

Effect of pH The **pH-value** affects the solubility of the protein as it changes the net charge of the "polyvalent ion", e.g. the solubility is at its lowest at the isoelectric point of the protein due to its net zero charge. Variation of the protein surface charge distribution allows the binding of different counter ions and hence alters protein-protein interactions. The buffers commonly preferred for adjusting the pH in crystallization experiments are the low ionic strength Good's buffers as HEPES (4-(2-hydroxyethyl)piperazin-1-ethansulfonic acid), MES (2-morpholinoethanesulfonic acid monohydrate), Tris, MOPS etc. (Good et al. 1966). These zwitterionic buffers are biologically and chemically non-reactive, non-toxic, and their pK_a is relatively insensitive to temperature and the ionic strength of the buffer (Harris 1989). As a wide range of pH values (4 – 10) are covered in a crystallization experiment, it is almost certain that your proteinase will be at its active pH in many of the setups. As mentioned before, addition of an inhibitor may be necessary to avoid self-degradation.

Many of the factors influencing protein solubility are **temperature depend-ent**, e.g. the free energy of the solution (ΔG), or the dielectric constant, which bears a reciprocal relationship to the temperature. In a high ionic strength medium, many proteins are less soluble at $20°C$ than at $4°C$, while in a low ionic strength medium the solubility decreases with decreasing temperature. Crystallization experiments at different temperatures therefore lead to different results. In addition, many proteinases are more active at higher temperatures, so that the long-term stability of the proteinase at $20°C$ should be tested.

An appropriate combination of the parameters listed in Table 1 leads to supersaturation in the protein solution, a metastable state which can be changed to a stable one by **crystal nucleus formation**. The nucleation often appears spontaneously, but can be controlled by the addition of a nucleus (**seeding**). A very highly supersaturated solution and/or contamination with dust or denatured protein molecules (e.g. as a result of lyophilization, long storage or high temperatures) produces many nuclei and many small crystals as a consequence. The surfaces of contact to the air or crystallization container can also act as nucleation centres, and should therefore be re-duced as much as possible (e.g. hanging drops to keep the drop spherical, avoidance of air bubbles). Mechanical vibrations (e.g. in a refrigerator) or other disturbances should also be avoided, as they may produce showers of microcrystals; on the other hand, mild mechanical shock such as streaking or seeding may be helpful in inducing controlled crystallization in just supersaturated solutions.

The **properties of protein crystals** differ from those of small molecules in that they contain 30 – 80% solvent; protein crystals are therefore very sen-sitive towards desiccation and mechanical stress. The molecules in a protein crystal are stabilized by only a small number of intermolecular contacts (such as hydrogen bonds and van der Waals interactions, salt bridges, di-pole interactions and stacking). This guarantees the native conformation of the proteins and preserves the activity of the individual molecules in the crystal; on the other hand, it means that protein crystals are weak and sen-sitive compared to small molecule crystals. As described in the case study below, this special feature of protein crystals allows the soaking of small substrates or inhibitors. Crystals suitable for X-ray diffraction should be of the order of 0.3 mm in each direction at least, as protein crystals interact with X-rays only very weakly, and the diffracted intensity is in general proportional to the volume of the crystal. Protein crystals do not show strong scattering of polarized light, which can be used as a tool to distinguish them from salt crystals (see below).

Materials

Fundamental equipment Most equipment necessary for crystallization - such as chemicals, protein concentration and dialysis apparatus, centrifuges, SDS- and IEF gel electrophoresis, UV/VIS-spectrophotometers, binocular microscopes (magnification at least x 50) with polarization filters, temperature controlled rooms or boxes (4, 20 and 30°C), small volume pipettes - can be found in most biochemical laboratories. A variety of specially designed receptacles for crystallization setups (Fig. 2) are available commercially, including the Cryschem MVD24 plates (C. Supper Company, Natick, Massachusetts 01760, USA), Linbro tissue culture boxes (ICN Biomedicals, Inc., Aurora, Ohio 44202, USA) with microscope glass cover slips (diameter 22 mm), Crystal Clear strips (Douglas Instruments Ltd., London SW11 4NB, U.K.), Q-plates (Hampton Research, 25431 Cabot Road, Suite 205, Laguna Hills, CA 92653-5527 USA, e-mail: xtalrox@aol.com), Petri dishes with benches (Labor Service Boll, Waltherstr. 19, München, Germany) and many others. The advantages and disadvantages of each of the plates will be described below. An initial setup of 24 crystallization experiments costs around $ 4 for each type of dish. For those planning on crystallizing a large number of proteins, pipetting robots such as the IMPAX Automatic Protein Crystallization System (Douglas Instruments, London, Great Britain), the BIOMEK Automated Laboratory Workstation (Beckman, Palo Alto, California) and the TECAN-RSP-system (Zinsser Analytic GmbH, Frankfurt, Germany) can be a useful, but expensive timesaver (each around $ 30,000).

Stock solutions Before starting your crystallization experiments, you should prepare stock solutions of precipitants (up to 4M for precipitating salts if possible, 50 % w/v solutions for PEG), additional salt solutions (1 M, if possible) and buffers from pH 4 to 10 (1 M, if possible), including 0.03% NaN3 in all solutions to avoid bacterial growth. The Good's buffers represent the buffer substances of choice, as they are of low ionic strength and therefore do not compete with the precipitant. Ready-to-use crystallization screening solutions and crystal handling equipment can be purchased from Hampton

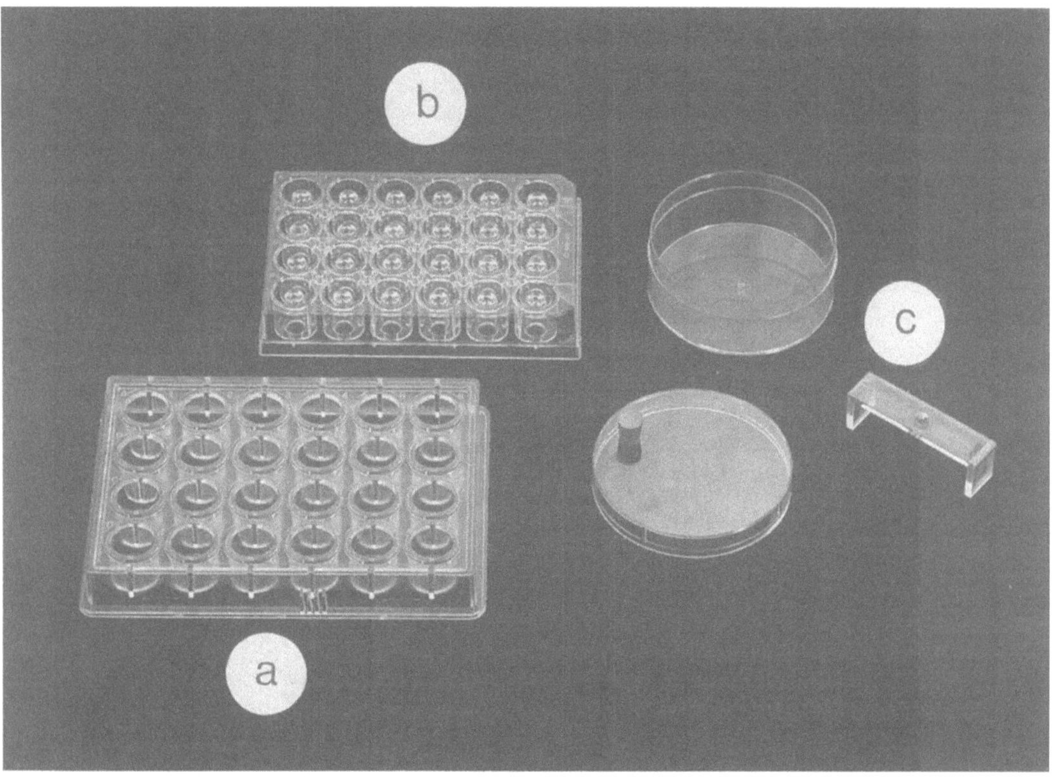

Fig. 2. Crystallization dishes. a Linbro tissue culture boxes (ICN Biomedicals, Inc., Aurora, Ohio 44202, USA) with microscope glass cover slides; b Cryschem MVD24 plates (C. Supper Company, Natick, Massachusetts 01760, USA); c Petri dishes with benches (Labor Service Boll, Waltherstr. 19, München, Germany)

Research, 25431 Cabot Road, Suite 205, Laguna Hills, CA 92653-5527 USA, e-mail: xtalrox@aol.com.

Procedure

Choice of Suitable Precipitation Conditions

SDS-(reduced and non-reduced) and IEF-electrophoresis should be carried out before starting the crystallization experiment in order to check the purity of the protein, the size, the number of peptide chains, the glycosylation or sequence heterogeneities, the state of degradation and the isoelectric point (pI). Knowledge of the pI is important for choosing the correct buffer for the protein: the pH-value must be different from the pI to allow optimal solubility and concentration of the protein.

Many protein purification protocols include at least one precipitation step, so that the behaviour of the protein is already known for this special condition. The solubility of your proteinase can be tested by the dropwise addition of solutions of some of the most common precipitating agents (phosphate, ammonium sulphate, citrate, PEG) to a drop of protein solution at controlled temperature and pH, and observing the concentration at which precipitation occurs. This information can help in choosing initial concentrations of precipitants in your first crystallization trial. If your proteinase precipitates upon dialysis against a low ionic strength buffer or water, it is worthwhile carrying out low-salt crystallization experiments.

As summarized in Figure 3, most chemicals of a well equipped biochemical laboratory may be tested as precipitant; the limits being set by the amount of protein available. The most commonly used precipitants are 1 – 3 M sodium-potassium-phosphate, 10 – 30% PEG 6000 and 1 – 4 M ammonium sulphate and 0.5 – 1.5 M sodium citrate at pH from 4 – 10, which we use in our labs as an initial screening. The factorial screening methods developed by Jancarik and Kim 1991 and Carter and Carter 1979 cover a broader range of possible crystallization conditions, at the expense of a fine sampling interval. If your protein requires e.g. calcium for stabilization, remember to exclude any buffers or precipitants that might form salt crystals (in such a case phosphate for example).

The increasing importance of structure based inhibitor design in the search for novel pharmaceuticals has fostered a staggering number of proteinase-inhibitor complex structure determinations: for instance, there were 28 structures of the HIV-1 proteinase deposited in the Brookhaven Protein Databank (PDB) at the time of writing (see appendix). It is important to keep in mind, however, that even small synthetic proteinase inhibitors may alter the overall structure of the target proteinase - not only in the active site - and may therefore require a renewed search for crystallization conditions for the new complex.

Preparation of the Protein

The proteinase used for crystallization experiments should be 99% pure (one band on SDS- and IEF-gel electrophoresis using silver staining) and free of any contaminating proteolytic activity. As most proteinases will degrade themselves under suitable conditions during the long time

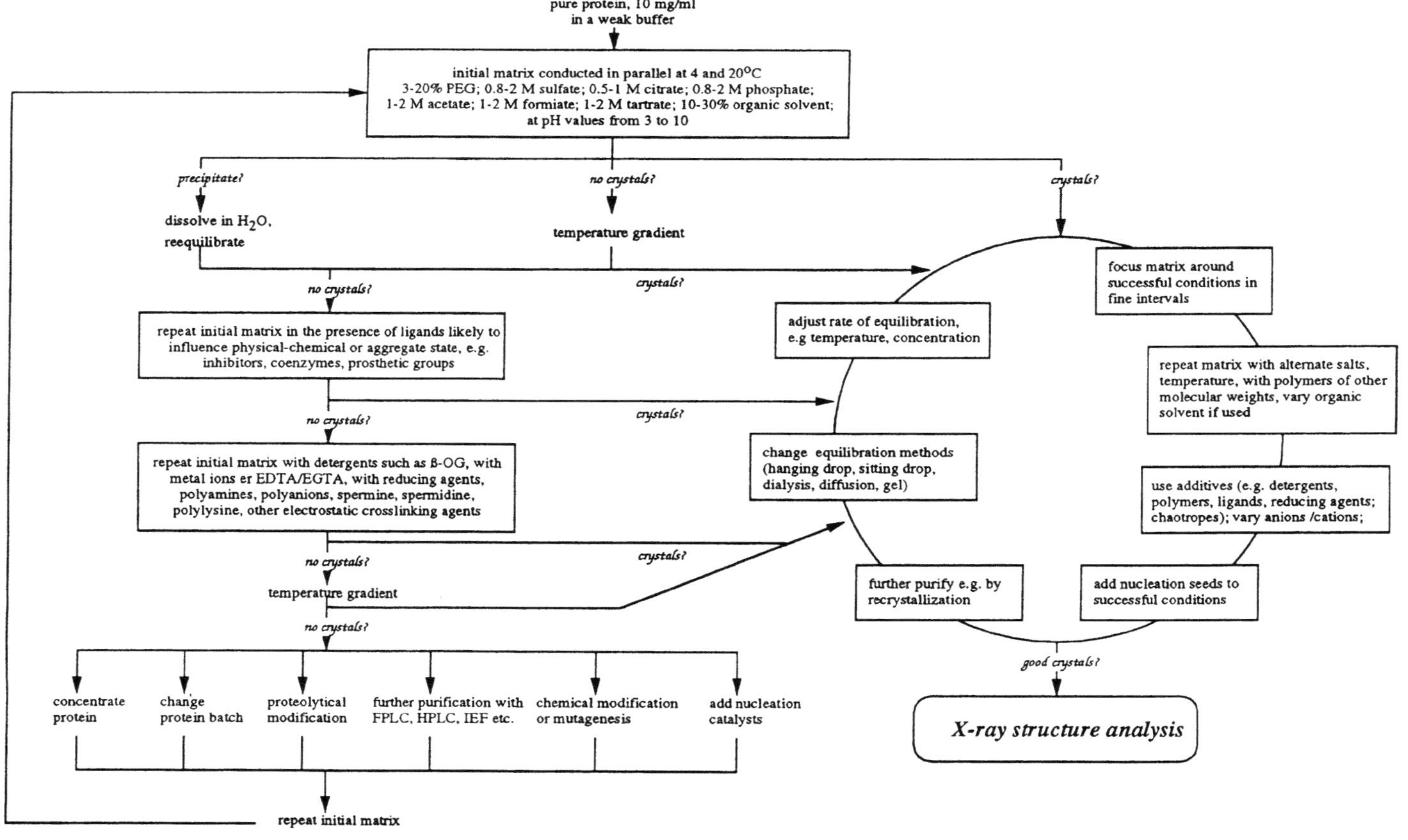

Fig. 3. Crystallization flowchart (after McPherson et al., 1995)

needed for crystallization, the setup should either contain a sufficient amount of an inhibitor specific to the proteinase, or should be carried out at conditions (pH, temperature, in absence of a cofactor, zymogen) where the proteinase is not active. With proteinaceous inhibitors, a 1.5-fold molar excess over the proteinase concentration is sufficient, while non-covalent low molecular weight inhibitors should be added to a 10-fold molar excess prior to the crystallization setup. It should be borne in mind that the inhibitor alone is also a crystallizable molecule; similarly, each component of a protein-protein inhibitor or a proteinase-cofactor complex can crystallize independently, as such complexes are stabilized largely by weak interactions that may be disrupted by the high salt concentrations prevalent in the crystallization experiment. If you have produced your protein by recombinant methods, it is possible to inactivate it through mutation of an active site residue.

20 mg of pure protein is an ideal amount to start crystallization trials. Mixing of protein samples from different preparation batches is not recommended, as it is often the case that only one batch will crystallize. For the same reason, the batch of the sample used for the crystallization experiment should be noted. If a large amount of protein is available from one batch, it should be frozen in approximately 200 µl aliquots (concentration 10 mg/ml, add 20% glyerol, if necessary) in Eppendorf tubes, to be thawed as necessary. All long-term storage methods (lyophilization, freezing, precipitation or similar methods) may inactivate or denature a part of the protein, decreasing the effective purity of the sample. In such cases, it is advisable to apply an additional purification step. Short term (overnight) storage of proteinases should be at 4°C.

All protein samples and crystallization buffers should contain 0.03% of NaN_3 to avoid bacterial growth. The protein should be dialyzed against a low ionic strength buffer (check the stability of your protein in this buffer! If necessary, add an inhibitor to the dialysis buffer, too), e.g. 10 mM Tris pH 7.5, using a dialysis membrane of appropriate molecular weight cutoff, or a desalting column (e.g. PD-10 columns, Pharmacia Biotech, Uppsala, Sweden). For small amounts of protein, dialysis buttons for 5 - 350 µl as offered by Hampton Research (25431 Cabot Road, Suite 205, Laguna Hills, CA 92653-5527 USA, e-mail: xtalrox@aol.com) or Tube-O-dialyzers (Novus Molecular Inc. 4660 La Jolla Village Drive, Suite 500, San Diego, CA 92122 USA, e-mail: novus@iinet.com) or Slide-A-Lyzer dialysis cassettes (Pierce, 3747 N. Meridian Rd. P.O. Box 117, Rockford, IL 61105, USA). It should then be concentrated to around 5-20 mg/ml, using e.g. centrifuge concentrators (supplied by e.g. PallGelman Sciences, 600 South Wagner Road, Ann Arbor, MI 48103-9019 USA, by Amicon/Millipore, and other firms). Do not forget to add stabilizing factors (e.g. calcium, reducing agents) or again inhibitors if your proteinase requires them. As some proteins do not stand such high concentrations at low salt conditions, you should observe your sample carefully. If precipitation does occur, either add a definite amount of NaCl or dilute the protein until precipitation disappears. As mentioned above, proteins which are unstable in low ionic strength media may crystallize at low-salt conditions. Check the concentration of your sample by measuring the absorbance at 280 nm (note that some buffers absorb light at 280 nm, take your buffer as the zero value) or by protein concentration assays such as Lowry, Biuret and Bradford assay. Immediately before the crystallization setup, the protein solution should be filtered to remove any denatured protein or dust particles, either using a syringe sterile filter (0.2 µl pore size), a Millipore Ultrafree centrifugation vial, or by high speed centrifugation.

Crystallization Experiment Setup

All crystallization experiments should be carried out at controlled temperatures of 4°, 20° or 30°C in a climatized room or incubator. A climatized room has the distinct advantage that crystal growth can be followed with minimal changes in the temperature; taking the experiments from a refrigerator to a microscope in a warm room should be avoided or kept to a minimum. Care should also be taken if your microscope is not equipped with a cold light source: the crystallization vessel will warm up during inspection, possibly dissolving any crystals present. The same holds for the warmth in your fingers!

There are various methods to bring the protein solution to the supersaturated phase. The simplest one is the **batch method**, where the protein and precipitant are mixed to weakly supersaturated conditions. A sophisticated application of the batch method is provided by the IMPAX pipetting robot (see above), which mixes protein and precipitant under a film of oil in microtiter plates, excluding air from the crystallization experiment.

Microdialysis setups are generally used for obtaining large crystals where the conditions are already known. The protein solution may be put into capillaries, plastic tubings or dialysis buttons, and sealed using a dialysis membrane of appropriate molecular weight cutoff. The receptacle is then placed in a solution of the required precipitant. The slow exchange allows better crystal growth than the batch setup. Microdialysis is particularly good for low salt crystallization experiments, where the receptacle is placed in a solution of decreasing ionic strength.

By far the most frequently used method is **vapour diffusion**, with either **hanging drop** or **sitting drop** experiments. A drop of protein solution is mixed with precipitant at subsaturation and then placed in a sealed vessel above a reservoir containing the same precipitant at higher concentration. Slow equilibration of water (several hours to days depending on the precipitant) takes place between the drop and the reservoir, leading to the supersaturated state. The method is not recommended for low-salt crystallizations, as the diffusion of solvent from the low concentration reservoir to the protein drop results in an increase in the drop volume and a dilution of the protein concentration.

A number of vessels for vapour diffusion crystallization have been developed over the last few years (Fig. 2), among them Cryschem plates (allow sitting drop only), Linbro or VDX plates (allow hanging drop only), and Q plates or plastic Petri dishes with benches (for both sitting and hanging drop). The Linbro/VDX, Cryschem and Q plates require little shelf space, unlike the Petri dishes. Cryschem plates are the fastest to set up. Some proteins tend to stick to plastic surfaces, especially in sitting drop experiments; in such cases, hanging drop experiments should be carried out using a siliconised glass cover slip (Linbro/VDX) or greased bench (Petri dish). With the bung in the lid of the Petri dishes, conditions in the reservoir can be changed relatively easily without disturbing the crystallization setup (see below).

Setting up of the Cryschem plates

Each of the 24 wells of the plate receives 500 µl of one of the precipitant solutions to be tested. 1-5 µl from the respective reservoir is mixed with 1-5 µl of the protein solution in the depression (the ratio can be varied depending on how much protein you have and how concentrated it is: for dilute protein solutions, take a higher protein : precipitant ratio). Try to keep the drop compact, as one that is smeared out runs the risk of drying out. Once all 24 drops have been set up, seal the plate with the self adhesive tape provided.

Setting up of the Linbro/VDX/Q plates

Each of the 24 wells of the plate receives 500 µl of one of the precipitant solutions to be tested. 1-5 µl from the reservoir is mixed with 1-5 µl of the protein solution on a siliconised cover slip (once again, take care to make a compact round drop), and this is then sealed over its respective reservoir using silicon grease or - in case of the Q plate - placed with the drop upwards (sitting drop) or downwards (hanging drop) over its repective reservoir. Make sure that you place the correct drop over the correct reservoir - your experiments will be irreproducible otherwise! Once all 24 drops have been set up, the lid of the Linbro plate must be fixed using a plasticene 'spacer' so that the cover slips do not come into contact with the lid. This procedure is not necessary with the VDX plates, the Q plates are sealed by a self adhesive tape.

Note: Be careful with handling the Q plates, do not drop or hit them. The glass slides are very close to the tape and may stick on it forever.

Each dish is filled with 3-5 ml of a precipitant solution. Each bench is siliconised by smearing it with a thin layer of silicon grease, and then polished. 1-5 µl from the reservoir is mixed with 1-5 µl of the protein solution on the top (sitting drop) or bottom (hanging drop) of a bench. Several experiments can be carried out on a single bench to test the optimum protein : precipitant ratio. The prepared bench is placed in the Petri dish, and the lid closed and sealed with tape to minimize evaporation.

Setting up of the Petri dishes

Results

The time that crystals need to grow varies tremendously from protein to protein and from crystal form to crystal form. Some crystals grow overnight, whilst some take several months to appear. It is also possible for crystals to form out of precipitate. The time taken for equilibration can be a few hours (with e.g. ammonium sulphate) or a matter of days (with e.g. PEG). You should therefore examine the drops with a binocular microscope (magnification up to 60x) immediately after making up the drops, the day after, and then at least every week thereafter (but not too often, as this can disturb the crystal growth). Some possible results of crystallization experiments are shown in Figure 4.

Troubleshooting

- **Clear solution**
  The precipitant concentration was not high enough. The precipitant concentration can be increased by adding a higher concentration of preci-

Fig. 4. Some possible results of crystallization experiments (drop diameter: 2 mm, photographed using polarised light). **a** – **g** shows the results of crystallization experiments with a protease/protease-inhibitor complex (human leukocyte elastase in complex with secretory leukocyte protease inhibitor) with 1 M sodium citrate as precipitant at increasing pH-values (pH 4.5 to 5.5). Crystals grow at pH 4.5 and 4.7 (**a** resp. **b**), on increasing pH from 4.8 to 5.0 (**c** and **d**) needle- and hedgehog-like forms appear, while at higher pH (**e** – **g**) only sphaerolytes are formed;**h** phase separation; **i** crystal of human leukocyte elastase; **k** microcrystals; **l** heavy precipitation; **m** salt crystals (ammonium sulphate)

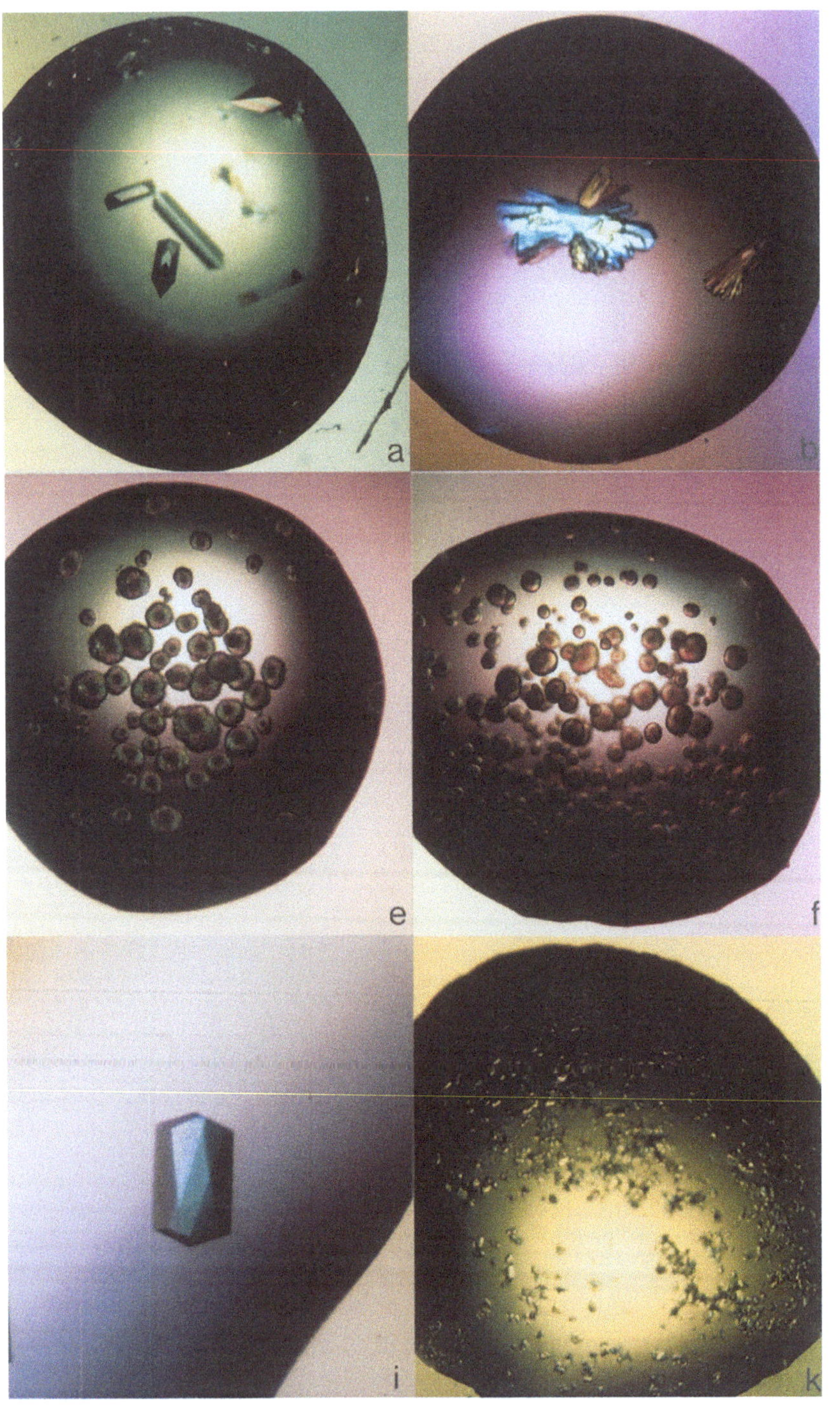

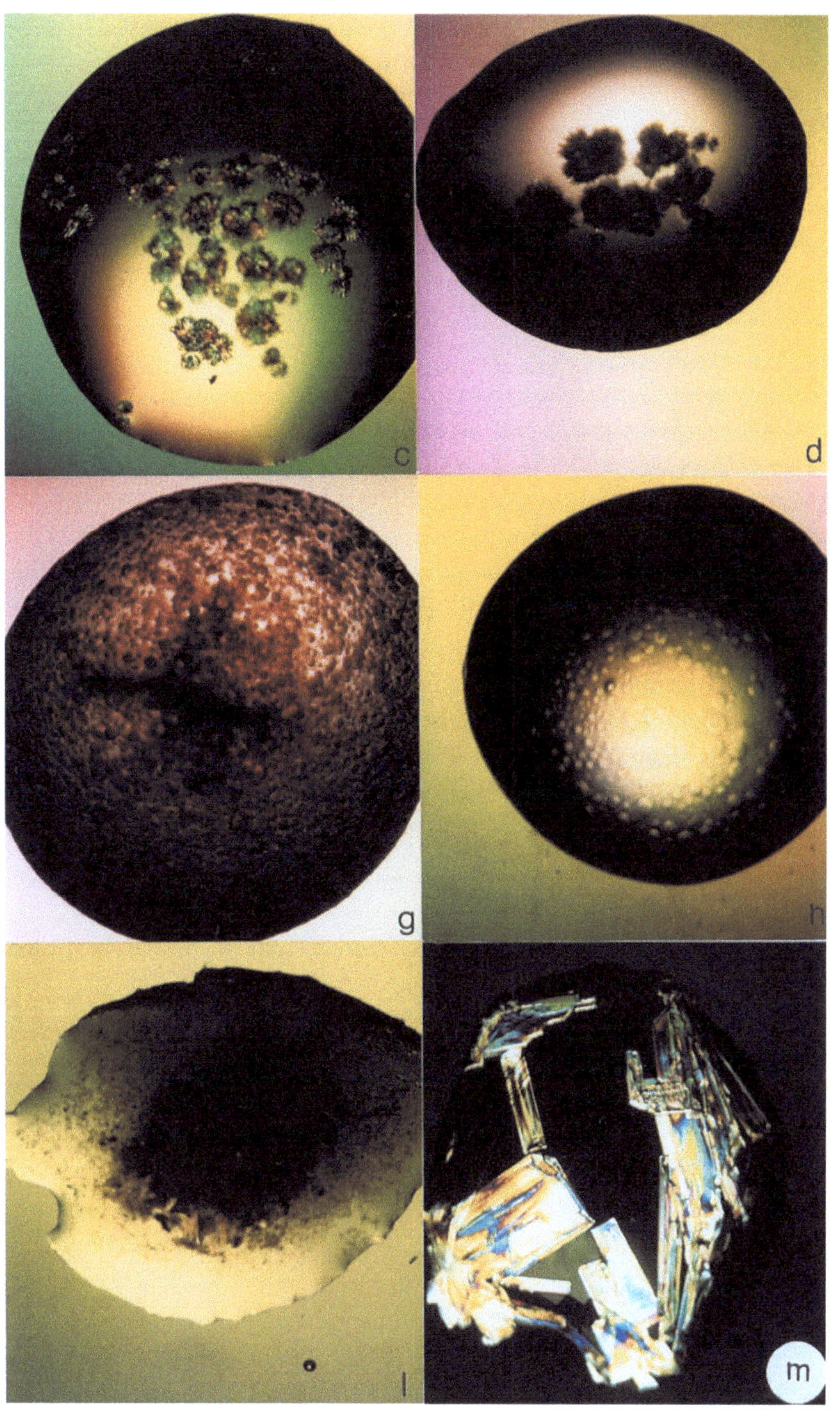

pitant (make sure not to change the pH if you are using volatile buffers / precipitants!), which is easily done using the Petri dish method - remove the bung, add the new solution, then reseal. Alternatively, set up new crystallization experiments with higher starting concentrations of protein and / or precipitant.

- **Precipitation**
  A broad range of precipitation is possible. Thick dark aggregates of protein that cannot be redissolved with water show irreversible denaturation, as do a yellow or brown colouring of the drop; repeating the experiment with fresh protein at a lower temperature may be worthwhile. Fluffy aggregation can often be redissolved using a little water or ammonia, and then re-equilibrated at a lower precipitant concentration. A closer examination of fine precipitate occasionally reveals small needles.
  If all your experiments show precipitation, try **low-salt crystallization**: dissolve the protein at high salt concentration, and dialyse or equilibrate it against decreasing salt conditions. Bear in mind that the volume of the protein solution will increase during this procedure if using vapour diffusion (a particular problem for hanging drop set ups).

- **Sphaerolytes**
  Needle cushions, hedgehogs and sphaerolytes (crystalline balls with no sharp edges, often very colourful under a polarized microscope) share a common cause: the crystallization has occurred too quickly, or the crystal growth has been disturbed mechanically or by impurities in the solution. Small changes in the crystallization conditions (e.g. in pH, temperature or concentration conditions) may lead to nice crystals.

- **Salt crystals**
  Probably the most disappointing moment during crystallization is the discovery of regularly shaped, nicely grown crystals that turn out to be crystallized precipitant or inhibitor. If you have any doubts, check your crystallization experiment buffer for any ingredient that may form insoluble compounds (e.g. calcium or magnesium with phosphate or tartrate buffer), and change the conditions if necessary. If your experiments have been standing for a long time, take a look into the reservoir solution; if similar crystals appear there as well, then your crystals are almost certainly not protein.
  Salt crystals can be distinguished from protein in a variety of ways. Salt crystals show strong colours when viewed under polarizing light. As they have a low solvent content, salt crystals are rather hard; they tend to crack into pieces when touched with a glass rod or metal needle, while protein

crystals are soft and smear like jelly. Special dye solutions are available that stain only protein crystals (e.g. Izit-stain, Hampton Research). If the crystal is washed several times in **harvesting buffer** (mother liquor with 10 – 20% higher concentration of precipitant) to remove any attached protein, it can then be dissolved in Laemmli buffer and loaded onto an SDS-PAGE. Finally, protein and salt crystals are easily distinguished according to their diffraction patterns.

- **Bad crystals**
  Small crystals and needles are often the result of the first screening. Small changes in the crystallization conditions (*e.g.* exchange of ammonium sulphate for sodium sulphate, different PEG molecular weight, different temperature) in a new crystallization setup may give better results. You can also try various additives, e.g. β-octyl-glucoside, reducing agents, crosslinkers etc. as described in the flowchart.
  Larger crystals can be obtained via **seeding techniques**, which also reduce the number of crystallization nuclei:

- **Microseeding**
  One or more crystals are crushed into tiny pieces with a thin glass needle, washed in mother liquor, and transferred to a fresh protein crystalliza-tion drop using a whisker or a glass loop. If this method is not possible (crystals do not stick on the whisker), vortex the crushed crystals and add at least the same volume of water (up to 100-fold, depending on the size of the crystals: small crystals dissolve very quickly). Take a small amount of this diluted solution (which now contains only a few crystal nuclei) by using a drawn out pasteur pipette (heat the pasteur pipette over a bunsen burner until it begins to melt, then take it out of the heat and pull quickly; after a minute or so, you can then break the drawn out part to get a 0.1 mm diameter pipette end) and transfer it to a freshly prepared crystal-lization drop with the same precipitant concentration as the initial mother liquor of the crystals. This method is especially useful in cases where nucleation of the supersaturated protein solution occurs only rarely or not at all.

- **Macroseeding**
  A protein crystal is washed twice in harvesting buffer to remove any other crystals, and then transferred to a diluted harvesting buffer. Within a few minutes, the crystal loses its sharp edges and smooth faces. Transfer of the crystal to a new protein crystallization drop allows further growth of the etched planes, resulting in larger crystals.

If the crystals remain small, or show twinning or are intergrown, the protein may contain an impurity or heterogeneity which can be removed by recrystallization. A large batch of protein is brought to crystallization conditions (e.g. by dialysis) and seeded with some small crystals. The resulting large number of microcrystals can be isolated by centrifugation, washed three times with harvesting buffer to remove protein other than that in the crystals, redissolved by lowering the precipitant concentration, and set up again for crystallization.

- **Good crystals**
  The best result is a protein drop containing one or a few individual crystals with sharp edges, defined shape, and larger than 0.2 mm in each direction. Before writing your paper, try to reproduce the crystals, especially if they needed a long time to grow, or if you work with a proteinase with little or no homology to proteins of known structure, as you may need the help of heavy metal derivatives (and that means more crystals) to solve the structure.
  In both our laboratories and others, it has often been observed that some proteinase crystals take a very long time (several months) to grow. In some cases, analysis of the crystals showed that they contained an autocatalytically cleaved molecule (see the thrombin case study below). In the case of proteinase inhibitor complex crystals, preferential crystallization of only one component may also occur. For a successful structure determination (and to be able to reproduce the crystals), it is imperative to know the identity of the protein in the crystals. It is therefore advisable to perform the **SDS-PAGE analysis** outlined above. **N-terminal sequencing** provides a more sensitive technique, and provides more detailed information. For a gas phase sequencer, you can take one large or three small crystals, which must be carefully washed using harvesting buffer. The crystal should then be transferred to methanol-wetted Prospin centrifuge cups (Applied Biosystems, Foster City, California 94404), and the polyvinylidenfluoride (PVDF)-membrane washed several times with water according to the manufacturers instructions to remove salt and buffer. The membrane bound protein is now ready for sequencing.

Finally, the crystallization conditions for proteinases can be greatly influenced by the addition of different inhibitors (see our thrombin case study below). Even low molecular weight inhibitors can alter the surface charge or hydrophobicity, or produce conformational rearrangements that allow the proteinase to crystallize with one inhibitor yet give poor or no crystals with another. The choice and variation of inhibitor can therefore be an important parameter in obtaining suitable crystals for your proteinase.

In a suitable harvesting buffer, many protein crystals are stable for years if **Storage** they are protected against drying out, temperature changes, mechanical stress, and light, which can cause cross linking reactions (as e.g. with PEG as precipitant [10]). If you have crystals, however, it is wise to contact a practicing crystallographer first, as some crystals do not like changing buffers and are best mounted straight from the crystallization drop. As the mounting of protein crystals and their measurement in X-ray beams is beyond the scope of this article, the interested reader is referred to a standard text such as Blundell and Johnson 1976 or McRee 1993.

In the rapidly developing field of rational drug design, experimental ver- **Soaking** ification of the binding modes of synthetic small molecule inhibitors to their target proteinase is imperative. Not only does the X-ray structure of a proteinase - ligand cocrystal provide a basis for understanding which chemical groups are important for affinity and selectivity, it suggests possible candidates for modification to *e.g.* improve binding or reduce toxicity; furthermore, experience shows that continuous experimental monitoring of proteinase - ligand interactions are essential for the efficient progress of the design process. To facilitate this cycle, soaking is the method of choice. Soaking is feasible if the target proteinase can be crystallized without an inhibitor, or in complex with a weakly binding small inhibitor.

Transfer a well formed crystal into a 10 µl drop of harvesting buffer (mother liquor with 10 – 20% higher concentration of precipitant) devoid of inhibitor, and keep the drop at a constant temperature in a suitable receptacle with a reservoir solution similar to the mother liquor. Observe the crystal for around 30 minutes to make sure that it withstands the harvesting buffer. Prepare an inhibitor solution of approximately 50 mM in the harvesting buffer (some inhibitors are only poorly soluble in the harvesting buffer, in this case they can first be dissolved in e.g. DMSO to around 100 mM and then diluted 1:1 with the harvesting buffer). Add 1 µl of this solution laterally to the drop containing the crystal. Be sure neither to touch the crystal during this procedure nor to bring the crystal immediately in contact with the inhibitor. The inhibitor should diffuse slowly through the drop towards the crystal. Observe the crystal again for 30 minutes (the crystal can crack if the inhibitor is too large or causes severe changes in the conformation of the protease), then add another µl of the inhibitor solution. Total exchange of inhibitor takes some hours up to one week (depending on its K_i and solubility); in this time, add another µl of the inhibitor solution once or twice. The crystal is now ready for mounting.

The concentrations given above are intended as a guideline; for a tight binding inhibitor, these can be reduced. This might be worthwhile if the inhibitor is poorly soluble in the crystallization buffer, or if the crystals crack on soaking. If the inhibitor continues to precipitate in the harvesting buffer, do not despair; often the small amount of dissolved inhibitor is enough to get reasonable occupancy of the proteinase active sites in the crystal.

Applications

Proteinase Crystallization Case Study: Thrombin

Thrombin has been the subject of intense structural studies, both as a result of its fundamental importance in the process of blood coagulation, and as a framework for antithrombotic drug design (Stubbs and Bode 1993). These structural studies illustrate several aspects of proteinase crystallization.

The first published structure of α-thrombin was that of human thrombin in complex with the covalent inhibitor D-Phe-Pro-Arg-chloromethyl ketone (PPACK) (Bode et al. 1989). Thrombin was incubated with a 10-fold molar excess of the inhibitor for 20 minutes at $20°C$ in 0.2 M phosphate buffer, concentrated to 10 mg/ml and dialysed against 2 mM MOPS pH 7 with 100 mM NaCl. Crystals were obtained by the microdialysis method with 16% w/v PEG 6000 in 0.2 M sodium-potassium buffer, pH 7. The crystals were orthorhombic, spacegroup $P2_12_12_1$, with cell constants a=87.74 Å, b=67.81 Å, c=61.07 Å, one molecule in the asymmetric unit, and diffracted to 1.9 Å.

The most potent natural inhibitor of thrombin is hirudin (subpicomolar K_i) from the European medicinal leech *Hirudo medicinalis*. Crystals of the 1:1 complex were grown by the vapour diffusion method with 28 – 30% PEG 4000, 0.2 M $MgCl_2$, 0.1 M sodium acetate buffer, pH 4.5 and 1 mM NaN_3 (Rydel et al. 1990). The crystals were tetragonal, spacegroup $P4_32_12$, with cell constants a=b=90.39 Å, c=132.97 Å, one molecule in the asymmetric unit, and diffracted to 2.3 Å.

The thrombin-hirudin complex revealed that it is a two component interaction, with extensive electrostatic attraction between the C-terminal 'tail' of hirudin with the basic 'fibrinogen recognition exosite' of thrombin. Skrzypczak-Jankun et al. (1991) succeeded in obtaining crystals of thrombin in complex with the tail peptide ('hirugen'). Human thrombin (3.03 mg/ml in 0.75 M NaCl) was incubated with hirugen in 50 mM phosphate buffer, pH 7.3, and concentrated to 3.7 mg/ml in 50 mM sodium phosphate buffer pH 7.3 with 375 mM NaCl and 0.5 mM NaN_3. Crystals were obtained using both sitting and hanging drop vapour diffusion against 0.1 M sodium phosphate and up to 30 % v/v PEG 8000. The crystals were monoclinic, spacegroup C2, with cell constants a=70.8 Å, b=71.7 Å, c=72.6 Å, $\beta=100.9°$, with one molecule in the asymmetric unit, and diffracted to 2.2 Å. The active site of thrombin is empty in these crystals, allowing the soaking in of low molecular weight inhibitors as a starting point for structure based drug design.

Several other crystal forms have also been reported for human thrombin in complex with active site inhibitors and hirudin C-terminal derivatives (Banner and Hadváry 1991, Stubbs et al. 1992, Priestle et al. 1993; authors' unpublished observations). Banner and Hadváry (1991) published the following procedure: a 10 mg/ml thrombin solution in 10 mM Tris/HCL pH 8.0 and 50 mM NaCl was inhibited with an active site inhibitor and [des-amino Asp55]hirudin 55-65, and equilibrated using vapour diffusion against 0.1 M Hepes pH 7.5, 0.2 M CaCl2 and 25 % PEG 6000. One crystal form was orthorhombic, space group $P2_12_12_1$, with cell constants a=104 Å, b=61 Å, c=119 Å, and the other tetragonal, space group $P4_32_12$, with cell constants a=b=90.8 Å, c=132.5 Å, both diffracted to 3.0 Å. It is worth noting that the latter spacegroup is the same as that for thrombin hirudin mentioned above (Rydel et al. 1990), suggesting that the electrostatic neutralisation provided by the C-terminal peptide may be an important factor in the crystallization. This same crystal form was obtained using N-acetylated chloromethylated fibrinopeptide A and N-acetylated hirudin(54-65) in complex with human α-thrombin (Stubbs et al. 1992). A 10-fold molar excess of fibrinopeptide A was added to thrombin, followed by a fourfold molar excess of hirudin(54-65), buffered in 3 mM MOPS, 50 mM NaCl and 200 mM $CaCl_2$, pH 7.0. 4.5 µl of protein solution at ~4mg/ml was added to 1.5 µl 0.2 M MES / 10% w/v PEG 6000 at pH 6.0 and equilibrated against 1.5 M sodium phosphate using hanging drop vapour diffusion. The procedure published by Priestle et al. 1993 is as follows: thrombin (4 mg/ml) was inhibited by a 10-fold excess of PPACK, then dialysed against 2 mM MOPS, pH 7.0, containing 0.1 M NaCl and 0.05% NaN_3, concentrated to 6 mg/ml, and a twofold molar excess of Tyr-63-sulphated hirudin(55-65) added. Vapour diffusion against 0.1 M sodium acetate buffer, pH 4.0 with 6 – 12% PEG 6000 resulted in orthorhombic crystals, spacegroup $P2_12_12$, with cell constants a=80.9 Å, b=107.5 Å, c=45.9 Å, one molecule in the asymmetric unit, and diffracted to 2.5 Å.

Crystals of bovine thrombin (Brandstetter et al. 1992) proved difficult to reproduce, and took a long time to grow. Analysis of the crystalline material revealed that the thrombin molecule was cleaved in an exposed surface loop. Well diffracting crystals could be obtained reproducibly only after treatment of bovine α-thrombin with porcine pancreatic elastase. After the final purification step, α-thrombin was incubated with 1/100 w/w porcine pancreatic elastase for 30 min at 20° C, pH 8. The elastase cleavage in the 149-insertion loop generates ε-thrombin, which is fully active and has the same number of amino acids as α-thrombin. Thrombin inhibitors (1 mg/

ml) were added to the ε-thrombin, together with 1 M ammonium sulphate buffered with 0.1 M potassium phosphate, and the protein concentrated to 10 mg/ml. Octahedral crystals grew at 20° C using vapour diffusion against 1.9 M ammonium sulphate. The crystals were tetragonal, space group $P4_22_12$, with cell constants a=b=88.5 Å, c=102.9 Å, one molecule in the asymmetric unit, and diffracted to 3.0 Å.

This case study shows the wide range of variations open to the proteinase biochemist to obtain crystals suitable for X-ray analysis. Important is a good supply of protein, a variety of possible inhibitors, careful analytical skills, and a high frustration threshold. The authors wish you luck and success in solving the structure of your proteinase!

References

Banner, D.W., Hadváry, P. (1991): Crystallographic analysis at 3.0 Å resolution of the binding to human thrombin of four active site directed inhibitors. J. Biol. Chem. 266, pp. 20085 - 20093

Barrett, A.J. (1994): Classification of peptidases. Meth. in Enzymol., 244, pp. 1 - 15

Bernal, J.D. & Crowfoot, D. (1934): X-ray photographs of crystalline pepsin. Nature 133, p. 794

Blundell, T.L. & Johnson, L.N. (1976): Protein crystallography. Academic Press, New York

Bode, W., Mayr, I., Baumann, U., Huber, R., Stone, S.R., Hofsteenge, J. (1989): The refined 1.9 Å crystal structure of human α-thrombin: interaction with D-Phe-Pro-Arg chloromethylketone and significance of the Tyr-Pro-Pro-Trp insertion segment. EMBO J. 8, pp. 3467 - 3475

Brandstetter, H., Turk, D., Hoeffken, H.W., Grosse, D., Stürzebecher, J., Martin, P.D., Edwards, B.F.P., Bode, W. (1992): Refined 2.3 Å X-ray crystal structure of bovine thrombin complexes formed with the benzamidine and arginine-based thrombin inhibitors NAPAP, 4-TAPAP and MQPA. J. Mol. Biol. 226, pp. 1085 - 1099

Carter, C.W., Jr. & Carter, C.W. (1979): Protein crystallization using incomplete factorial experiments. Journal of Biological Chemistry 254, pp. 12219 - 12223

Ducruix, A. & Giegé, R.(ed.) (1992): Crystallization of nucleic acids and proteins. A practical approach. Oxford University Press, Oxford, New York, Tokyo

Finzel, B.C (1993): Software for macromolecular crystallography: a users overview. Curr. Opin. Struct. Biol., 3, pp. 741 - 474

Good, N.E., Winget, G.D., Winter, W., Connolly, T.N., Izawa, S., Singh, R.M.M. (1966): Hydrogen ion buffers for biological research. Biochemistry 5, pp. 467 - 477

Harris, E.L.V. (1989) in: Protein Purification methods - a practical approach (ed. by E.L.V. Harris & S. Angal), pp. 1 - 64

Helliwell, J.R. (1992): Macromolecular crystallography with synchrotron radiation. Cambridge University Press, Cambridge

Hofmeister, F. (1888): Zur Lehre von der Wirkung der Salze: Über Regelmässigkeiten in der eiweissfällenden Wirkung der Salze und ihre Beziehung zum physiologischen Verhalten derselben. Arch. Exp. Pathologie und Pharmakologie (Leipzig) 24, p. 247

Jancarik, J. & Kim, S.-H. (1991): Sparse matrix sampling: A screening method for crystallization of proteins. Journal of Applied Crystallography 24, pp. 409 - 411

Kottke, T. & Stalke, D. (1994): Crystal handling at low temperatures. J. Appl. Crystallogr., 26, pp. 615 - 619

McPherson, A. (1982): Preparation and analysis of protein crystals. John Wiley & Sons, New York.

McPherson, A., Malkin, A.J., Kuznetsov, Y.G. (1995): The science of macromolecular crystallization. Structure, 3, pp. 759 - 768

McRee, D.E. (1993): Practical protein crystallography. Academic Press, San Diego

Pflugrath, J.W. (1992): Development in X-ray detectors. Curr. Opin. Struct. Biol., 2, pp. 811 - 815

Price, S.R. & Nagai, K. (1995): Protein engineering as a tool for crystallography. Curr. Opin. Biotech. 6, pp. 425-430

Priestle, J.P., Rahuel, J., Rink, H., Tones, M., Grütter, M.G. (1993): Changes in interactions in complexes of hirudin derivatives and human α-thrombin due to different crystal forms. Protein Science 2, pp. 1630 - 1642

Roussell, A., Serre, L., Frey, M., Fontecilla-Camps, J.C. (1990): Rapid access to an updated biological macromolecule crystallization database through artificial intelligence. J. Crystal Growth, 106, p. 405

Rydel, T.J., Ravichandran, K.G., Tulinsky, A., Bode, W., Huber, R., Roitsch, C., Fenton, J-W.II (1990): The structure of a complex of recombinant hirudin and human α-thrombin. Science 249, pp. 277-280

Scopes, R.K. (1982): Protein purification, principles and practice. Springer Verlag, New York

Skelly JV & Madden B (1996): Overexpression, isolation, and crystallization of proteins. In: Jones C, Mulloy B, Sanderson M (Eds) Methods in Molecular Biology, Vol 56. pp:23-53

Skrzypczak-Jankun, E., Carperos, V.E., Ravichandran, K.G., Tulinsky, A., Westbrook, M., Maraganore, J.M. (1991): Structure of the hirugen and hirulog 1 complexes of α-thrombin. J. Mol. Biol. 221, pp. 1379 - 1393

Stubbs, M.T. & Bode, W. (1993): Crystal structures of thrombin and thrombin complexes as a framework for antithrombotic drug design. Perspectives in Drug Discovery and Design 1, pp. 431 - 452

Stubbs, M.T., Oschkinat, H., Mayr, I., Huber, R., Angliker, H., Stone, S.R., Bode, W. (1992): The interaction of thrombin with fibrinogen - a structural basis for its specificity. Eur. J. Biochem. 206, pp. 187 - 195

von Hippel, P.H. & Schleich, T. (1969) in: Structure and stability of biological macromolecules (ed. by S.N. Timasheff and G.D. Fashman), pp. 417 - 574, Dekker, New York

Wiegand, G. (1990): How do you get large protein crystals? In: Tschesche, H. (ed.): Modern Methods in Protein- and Nucleic Acid Research. De Gruyter, Berlin New York.

Abbreviations

Tris	Tris(hydroxymethyl)aminomethane
PEG	Polyethylene glycol
NaN₃	Sodium azide
pI	isoelectric point
SDS	sodium dodecylsulfate
MOPS	3-morpholinopropane sulphonic acid
IEF	isoelectric focusing
PAGE	polyacrylamide gel electrophoresis
MPD	2-methyl-2,4-pentanediol
DMSO	Dimethylsulfoxide

Chapter 10

Basic Kinetic Mechanisms of Proteolytic Enzymes

L. POLGÁR

Introduction

Peptidases, like other enzymes, generally follow Michaelis–Menten kinetics, and their reactions can be characterized by k_{cat}, K_m and particularly k_{cat}/K_m. Each of these parameters may involve several steps of the catalysis. The exploration of the elementary steps, their relationship to one another and the overall parameters is essential for the understanding of the catalytic mechanism. This chapter concerns the basic kinetic equations which help elucidate the most important features of the peptidase reactions, such as acylation and deacylation, formation of intermediates, the nature of the rate-limiting step, isomerization of the enzyme–substrate complex, and the ionizing groups that control these processes.

Mechanisms of Peptide Bond Cleavage

Four Classes of Peptidases

Four well–established mechanisms are known concerning the hydrolysis of peptide bond, each associated with a particular enzyme group, namely the serine, cysteine, aspartic and metallo peptidases. Two basic types of peptide bond hydrolysis may be distinguished. One involves the direct attack of a water molecule on the carbonyl carbon of the substrate, which leads to the formation of a tetrahedral intermediate that decomposes in the next step, providing the two hydrolysis products. This single addition–elimination mechanism holds in the reactions of the aspartic and metallo peptidases. The other mechanism, the double addition–elimination (Fig. 1), operates in

L. Polgár , Hungarian Academy of Sciences, Biological Research Center, Institute of Enzymology, P.O. Box 7, Budapest, 1518, Hungary (*phone* 36-1-466-8858; *fax* 36-1-166-466-5465; *e-mail* polgar@hanga.enzim.hu)

the catalyses of serine and cysteine peptidases. In this mechanism a nucleophile of the enzyme, such as the serine oxygen or the sulfur atom of a cysteine residue, rather than a water molecule, reacts with the substrate carbonyl carbon, leading through a tetrahedral intermediate to an acyl–enzyme and amino product. The acyl–enzyme, a covalent intermediate, is then hydrolyzed through the reverse reaction pathway of acylation, but in the second addition–elimination a water molecule is the attacking nucleophile.

The attack of nucleophile is always facilitated by an enzymic group (imidazole or carboxylate) by accepting the proton from the nucleophile (general base catalysis) concomitantly with the formation of the tetrahedral intermediate, and by transferring the accepted proton to the leaving peptide nitrogen while the tetrahedral intermediate breaks down (general acid catalysis). A more detailed discussion of the similarities and differences regarding the mechanisms of the four peptidase groups is found elsewhere (Polgár 1989; 1990).

Kinetic Equations for Simple Peptidase Reactions

Single Displacement Reactions

The reactions of aspartic and metallo peptidases generally obey the simple Michaelis–Menten kinetics which involves the formation of an enzyme-substrate complex or Michaelis complex (ES) held together by physical forces (eqn. 1). This rapid and reversible process is followed by the irreversible chemical reaction characterized by the first–order rate constant k_{cat}. P_1

No acyl–enzyme intermediate

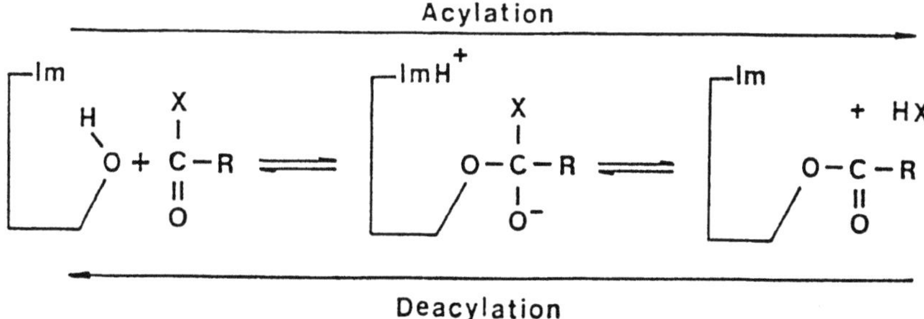

Fig. 1. Scheme of the mechanism of action of serine peptidases. Im represents an imidazole group and X stands for the leaving group in acylation and for the OH group of a water molecule in deacylation. The leaving group may be an amino compound with peptides or alcohol with ester substrates.

and P_2 stand for the amino and acyl products of the substrate, respectively. The dissociation constant of the enzyme–substrate complex is defined by eqn. 2, and the rate of the enzyme reaction is shown by eqn. 3, where $[E_o]$ is the total enzyme concentration ($[E] + [ES]$), and $V = k_{cat}[E_o]$ stands for the maximal rate that is obtained when the enzyme is saturated with substrate.

$$E + S \underset{}{\overset{K_s}{\rightleftharpoons}} ES \xrightarrow{k_{cat}} E + P_1 + P_2 \tag{1}$$

$$K_s = [E][S]/[ES] \tag{2}$$

$$v = k_{cat}[ES] = k_{cat}[E_0][S]/(K_s + [S]) = V[S]/(K_s + [S]) \tag{3}$$

The Michaelis–Menten mechanism is limited to the fast preequilibrium formation of ES, i. e. when k_{cat} is much smaller than k_{-1} of eqn. 4, an extended form of the Michaelis–Menten equation. The general forms represented by eqns. 4 and 5 (Briggs–Haldane kinetics) permit that k_{cat} be comparable to $k-1$. Here K_s is replaced by K_m which represents the Michaelis constant, a pseudoequilibrium constant (eqn. 6) that may be defined as the substrate concentration at half-maximum rate. Eqn. 6 simplifies to $K_m = k_{-1}/k_1 = K_s$ when $k_{cat} \ll k_{-1}$, otherwise $K_s < K_m$.

$$E + S \underset{k_{-1}}{\overset{k_1}{\rightleftharpoons}} ES \xrightarrow{k_{cat}} E + P_1 + P_2 \tag{4}$$

$$v = V[S]/(K_m + [S]) = k_{cat}[E_0][S]/(K_m + [S]) \tag{5}$$

$$K_m = (k_{-1} + k_{cat})/k_1 \tag{6}$$

Double Displacement Reactions

Acyl–enzyme intermediate Serine and cysteine peptidases have, in addition to the Michaelis complex, a covalent acyl–enzyme intermediate (EA), as shown by eqn. 7, where k_2 and k_3 represent the first-order acylation and deacylation rate constants, respectively. The related Michaelis–Menten parameters (k_{cat} and K_m) can be expressed as shown by eqns. 8 and 9.

$$E + S \underset{k_{-1}}{\overset{k_1}{\rightleftharpoons}} ES \overset{k_2}{\underset{P_1}{\searrow}} EA \overset{k_3}{\underset{P_2}{\searrow}} E \tag{7}$$

$$k_{cat} = V/[E_0] = k_2k_3/(k_2 + k_3) \tag{8}$$

$$K_m = K_sk_3/(k_2 + k_3) \tag{9}$$

It follows from eqn. 9 that $K_m < K_s$. When $k_2 \ll k_3$, eqns. 8 and 9 simplify to eqns. 10 and 11, respectively. Under these conditions the acyl–enzyme does not accumulate during the reaction. This mechanism holds, for example, in the hydrolysis of amide substrates catalyzed by the serine peptidases trypsin or chymotrypsin (Polgár 1987; 1989).

$$k_{cat} = k_2 \tag{10}$$

$$K_m = K_s \tag{11}$$

When $k_3 << k_2$, eqns. 8 and 9 are reduced to eqns. 12 and 13, respectively. Under these conditions accumulation of the acyl–enzyme is detectable during the catalysis. Reactions conforming these equations also occur in practice, as in the hydrolysis of activated esters (e. g. nitrophenyl esters) catalyzed by chymotrypsin (Polgár 1987; 1989). Specifically, the fast liberation of the nitrophenolate ion, i. e. the "burst" reaction, has indicated the formation of the acyl–enzyme intermediate, which is followed by a slower steady state reaction corresponding to the hydrolysis of the acyl–enzyme.

$$k_{cat} = k_3 \tag{12}$$

$$K_m = K_sk_3/k_2 \tag{13}$$

Dividing eqn. 8 by eqn. 9, eqn. 14 is obtained. The k_{cat}/K_m or k_2/K_s is the second–order acylation rate constant and characterized by eqn. 15.

$$k_{cat}/K_m = k_2/K_s \tag{14}$$

$$E + S \xrightarrow{k_{cat}/K_m} EA + P_1 \tag{15}$$

Meaning of the Kinetic Parameters

In the catalysis of aspartic and metallo peptidases, where a covalent intermediate is not formed, the simple Michaelis–Menten mechanism often prevails. In this case k_{cat} represents the first–order rate constant of the chemical conversion of the enzyme–substrate complex into products, and K_m is commonly equal to K_s. On the other hand, these parameters are less meaningful in the serine and cysteine peptidase reactions because the complexities of the two parameters (eqns. 8 and 9) do not permit to assign the constants to any single elementary process, except in special cases as shown by eqns. 10–12. Accordingly, these macroscopic constants in the absence of additional data, do not bear any straightforward mechanistic information. They may be useful, however, when the reactions of a particular peptidase with two or more substrates are to be compared.

k_{cat}/K_m The k_{cat}/K_m is an apparent second–order rate constant that depends on the concentrations of the free enzyme and free substrate. It determines the specificity for competing substrates, and therefore k_{cat}/K_m is often referred to as the specificity rate constant, which involves binding. It is a macroscopic constant that may approach a true microscopic rate constant when the rate–limiting step in the catalysis is the association of the enzyme and substrate (k_1 in eqns. 4 and 7).

An additional factor that can be rate–limiting for k_{cat}/K_m is the conversion of the ES into ES' complex, as illustrated by eqns. 16 and 17 for the single and double displacement reactions, respectively. Such conformational changes have been found with several peptidases, like HIV-1 protease, an aspartic peptidase (Hyland et al., 1991; Polgár et al., 1994) and the serine peptidases, subtilisin (Matta and Andracki 1988) and prolyl oligopeptidase (Polgár, 1992).

$$E + S \rightleftharpoons ES \rightleftharpoons ES' \longrightarrow E + P_1 + P_2 \tag{16}$$

$$E + S \rightleftharpoons ES \rightleftharpoons ES' \underset{\searrow P_1}{\longrightarrow} EA \longrightarrow E + P_2 \tag{17}$$

Distinction Between Rate–Limiting Chemical and Physical Steps

The rate-limiting step for the complex process characterized by k_{cat}/K_m is mostly the chemical reaction involving the formation and decomposition of the tetrahedral intermediate catalyzed by a general base and a general acid, respectively. General acid/base catalysis proceeds slower in deuterium oxide by a factor of 2–3 (Polgár, 1987; 1989). Therefore, rate-limiting conformational change and rate-limiting association of enzyme and substrate, as these processes do not exhibit kinetic deuterium isotope effects, may be distinguished from the rate-limiting chemical reaction which does have a solvent isotope effect. In the absence of kinetic deuterium isotope effects, conformational change is expected to be rate–limiting if k_{cat}/K_m is significantly lower than the diffusion–controlled rate constant (10^7–10^8 $M^{-1}s^{-1}$). On the other hand, diffusion may be partly or completely rate–determining if k_{cat}/K_m approaches the diffusion limit.

The k_{cat}/K_m has the advantage over k_{cat} that it is not affected by non–pro- **Non–productive** ductive binding. Peptidases may have alternative binding modes for a par- **binding** ticular substrate, resulting in active (ES) and inactive (ES_i) complexes (eqn. 18). Non–productive binding proportionally decreases k_{cat} and K_m since a fraction of the enzyme forms an inactive complex (ES_i), and only the rest takes part in the catalysis.

$$ES_i \xrightleftharpoons{K_i} E + S \xrightleftharpoons{K_s} ES \xrightarrow{k_{cat}} E + P_1 + P_2 \tag{18}$$

$$1/v = 1/V + K_m/V[S] \tag{19}$$

Estimation of the Kinetic Parameters

Linear and Nonlinear Regression Analysis

The majority of the kinetic data of the peptidase literature concerns k_{cat} and K_m. Until recently the linearized forms of eqn. 5, such as the Lineweaver-Burk plot (eqn. 19), have been used most frequently. According to eqn. 19 the plot of $1/v$ against $1/[S]$ yields a straight line which is characterized by the slope of K_m/V and the intercept of $1/V$ on the ordinate. Unfortunately, the double reciprocal plot introduces statistical bias unless subjected to weighted regression analysis. The ready availability of personal computers in laboratories renders it possible to analyze the original nonlinear data set more precisely by nonlinear regression. Computer programs performing nonlinear regression analysis are available from various commercial sources (e. g. Enzfitter, Elsevier Biosoft; GraFit, Erithacus Software; Sigma-Plot, Jandel Scientific). These programs offer a wide choice of kinetic equations, including the Michaelis equation, and perform linear and nonlinear regression data analyses even with the user's own equations.

Initial Rate Measurements

When measuring initial rates it is desirable to keep the substrate consumption below 5%. This ensures the linearity of the reaction at a practically constant substrate concentration. Most k_{cat} and K_m data of the literature have been acquired by initial rate measurements. Initial rates are calculated from the decrease in substrate concentration per second ($\Delta[S]/s$) or from the increase of product concentration per second ($\Delta[P]/s$). The substrate or product concentration can readily be determined by dividing the changes of absorbance (ΔA) by the molar extinction coefficient for the substrate hydrolysis ($\Delta\varepsilon$). If $\Delta\varepsilon$ is not known, it can be determined from the complete hydrolysis of a known concentration of substrate. In the case of fluorescence measurements the substrate concentration can be calculated in a similar manner. From the initial rate data the kinetic parameters may be obtained by nonlinear regression analysis, using eqn. 5.

Because peptidases often contain a substantial amount of inactive material as a result of oxidation or autolysis, the active enzyme concentration $[E_o]$ should be used to obtain precise rate constants. This operative concentration can be determined by active site titration (Polgár 1989; Knight 1995a).

Determination of k_{cat} and K_m Requires $[S_o] > K_m$

When determining k_{cat} and K_m care must be taken that the substrate concentration be higher than the K_m, preferably in the range of $(0.2–6)Km$. Poor solubility of the substrate may pose a problem, and there are many examples in the literature when the parameters are calculated from substrate concentrations below K_m. Such data, of course, can be considered as approximate values only.

Problems with Organic Solvents

The reactions are often conducted at high concentrations of organic solvents to overcome solubility problems with substrates dissolving poorly in the reaction medium. The sensitivities of peptidases to organic solvents differ considerably. For example, papain is active at very high concentrations (20–30%) of organic solvents, while prolyl oligopeptidase tends to be inhibited by as low as 0.1% acetonitrile, dimethylformamide or dioxane, and the inhibition by 1% solvent may be more than 50%. Nevertheless, experiments have been carried out in the presence of 20% dioxane, where the enzyme possesses only very low activity. If the substrate is dissolved in organic solvent, care must be taken to keep the solvent concentration constant when the K_m is determined with initial rate measurements using increasing concentrations of substrates. Otherwise, the inhibition by the varied concentration of solvent can cause serious errors.

Integrated Michaelis–Menten Equation

The k_{cat} and K_m may be calculated from a single progress curve by using the integrated Michaelis–Menten equation (eqn. 20), where [P] stands for the concentration of the product at t time. Unfortunately, most of the computer programs that are commercially available require the explicit form of [P] to be expressed as a function of t, which is not possible with eqn. 20. However, the inverse formula, i. e. t being the explicit function of [P], can be obtained by dividing eqn. 20 by V, and this form may provide approximate values for k_{cat} and K_m. The absorbance or fluorescence data column obtained from the progress curve can readily be transformed, for example by the GraFit program (Leatherbarrow 1990), to the values required for plotting. Thus [P] = $A[S_o]/(A_\infty - A_o)$ can be calculated by using the Rescale command from the Manipulate menu. To this end, it is essential to determine exactly the absorbance (A) at the beginning (A_o) and the end (A_∞) of the reaction. Problems with the integrated Michaelis–Menten equation have recently been discussed (Duggleby 1995).

An additional approach utilizes the linearized form of the integrated Michaelis–Menten equation (eqn. 21), where $(1/t)\ln([S_o]/([S_o]-[P]))$ is plotted versus [P]/t (Fig. 2). The slope of the curve is $-1/K_m$, and the intercept on the ordinate is V/K_m, which gives $V = k_{cat}E_o$ when multiplied by K_m. The points relating to the lowest and highest substrate concentrations are often less accurate and can be neglected if sufficient points remain below and above the Km as seen in Fig. 2. Using either eqn. 20 or eqn. 21 we have obtained identical Michaelis-Menten parameters. Of course, it is more convenient to use eqn. 20, but the points causing division by zero should be removed from the data set, as for example [P] = $[S_o]$. This is usually not a problem with the calculation of the parameters, but rather with the drawing of the curve if the "finish" of the [P] axis is equal or greater than $[S_o]$.

$$Vt = [P] + K_m(\ln([S_0]/([S_0] - [P]))) \qquad (20)$$

$$(1/t)\ln([S_0]/([S_0] - [P])) = V/K_m - (1/K_m)([P]/t) \qquad (21)$$

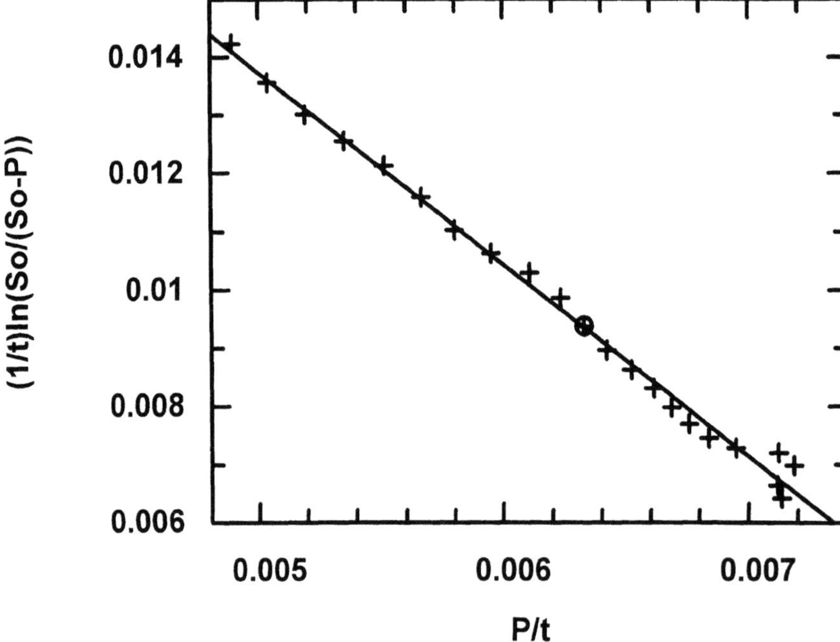

Fig. 2. A plot of the linearized form of the integrated Michaelis–Menten equation. The hydrolysis of aminobenzoyl–Thr–Ile–Nle–Phe(NO$_2$)–Glu–Arg substrate (1.28 µM) by HIV-1 protease (3.51 nM) was monitored with a Jasco FP 777 spectrofluorometer using 337 nm and 420 nm excitation and emission wavelength, respectively, at pH 5.41 in the presence of 2 M NaCl. The substrate concentrations used for the calculation extend from 0.04 µM to 0.96 µM. The point related to the K_m value is represented by an open circle. The following parameters were obtained from the data: K_m =0.30 µM and k_{cat} = 2.61 S^{-1}.

Advantages and Limitations of the Progress Curve and Initial Rate Methods

- The progress curve for determination of k_{cat} and K_m can be measured in a single run at a substrate concentration of 4–8 times as high as the K_m, whereas initial rates requires about 10 runs at variable substrate concentrations, say from 0.2 K_m to 6 K_m.

- A single run of initial rate measurements requires lower concentration of enzyme by about one order of magnitude than the determination of the progress curve. More enzyme and substrate are needed, however, for measuring a complete data set of initial rates.

- A considerable advantage of initial rate measurements over the progress curve method is that product inhibition does not interfere with the determination conducted only at the beginning (5% or less) of the reaction.

- The initial rate measurements cannot be used when the assay is not sensitive enough to monitor only 5–10 % of the reaction at low substrate concentrations. Therefore, the progress curve method may be invaluable for reactions having very low K_m.

- Only irreversible reactions can be analyzed with the progress curve method. Fortunately, the peptide bond hydrolysis usually meets this requirement.

Determination of k_{cat}/K_m when $K_m \gg [S_0]$

The specificity rate constant, k_{cat}/K_m, is generally calculated from k_{cat} and K_m. However, these parameters cannot always be determined because of solubility reasons. Furthermore, k_{cat} and K_m are less accurate than their ratio that can be directly determined under first–order conditions, where $K_m \gg [S_0]$. In this case eqn. 22 is obeyed, which has been derived from eqn. 5. The pseudo first–order rate constant ($k_{cat}/K_m[E_0]$ in eqn. 22) divided by $[E_0]$ will provide k_{cat}/K_m. When the decrease in absorbance is monitored, as in the hydrolysis of peptide substrates containing 4–nitrophenylalanine at P1' position, eqn. 23 can be used, where t is the time and k represents the first–order rate constant for a single exponential decay. If the absorbance or fluorescence increases with product formation, eqn. 24 applies. Note that $[S_0] = [P_\infty]$. When working with first–order processes, it is not necessary to follow more than 3–4 half-lives ($t_{1/2} = 0.693/k$) of the reaction to obtain a sufficiently accurate rate constant. This is very helpful with slow reactions. An additional advantage may be that the first–order rate constant is independent of substrate concentration. Therefore, the knowledge of the precise substrate concentration is not an absolute requirement.

If the reaction is very slow, k_{cat}/K_m can be estimated by measuring the initial rate (v_0) of eqn. 25, provided that $K_m \gg [S_0]$. However, in this case the knowledge of the exact concentration of the substrate is essential.

$$v = (k_{cat}/K_m)[E_0][S] \tag{22}$$

$$[S] = [S_0] \exp(-kt) \tag{23}$$

$$[P] = [P_{\langle\infty\rangle}](1 - \exp(-kt)) \tag{24}$$

$$k_{cat}/K_m = v_0/([E_0][S_0]) \tag{25}$$

$$v = V[S]/(K_s + [S] + [S]^2/K_{2s}) \tag{26}$$

Substrate Inhibition

Determination of the Michaelis–Menten parameters may be perturbed by substrate inhibition, when the binding of two substrate molecules to the same enzyme molecule generates an inactive complex. In such a case eqn. 26 applies, where K_{2s} is the second dissociation constant of the enzyme–substrate complex. If [S] is low, then $v = V[S]/K_s = [S][E_o]k_{cat}/K_s$, and the determination of the specificity rate constant is not affected by the substrate inhibition. As [S] increases, a maximum value of v is attained followed by a decrease. It is worthy of note that a similar phenomenon may be observed when initial rates are measured with increasing amounts of substrate dissolved in organic solvent, so that the inhibitory effect of the increasing solvent concentration on the peptidase activity is disregarded.

Effects of pH on Kinetic Parameters

The enzymic rate constants generally alter with pH (for reviews see Fersht 1985; Dixon 1992; Brocklehurst 1994). In the simplest case they depend on the ionization of one enzymic group that is involved in the catalysis. Eqns. 27 and 28 show the pH-dependence of parameter k (k_{cat}, k_2, k_3, k_{cat}/K_m, K_m) which follows a simple sigmoid curve characterized by the pK_a of a kinetically important group in the basic and acidic form, respectively. The k(limit) stands for the pH-independent, maximum value of the constant. The participation of a histidine residue in the reactions of serine peptidases has been first suggested by such studies (Polgár 1987;1989). For example, deacylation of chymotrypsin obeys eqn. 27 and shows the catalytic importance of a base having a pK_a of about 7. On the other hand, acylation of chymotrypsin depends on the ionizations of two groups, one base and one acid. This gives rise to a bell-shaped pH-dependence as described by eqn. 29, where pK_1 and pK_2 refer to the ionizations of the catalytically competent base and acid, respectively. In the chymotrypsin reaction pK_1

reflects the ionization of the essential histidine residue, whereas pK_2 is responsible for a group that maintains the active conformation of the protein.

$$k = k(\text{limit})\left[1/\left(1 + 10^{(pK_a - pH)}\right)\right] \tag{27}$$

$$k = k(\text{limit})\left[1/\left(1 + 10^{(pH - pK_a)}\right)\right] \tag{28}$$

$$k = k(\text{limit})\left[1/\left(1 + 10^{(pK_1 - pH)} + 10^{(pH - pK_2)}\right)\right] \tag{29}$$

The pK_a values extracted from the pH–dependence of k_{cat}/K_m reflect the ionizing groups that are encountered in the free enzyme and free substrate since k_{cat}/K_m concerns the process $E + S \rightarrow ES$. The same holds for the pH–dependence of $1/K_m$. On the other hand, the pH–dependencies of k_{cat} and K_m follow the pK_a of the enzyme–substrate complex $ES \rightarrow E + P$ and $ES \rightarrow E + S$, respectively. If an acyl–enzyme intermediate is accumulated on the reaction pathway, the pK_a extracted from the pH–k_3 profile reflects the ionization of the acyl–enzyme.

The pK_a of the catalytically competent group of the free enzyme should be independent of the reacting substrate. This is usually found with the pH–dependence of k_{cat}/K_m, but some exceptions may be noted as follows.   **Kinetic pK_a**

- The rate–limiting step changes with pH. Specifically, if the diffusion is rate–limiting at high pH where a catalytically competent base is operative, and the chemical reaction is rate–limiting at low pH where the base is largely protonated, then an apparent shift in the true pK_a of the catalytic group ensues.

- Overlapping ionization between the catalytic group and some kinetically influential enzymic group(s). Such additional group(s) with pK_a value(s) not far from that of the catalytic group create(s) an electrostatic environment that changes with pH. This causes the enzyme to react with different substrates with distinct apparent pK_a values, in particular, when the substrates have different charge distributions.

The pK_a values discussed with the above two instances do not represent true ionization constants. They are affected by rate constants that differently depend on pH, and such apparent pK_as can be classified as kinetic pK_as.

Effects of Ionic Strength on pK$_a$

The pK$_a$ of a catalytic group may be perturbed by surface charges and the perturbation may be modified by changes in the ionic strength of the medium. The shift from the true pK$_a$ is most marked at low ionic strength (below 0.1 M), and may be depressed at high ionic strength (above 1 M). Therefore, the determination of pH–dependence curves should be performed at constant ionic strength. Buffers of constant ionic strength for studying pH–dependent processes have been described (Ellis and Morrison 1982).

Effects of Temperature on Kinetic Parameters

In contrast to most enzymes, peptidases are able to react with various substrates. Mechanistic differences in the reactions of one peptidase with two or more substrates may be explored by studying the temperature dependencies of the reactions. Different substrates cleaved with similar rate constants i. e. with similar free energies of activation (ΔG^*) as shown by eqn. 30, may display very different activation enthalpies (ΔH^*) and entropies (ΔS^*), indicating some mechanistic difference between the reactions. In eqn. 30 k_B is the Boltzmann constant (1.381×10^{-23} J/K), T is the absolute temperature, h is the Planck constant (6.626×10^{-34} Js) and R is the gas constant (8.314 J/mol K).

$$k = (k_B T/h) \exp(-\langle\Delta\rangle G^*/RT) \tag{30}$$

$$E = \langle\Delta\rangle H^* + RT \tag{31}$$

$$\ln(k/T) = \ln(R/N_A h) + \langle\Delta\rangle S^*/R - \langle\Delta\rangle H^*/RT \tag{32}$$

$$\langle\Delta\rangle G^* = \langle\Delta\rangle H^* - T\langle\Delta\rangle S^* \tag{33}$$

$$\langle\Delta\rangle G^* = RT \ln(k_B T/hk) = 2478.9 \ln(620.79 \times 10^{10} \times 1/k) \tag{34}$$

A number of studies have been concerned with the determination of activation energy (E) by plotting lnk against 1/T (Arrhenius plot). Linear fit provides the slope = –E/R. The activation energy acquired from the Arrhenius plot is related to the activation enthalpy by eqn. 31. More information is obtained from the Eyring plot (eqn. 32) which allows to calculate both ΔH^* and ΔS^*. In eqn. 32 N_A stands for the Avogadro number (6.022×10^{23}/mol). Linear regression analysis gives $\Delta H^* = -(\text{slope})8.314$ J/

mol and $\Delta S^* = (\text{intercept} - \ln(R/N_A h))R = (\text{intercept} - 23.76)8.314$ J/mol K. The free energy of activation can be calculated from eqn. 33 or directly from the rate constant (eqn. 34), given in J/mol at 298.16 K.

The transition state of a normal chemical reaction is usually more ordered than its ground state, and thus the reaction is associated with negative entropy of activation. Therefore, a large positive entropy of activation may suggest rate–limiting reorganization of the water molecules in the active site cleft and/or around the substrate while the enzyme–substrate complex is formed (see for example Matta and Andracki 1988).

Assays for Peptidases

For studying the kinetic mechanisms of peptidases, synthetic chromogenic and fluorogenic peptide derivatives are the method of choice. A great advantage of these compounds is that their hydrolysis can be continuously monitored with a spectrophotometer or spectrofluorometer, so that a large number of points and thus more accurate data can be acquired. Even the initial rates when recorded continuously are more precise compared with those determined at a single time, though the latter procedure is widely used in practice.

Chromogenic Substrates

A wide variety of chromogenic substrates have been used for the kinetic investigations of peptidases. In the early studies of the mechanism of chymotrypsin action, p–nitrophenyl esters played a key role, showing the presence of an acyl–enzyme intermediate by a "burst" reaction (Polgár 1987; 1989). Since the p-nitrophenolate ion is a good leaving group, it can be released from the tetrahedral intermediate without the aid of general acid catalysis. This implies that the hydrolysis of nitrophenyl esters proceeds by a simpler mechanism than the cleavage of the peptide bond. This is consistent with the similar rate constants found for the hydrolysis rates of D– and L–amino acid nitrophenyl esters, whereas the corresponding alkyl esters, which are less reactive compounds, exhibit several orders of magnitude difference in k_{cat}/K_m (Polgár and Fejes 1979). Some advantages and limitations of the use of nitrophenyl ester substrates are mentioned below.

Nitrophenyl esters

- As activated esters, nitrophenyl esters are very sensitive substrates of peptidases.

- The molar extinction coefficient of the p-nitrophenolate ion is high ($18,300$ $M^{-1}cm^{-1}$ at 400 nm), but changes considerably with pH around neutrality as its pK_a is 7.15.

- Because of their high reactivity, nitrophenyl esters are hydrolyzed in neutral buffers in the absence of enzyme, and even more in the alkaline pH region, so that above pH 8 they cannot practically be used. Therefore, the enzymic reactions should be corrected for buffer or "spontaneous" hydrolysis.

- Nitrophenyl esters have primarily been used as substrates for serine peptidases.

Thiolesters Like the nitrophenyl esters, thiolester substrates also have a good leaving group and thus exhibit high reactivities toward serine and metallo peptidases (Powers and Kam 1995). Cleavage of the thiolester bond can continuously be monitored if a thiol reagent such as 4,4'–dithiodipyridine or 5,5'–dithiobis(2–nitrobenzoic acid) is included in the assay mixture to produce a chromogenic compound. In contrast to the nitrophenyl esters, thiolesters have low background hydrolysis rates. However, the use of thiolester substrates has been limited by the commercial availability of only a small number of compounds.

Nitroanilides Nitroanilides possessing an amide bond resemble more the natural peptide substrates than do nitrophenyl esters or thiolesters. Hence, they are less reactive, but more stable, and more specific compared with the activated esters. Being more adequate for studying the peptide bond cleavage, 4–nitroanilides are the most widely used substrates of the present–day investigations of serine and cysteine peptidases. The hydrolysis of 4–nitroanilides can be monitored at 410 nm, and $\Delta\varepsilon = 8800$ $M^{-1}cm^{-1}$ is used to calculate the concentration of the released product.

Substrates Containing Amino Acid Residues on both Sides of the Scissile Peptide Bond

Nitrophenyl ester and nitroanilide substrates contain amino acids only on the left–hand side of the bond to be cleaved (acyl side, P1..Pn residues). Aspartic and metallo peptidases, however, require substrates that possess amino acids on the amino side (P1'...Pn') as well. To study such substrates, peptides containing 4–nitrophenylalanine, $Phe(NO_2)$, has been introduced into the P1 or P1' position. Fortunately, most aspartic and metallo peptidases prefer an aromatic residue at position P1 or P1', or both. When a substrate with $Phe(NO_2)$ at P1 is hydrolyzed, an increase in absorbance at 310 nm can be observed (Inouye and Fruton 1967). When the chromogenic residue is located at P1', the absorbance decreases as the reaction proceeds. An example of using such a substrate is given below.

Determination of k_{cat}/K_m for the HIV–1 Protease Reaction

- Enzyme: Human immunodeficiency virus type–1 protease (HIV–1) is commercially available (for example Bachem, Bubendorf, Switzerland). The stock solution contains 0.3 mg/ml peptidase or less in 50 mM phosphate buffer, pH 7.5, 1 mM EDTA, 1 mM dithioerythritol (DTE), 0.1 M NaCl and 10% glycerol, and stored at –80 °C.

- Substrate: Lys–Ala–Arg–Val–Leu*$Phe(NO_2)$–Glu–Ala–Nle or its derivative amidated at the C-terminal norleucine (Nle). The * stands for the scissile bond. It is commercially available (Bachem) or can be prepared by solid phase synthesis. The stock solution contains 2 mg compound per ml of 1 mM HCl, and stored at –18 °C.

- Buffer: 50 mM acetate buffer, pH 4.0, containing 1 mM EDTA, 1 mM DTE, 5 % glycerol.

- A sample of 10 μl substrate is added to 985 μl buffer equilibrated in the spectrophotometer cell. The reaction is started with the addition of 5 μl enzyme, diluted appropriately to give a concentration of 40–150 nM in the cell and the reaction is followed at 300 nm (Richards et al. 1990) by a sensitive spectrophotometer that can detect 0.0001 absorbance.

- The first–order rate constant is calculated by nonlinear regression analysis using an appropriate software (e. g. GraFit) and eqn. 23. The first–order rate constant divided by the enzyme concentration provides k_{cat}/K_m.

Note: The ionic strength and pH considerably affect the rate constant (Polgár et al. 1994). Increasing ionic strength increases k_{cat}/K_m but decreases K_m, weakening the first–order assay condition.

Assays with Fluorogenic Substrates

Two major groups of fluorogenic substrates may be distinguished. One is related to the nitroanilides by containing aromatic amines, such as 2–naphthylamine or 7–amino–4–methylcoumarin. The other group of substrates contain a fluorophore and a quencher which are separated by the scissile bond. Upon hydrolysis the quenching is relieved, allowing to monitor the increasing fluorescence. Such quenched fluorescent substrates can be used with all four groups of peptidases, and they are particularly invaluable in the studies of aspartic and metallo peptidases. A review on fluorogenic substrates has recently been published (Knight 1995b).

Naphthylamides and 7–amino–4–methylcoumarin

Peptidyl 2–naphthylamides are highly sensitive substrates and their reactions can be followed at 340 nm and 410 nm excitation and emission wavelengths, respectively. Substrate concentrations as low as 0.1 µM may be detected so that most reactions can be monitored under first–order conditions. The 7–amino–4–methylcoumarin derivatives are even more sensitive. The excitation and emission wavelengths used for measuring their hydrolysis are 370 nm and 460 nm, respectively.

The coumaryl substrates are more expensive than the naphthylamides, but the latter compounds concern about the carcinogenic nature of 2–naphthylamine. An additional drawback of 2-naphthylamine is its photosensitive nature. It is therefore important to use a small excitation slit width. With the Jasco FP 777 spectrofluorometer, for example, the hydrolysis of naphthylamides can be readily measured at 1.5 nm band width, but not at 5 nm. No such problem emerges with the corresponding coumarylamides.

Quenched fluorescent substrates

Several quencher–fluorophore pairs have been developed to monitor the hydrolysis of oligopeptide substrates. Thus, the tryptophan fluorescence can be quenched by the dansyl (5–dimethylaminonaphthalene–1–sulfonyl) group or by the more effective 2,4–dinitrophenyl (Dnp) group. The Dnp also quenches the fluorescence of o–aminobenzoyl (Abz) group. Likewise, the 4–nitrophenylalanine (Phe(NO$_2$)) and 3–nitrotyrosine can form quenched fluorescence substrates with the Abz group. Commonly, the N-terminus of the peptide substrate is labeled with one member of the pair (e. g. dansyl or Abz), and a hydrophobic residue on the C–terminal side of the scissile bond is replaced by an aromatic quencher (e. g. Phe(NO$_2$)). When designing a quenched fluorescent substrate, it is important to choose a quencher–fluorophore pair which renders it possible to prepare the peptide by solid phase synthesis, such as dansyl–tryptophan (Stöcker et al. 1990) or Abz–Phe(NO$_2$).

General Comments

- Fluorescence measurements depend considerably upon temperature so that an efficient thermostated cell holder should be employed, and sufficient time should be allowed for equilibration.

- The fluorescence data is obtained in arbitrary units, which can be directly used for the determination of first–order rate constants. For the calculation of the Michaelis parameters, the fluorescence data should be converted to substrate concentrations as described above.

- Careful calibration regarding the linearity of fluorescence with substrate concentration is essential, in particular with quenched substrates, since light absorption by the quenching group can lead to misleading results at substrate concentrations higher than about 10 μM. One method of calibration involves zeroing the fluorometer with the buffer in the cell, and the residual fluorescence of substrate at increasing concentration is measured. The difference between the extrapolated initial linear portion and the experimental points at higher substrate concentrations shows the extent of self–quenching, which should be less than 50 % to avoid large corrections associated with loss of accuracy.

- With a new fluorescent oligopeptide the hydrolysis products should be characterized by HPLC analysis, as not only a single peptide bond may be cleaved.

References

Brocklehurst K (1994) A sound basis for pH–dependent kinetic studies on enzymes. Protein Engineering 7:291–199

Dixon HBF (1992) How groups of protein titrate – a new approach. Essays in Biochemistry 27:161–176

Duggleby RG (1995) Analysis of enzyme progress curves by nonlinear regression. In: Purich DL (ed) Enzyme Kinetics and Mechanism. Methods Enzymol, vol 249. Academic Press, London, pp 61–90

Ellis KJ, Morrison JF (1982) Buffers of constant ionic strength for studying pH–dependent processes. In: Purich DL (ed) Enzyme Kinetics and Mechanism. Methods Enzymol, vol 87. Academic Press, London, pp 404–426

Fersht A (1985) Enzyme Structure and Mechanism, 2nd edition, Freeman WH and Company, New York

Hyland LJ, Tomaszek JA Jr, Meek TD (1991) Human immunodeficiency virus–1 protease. 2. Use of pH rate studies and solvent kinetic isotope effects to elucidate details of chemical mechanism. Biochemistry 30:8454–8463

Inouye K, Fruton JS (1967) Studies on the specificity of pepsin. Biochemistry 6:1765–1777

Knight CG (1995a) Active site titration of peptidases. In: Barrett AJ (ed) Proteolytic Enzymes: Aspartic and Metallo Peptidases. Methods Enzymol, vol 248. Academic Press, London, pp 85–101

Knight CG (1995b) Fluorimetric assays of proteolytic enzymes. In: Barrett AJ (ed) Proteolytic Enzymes: Aspartic and Metallo Peptidases. Methods Enzymol, vol 248. Academic Press, London, pp 18–32

Leatherbarrow RJ (1990) GraFit, Version 2.0, Erithacus Software Ltd, Staines, UK

Matta MS, Andracki ME (1988) Acylation of subtilisin A by aryl esters: contribution to rate limitation by a physical step preceding general acid–base catalysis. Biochemistry 27:8000–8007

Polgár L (1987) Structure and function of serine proteases. In Neuberger A, Brocklehurst K (eds) Hydrolytic Enzymes. New Comp. Biochem. vol 16. Elsevier, Amsterdam, pp 159–200

Polgár L (1989) Mechanisms of Protease Action., CRC Press, Inc., Boca Raton, FL

Polgár L (1990) Common feature of the four types of protease mechanism. Biol Chem Hoppe-Seyler 371:327–331

Polgár L (1992) Prolyl endopeptidase catalysis. A physical rather than a chemical step is rate–limiting. Biochem J 283:647–648

Polgár L, Fejes J (1979) Mechanism-controlled stereospecificity. Acylation of subtilisin with enantiomeric alkyl and nitrophenyl ester substrates. Eur J Biochem 102:531–536

Polgár L, Szeltner Z, Boros I (1994) Substrate-dependent mechanisms in the catalysis of human immunodeficiency virus protease. Biochemistry 33:9351–9357

Powers JC, Kam C–M (1995) Peptide thioester substrates for serine peptidases and metalloendopeptidases. In: Barrett AJ (ed) Proteolytic Enzymes: Aspartic and Metallo Peptidases. Methods Enzymol, vol 248. Academic Press, London, pp 3–18

Richards AD, Phylip LH, Farmerie WG, Scarborough PE, Alvarez A, Dunn BM, Hire & P–H, Kovalinka J, Strop P, Pavlickova L, Kostka V, Kay J (1990) Sensitive, soluble chromogenic substrates for HIV–1 proteinase. J Biol Chem 265:7733–7736

Stöcker W, Ng M, Auld DS (1990) Fluorescent oligopeptide substrates for kinetic characterization of the specificity of Astacus proteinase. Biochemistry 29:10418–10425

Chapter 11

Kinetic Analysis of Protease Inhibition by Synthetic Inhibitors

C. GRAHAM KNIGHT

Introduction

Inhibitors of proteolytic enzymes are molecules that bind at or near the active site to produce a decrease in catalytic activity. If the binding is reversible,

$$E + I \rightleftharpoons^{K_i} EI \tag{1}$$

the strength of the interaction is defined by K_i, the inhibition constant. K_i is the dissociation constant of the enzyme-inhibitor complex,

$$K_i = E \cdot I / EI \tag{2}$$

where italics indicate concentrations.

By contrast, irreversible inhibitors interact with the enzyme to form a stable covalent bond

$$E + I \xrightarrow{k_{2(app)}} E - I \tag{3}$$

and the interaction is characterised by an apparent second-order inactivation rate constant, $k_{2(app)}$.

A special class of inactivators, the mechanism-based inhibitors, depart from 1:1 stoichiometry. These substrate-like molecules are transformed by the enzyme to reactive species that may modify an active site residue or be quenched by the solvent (Silverman 1995). They have been described only rarely for proteolytic enzymes (for a recent example, see Groutas et al. 1997) and will not be discussed here.

C. Graham Knight, University of Cambridge, Department of Biochemistry, Tennis Court Road, Cambridge, CB2 1QW, UK (*phone* +44 (0)1223 766110; *fax* +44 (0)1223 333345; *e-mail* cgk21@mole.bio.cam.ac.uk)

Discrimination between reversible and irreversible inhibition is not always easy (Bieth 1995, Szedlacsek and Duggleby 1995), for when $K_i \ll E_0$, the concentration of active enzyme in the assay, the inhibitor will titrate the enzyme and so appear to behave irreversibly. Thus, it is very useful to have some idea of the value of E_0. For well-characterised enzymes, a suitable method of active site titration may be available (Knight 1995a). In other cases, the molarity of a homogeneous enzyme preparation may be estimated from the protein concentration P_0 (in g/L) and M_r ($E_0 \approx P_0/M_r$), although this will almost certainly be an overestimate. For reversible inhibitors, the importance of E_0 lies in its relation to K_i. If $E_0/K_i > 0.01$, a significant fraction of the total inhibitor becomes bound by the enzyme and the usual methods of kinetic analysis break down (Bieth 1995).

Subprotocol 1
Assay Methods

Introduction

Any method of kinetic analysis will be compromised if the results themselves are unreliable. Many different assay methods for proteolytic enzymes are available. They have been reviewed in two volumes of Methods in Enzymology (Barrett 1994, Barrett 1995), in Chapters 2 and 3 of this volume, and elsewhere (Sarath 1989). For kinetic studies, the procedure following is recommended.

Procedure

1. Select a fluorogenic or chromogenic substrate and monitor the release of product continuously. This allows time-dependent behaviour to be characterised.

2. Do not exceed 10% hydrolysis (5% is better). This minimises the effects of substrate depletion and product inhibition.

3. Use a recording spectrophotometer or fluorimeter equipped with a stirrer and thermostated cuvette holder, and ideally under computer control. Pre-equilibrate the buffer in a separate water bath.

4. Zero the instrument with substrate in the cuvette and calibrate the detector with product so that the full scale deflection on the recording device corresponds to 10% (or less) hydrolysis.

5. For 1 cm² cuvettes a total assay volume of 2.50 ml is convenient and individual additions are made in <50 µl. Check the calibration of the micropipettes frequently.

6. For tight binding reversible inhibitors and rapidly reacting covalent inhibitors, decrease E_0 by using a highly sensitive fluorometric assay (Knight 1995b, Auld, this volume).

7. Assays using quenched fluorescent (QF) peptides (Knight 1995b) are very temperature-dependent and these substrates must be used at low concentrations to minimise light absorption effects. A QF substrate may appear to be useless at 150 µM, but be highly sensitive at 10 or even 2 µM (Knight et al. 1992).

Subprotocol 2
Substrate Kinetics

Before doing any inhibition experiments, it is necessary to characterise the effects of substrate concentration on the rate of product release in the absence of inhibitor. When initial rates are measured, S may be equated to the initial substrate concentration. The simplest model (Fersht 1985) is

$$\text{E} + \text{S} \xrightleftharpoons{K_m} \text{ES} \xrightarrow{k_{cat}} \text{E} + \text{P} \tag{4}$$

where K_m is the apparent dissociation constant of the enzyme-substrate complex ES and k_{cat} is the first-order rate constant for the breakdown of ES to products. The value of ES cannot exceed E_0, so as S increases the reaction velocity v approaches $k_{cat}E_0$, the maximum velocity, V. Thus, according to this scheme, v varies with S as follows

$$v = k_{cat}E_0 S/(K_m + S) \tag{5}$$

Procedure

1. Set up the assays as recommended in Subprotocol 1.

2. Keep E_0 constant and measure initial velocities, v, at several different substrate concentrations, ideally in the range $0.2K_m < S < 5K_m$, although this may not be possible due to insolubility or absorbance effects.

3. Determine V and K_m by fitting the data directly to eqn.(5) by non-linear regression analysis using ENZFITTER (Biosoft) or GRAFIT (Erithacus Software) (both are available from Sigma/Aldrich) or equivalent software.

4. Other valid methods for the determination of V and K_m are discussed by Cornish-Bowden (1995a, 1995b). Leave the Lineweaver-Burk plot of $1/v$ versus $1/S$ to rest in peace, as it makes very poor data look satisfactory.

5. When the computed value of K_m is greater than the highest S, both V and K_m will be unreliable and the experiments should be extended to higher values of S if possible.

6. If v varies linearly with S over the range of substrate concentrations available, then $S \ll K_m$ and individual values of k_{cat} and K_m cannot be calculated, although the value of the specificity constant, k_{cat}/K_m, is well defined (Fersht, 1985) and can be calculated from the equation $k_{cat}/K_m = v/E_0S$.

Subprotocol 3
What is the Mechanism of Inhibition?

When working with a previously uncharacterised inhibitor, some simple initial experiments can clarify whether or not the inhibition is reversible, and may reveal complex behaviour such as tight or slow binding. Even if the inhibitor is well-known, these experiments are valuable as your enzyme may show unexpected behaviour.

Procedure

A preliminary experiment

1. Set up the assay so that no more than 10% (or less) of the substrate is consumed over 10 – 20 min. Start the reaction and measure the initial uninhibited rate, v_0, over 3 or 4 min. Then add a small volume of inhibitor solution sufficient to decrease the rate by about 50%.

2. A suitable inhibitor concentration may be found by initial experiments spanning four orders of magnitude, for example, 1, 10, 100 and 1000 nM after dilution into the assay.

3. If a new constant rate of hydrolysis, v_i, is reached within seconds, the inhibition is rapidly reversible.

4. If the rate decreases to a new steady-state over a period of minutes, the inhibitor is showing slow, reversible binding.

5. Irreversible inhibition is implied if the rate continues to fall and appears to approach zero.

1. Mix the enzyme and inhibitor at concentrations 50-fold greater than those used in the preliminary experiment.

2. After 10 and 30 min, dilute a portion of the mixture 50-fold into substrate and observe the rate of product release. Be alert for unexpected behaviour.

3. If the E-I interaction is rapidly reversible, the value of v_i seen previously will be attained immediately. A slow-binding reversible inhibitor will show a lag period before v_i reaches a constant value.

4. If the reaction is irreversible no activity may be seen.

5. Check for enzyme denaturation by incubating controls diluted with buffer alone (see also Subprotocol 6).

Confirmation of suspected mechanism

1. Measure the initial uninhibited rate, v_0. Then make experiments in which I_0 is varied to produce from 5 to 75% inhibition.

2. If you suspect slow, tight-binding or irreversible inactivation, allow the enzyme and inhibitor to equilibrate in the cuvette for 10 or 30 min before adding the substrate.

3. Plot a dose-response curve of v_i/v_0 versus I_0, the total inhibitor concentration.

4. If the plots are concave and inhibition is significant only at $I_0 \gg E_0$, the inhibition is freely reversible and the data may be used to make a preliminary estimate of K_i.

5. If the initial portion of the plot is linear and extrapolates to intersect the baseline at $I_0 \approx E_0$, either reversible tight-binding or irreversible inhibition is indicated.

Making a dose-response curve

Although these preliminary experiments may suggest the type of inhibition, more careful studies are required to determine accurate values of K_i and to characterise the rates of time-dependent inhibition. Methods for the analysis of the more commonly encountered mechanisms are outlined below. More comprehensive coverage and suggestions for the analysis of more complex phenomena will be found in the references cited.

> ## Subprotocol 4
> ## Mechanisms of Reversible Inhibition

Let us assume that the inhibition is freely reversible at $I_0 \gg E_0$. The possible modes of interaction in enzyme-substrate-inhibitor mixtures are illustrated in Fig. 1. This scheme provides the basis for the quantitative description of inhibition in terms of the individual rate and equilibrium constants. The experimenter, however, is faced with the practical problem of interpreting a set of I_0 and v_i data to provide valid equilibrium constants.

Checking for non-linear (partial) inhibition   Although it is tempting to assume a simple mechanism of inhibition, this should be confirmed. From the scheme shown in Fig.1, Yoshino (1987) derived an equation relating the fractional velocity, $v_i/(v_0 - v_i)$, to $1/I_0$

$$v_i/(v_0 - v_i) = \frac{(1 + S/K_m)(\beta + K_{iu}/I_0)}{(1 - \beta)S/K_m + (K_{iu}/K_{ic} - \beta)} \tag{6}$$

where $\beta = k_{cat}'/k_{cat}$, K_{ic} is the competitive inhibition constant and K_{iu} is the uncompetitive inhibition constant (see Fig. 1). Eqn.(6) predicts a straight line relationship between $v_i/(v_0 - v_i)$ and $1/I_0$ for all mechanisms of inhibition. The advantages of the Yoshino plot have been discussed by Whiteley (1997).

- If $\beta = 0$, as is the case for linear (complete) inhibitors, all such plots pass through the original for all mechanisms.

- When $1 > \beta > 0$ the plot intersects the baseline at $-\beta/K_{iu}$.

- The characterisation of complex inhibition requires many experiments spanning as wide a range of S as possible (Yoshino 1987, Cornish-Bowden 1995a).

- For linear inhibitors with $\beta = 0$, it is convenient to analyse the dose-response data initially by Dixon ($1/v_i$ versus I_0) or Cornish-Bowden (S/v_i versus I_0) plots (Cornish-Bowden 1995).

- Competitive inhibition is easily demonstrated. Dose-response experiments at two well-separated substrate concentrations will be sufficient to show that the plots of S/v_i versus I_0 are parallel.

Many of the advantages of the Yoshino (1987) plot are shared by the combination plot devised by Chan (1995). For an illustration of the method, see Sreedharan et al. (1996).

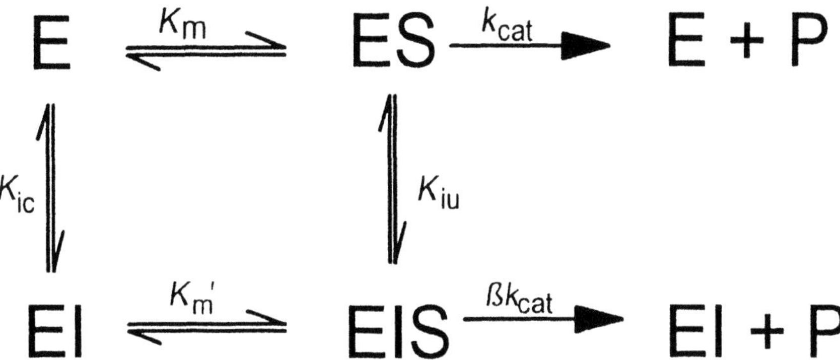

Fig. 1. Inhibitory binding modes. Inhibitor binding to free enzyme E is characterised by the competitive inhibition constant, K_{ic}. The binding of I to ES depends on K_{iu}, the uncompetitive inhibition constant. As the various enzyme forms are in mutual equilibrium, $K_m' = K_m \cdot K_{iu}/K_{ic}$. EIS is assumed to break down at a rate not exceeding that of the uninhibited reaction ($0 < \text{ß} < 1$).

Synthetic inhibitors of proteolytic enzymes are commonly structural ana- **Competitive** logues of good substrates, and features that increase k_{cat}/K_m also increase **inhibition** inhibitory potency. As might be expected, the binding of substrates and such inhibitors is mutually exclusive and is termed pure competitive inhibition. The discussion from now on will be restricted to this class of inhibitor. More complex mechanisms are discussed in the papers cited. In the presence of a competitive inhibitor, the velocity becomes

$$v_i = k_{cat}E_0S/[K_m(1 + I_0/K_i) + S] \tag{7}$$

Combining equations (5) and (7), we obtain

$$v_0/v_i = 1 + I_0/K_{i(app)} \tag{8}$$

where the apparent inhibition constant $K_{i(app)} = K_i(1 + S/K_m)$ reflects the competitive effect of substrate. Although eqn.(8) can be used to obtain a preliminary estimate of $K_{i(app)}$, it is unsuitable for kinetic studies (Leatherbarrow 1990).

Procedure

1. Determine a dose-response curve of v_i/v_0 versus I_0 as described in Subprotocol 3, using at least six values of I_0.

2. With continuous rate assays, it is convenient to measure v_0 and v_i in the same experiment, as in Subprotocol 2, and to treat the data as v_i/v_0.

3. Analyse the data by non-linear regression, to obtain the apparent inhibition constant, $K_{i(app)}$, using ENZFITTER and the Morrison (1969) equation:

$$v_i/v_0 = \{E_0 - I_0 - K_{i(app)} + [(I_0 - E_0 + K_{i(app)})^2 + 4E_0K_{i(app)}]^{1/2}\}/2E_0 \qquad (9)$$

4. The E_0 term appears in eqn.(9) because the Morrison equation is quite general and, as discussed in Subprotocol 5, applies to both loose and tight binding inhibitors.

5. When correctly entered into ENZFITTER, the equation editor screen for eqn.(9) is as shown in Fig. 2.

6. ENZFITTER will prompt for estimates of E_0 and $K_{i(app)}$. When $E_0/K_{i(app)} < 0.01$, enter E_0 as 1% of the lowest I_0. Do not enter zero.

7. Use eqn.(8) to estimate $K_{i(app)}$, by dividing any I_0 by the corresponding (v_0/v_i)-1. For example, if $v_0 = 67$ nM.min^{-1} and $v_i = 29$ nM.min^{-1} when $I_0 = 32$ nM, then $K_{i(app)} \approx 24$ nM.

```
============================Equation Name============
Competitive inhibition eqn.(9)

=======Y Axis Name======      =======X Axis Name======    ==VAR==
Relative velocity, vi/vo      [Inhibitor]                 Io

======Variable Name======  ==VAR==       ======Variable Name======  ==VAR==
1 Inhibition constant      Kiapp    5
2                                   6
3                                   7
4                                   8

=====Prompted Constant=====  ==VAR==      =====Prompted Constant=====  ==VAR==
1 Enzyme concentration     Eo       3
2                                   4

============================Equation Definition============
Temp1
Temp2

Y =   (Eo-Io-Kiapp+SQRT(SQR(Io-Eo+Kiapp)+4*Eo*Kiapp))/(2*Eo)
```

Fig. 2. ENZFITTER equation editor screen for the use of eqn. (9) to analyse classical competitive inhibition ($I_0 \gg E_0$).

8. Plot the data using ENZFITTER and check that the residuals (calculated v_i/v_0 - observed v_i/v_0) are evenly distributed. If not, you may be wrong in assuming linear inhibition or there may be systematic errors in E_0 or I_0. The diagnostic value of residuals has been discussed by Ellis and Duggleby (1978) and Cornish-Bowden (1995b). For an illustration of the power of residual analysis, see Cardenas and Cornish-Bowden (1993).

You may feel uncomfortable at relying on software to calculate $K_{i(app)}$, but it is well-established that non-linear regression analysis avoids the distortion of the error distribution introduced by linearisation (Leatherbarrow 1990). Make your Dixon plots if you wish, but only to obtain initial estimates of $K_{i(app)}$.

Subprotocol 5
Tight-Binding Competitive Inhibitors

From a kinetic standpoint, tight-binding inhibitors are those with $K_{i(app)}$ values such that under standard assay conditions $E_0/K_{i(app)} > 0.01$. In practice, the enzyme-inhibitor interaction only approaches 1:1 stoichiometry when $E_0/K_{i(app)} > 10$. Tight-binding does not imply some special property of the inhibitor (Szedlacsek and Duggleby 1995), but reflects a combination of experimental conditions (E_0) and thermodynamic properties ($K_{i(app)}$). Thus use of a poor substrate may lead to tight-binding behaviour by an inhibitor of only modest potency (Bieth 1995). The assay of tight-binding inhibitors is straightforward in principle, but complications arise if insufficient time is allowed for the system to reach equilibrium, as discussed in Subprotocol 6. Also, the amount of inhibitor bound by the enzyme cannot be ignored, so that the value of E_0 enters explicitly into the calculation of $K_{i(app)}$.

Procedure

1. Check that the system is at equilibrium. Preincubate the enzyme and inhibitor in assay buffer (2.45 ml) for various times before adding the substrate (50 µl). The inhibited velocity v_i should be independent of the incubation time. Test for enzyme denaturation by incubating the enzyme in buffer alone.

2. Determine a dose-response curve of v_i/v_o versus I_o, as described in Subprotocol 3, using preincubated mixtures. Be careful to measure steady-state values of v_i.

3. Fit the data to the Morrison (1969) equation, but treat E_0 as an unknown variable, of comparable magnitude to I_0. The ENZFITTER equation editor screen for tight-binding inhibition is shown in Fig. 3.

4. As an example, the tight binding of pepstatin methyl ester to cathepsin D is shown in Fig. 4, which illustrates too the usefulness of tight-binding inhibitors as active-site titrants for proteases (Knight 1995a).

5. The value of $K_{i(app)}$ will be poorly defined if the plot shows an extended linear portion representing stoichiometric inhibition, although E_0 will be known with accuracy. If this behaviour is seen, decrease E_0 and I_0 in order to increase the degree of dissociation.

6. When E_0 is known by active-site titration, it may be treated as a constant and a more accurate estimate of $K_{i(app)}$ obtained by fitting the data as in Fig.2.

```
==============================Equation Name==============================
Tight-binding competitive inhibition eqn.(9)

=========Y Axis Name=========   ==========X Axis Name==========  ==VAR==
Relative velocity, vi/vo        [Inhibitor]                      Io

========Variable Name========  ==VAR==   =========Variable Name=========  ==VAR==
1 Inhibition constant          Kiapp    5
2 Enzyme concentration         Eo       6
3                                        7
4                                        8

======Prompted Constant======  ==VAR==   ======Prompted Constant======  ==VAR==
1                                        3
2                                        4

===========================Equation Definition===========================
Temp1
Temp2

Y =   (Eo-Io-Kiapp+SQRT(SQR(Io-Eo+Kiapp)+4*Eo*Kiapp))/(2*Eo)
```

Fig. 3. ENZFITTER equation editor screen for the use of eqn.(9) to analyse tight-binding inhibition ($I_0 \approx E_0$).

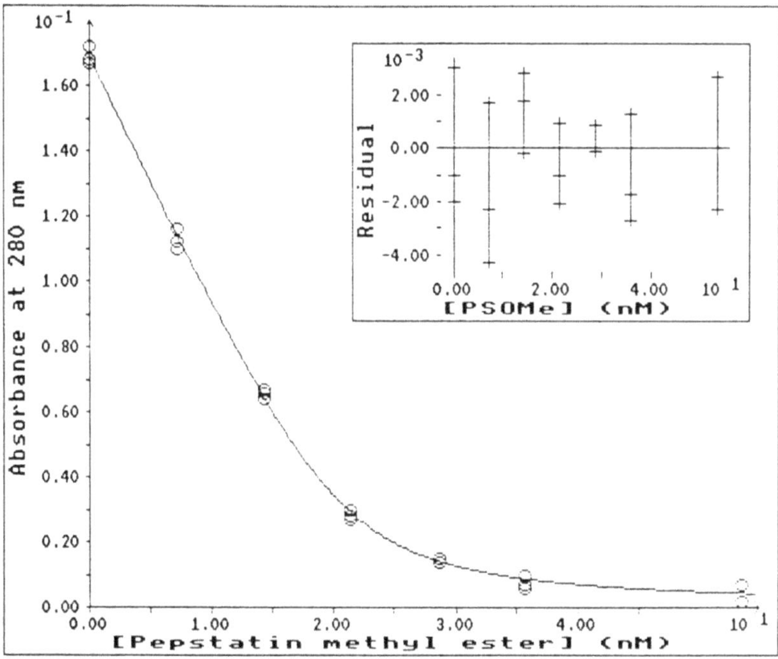

Fig. 4. Inhibition of cathepsin D by pepstatin methyl ester. Data were fitted to eqn.(9) using ENZFITTER to give $K_{i(app)} = 8.7 \pm 0.8 \times 10^{-10}$ M and $E_0 = 20.8 \pm 0.4 \times 10^{-9}$ M. Note the even distribution of the residuals. Assays were made in triplicate and measured the release of trichloroacetic acid-soluble peptides from haemoglobin during 10 min at 37°C.

Subprotocol 6
Slow and Slow, Tight-Binding Inhibitors

The use of stopped assays, as in Fig. 4, can mask more complex kinetic behaviour, as the enzyme and inhibitor may approach equilibrium only slowly after mixing (Cha 1975, Williams and Morrison 1979, Morrison and Walsh 1987). In some cases this slowness reflects a conformational change in the enzyme-inhibitor complex, but it can also arise directly from the low concentrations necessary to study very tight interactions (Cha 1975). At equilibrium

$$E + I \underset{k_{off}}{\overset{k_{on}}{\rightleftharpoons}} EI \qquad (10)$$

EI dissociates and reassociates at the same rate, and K_i is equal to the ratio of the rate constants, $K_i = k_{off}/k_{on}$. When $I_0 \gg E_0$, the approach to equilibrium is characterised by the pseudo-first-order rate constant, k_{obs}

$$k_{obs} = k_{on}I_0 + k_{off} \qquad (11)$$

Consider 10^{-10} M enzyme treated with 10^{-9} M inhibitor. If $K_i = 10^{-10}$ M, and $k_{on} = 10^7$ $M^{-1}.sec^{-1}$, a value seen in many enzyme-substrate interactions (Fersht 1985), it follows that $k_{off} = 10^{-3}$ sec^{-1}. Eqn.(11) predicts that approach to equilibrium will occur with a half-time of 63 sec ($0.693/k_{obs}$), requiring about 7 min for completion. This slow-binding behaviour is illustrated in Fig. 5.

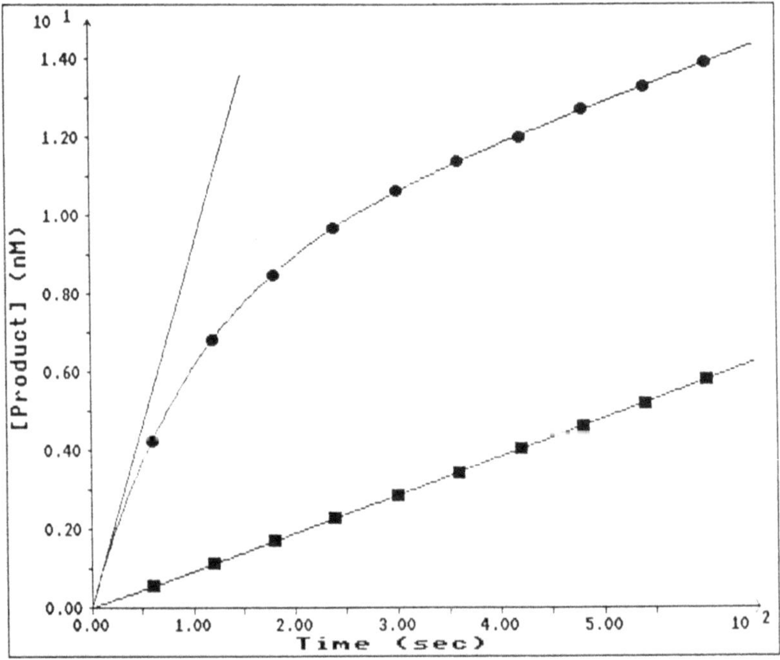

Fig. 5. Simulated slow binding. The lines were calculated with $k_{on} = 10^7$ $M^{-1}.s^{-1}$, $k_{off} = 10^{-3}$ s^{-1} ($K_i = 10^{-10}$ M), $I_0 = 10^{-9}$ M, $E_0 = 10^{-10}$ M, $k_{cat} = 10$ s^{-1}, $S = 10^{-5}$ M, $K_m = 10^{-4}$ M. Reaction started with ● enzyme; ■ substrate. The straight line shows the uninhibited reaction.

As synthetic proteinase inhibitors with dissociation constants in the pico-molar range become more common, time-dependent binding is likely to be encountered more frequently. The aim of kinetic analysis is to determine the rate constants for the individual steps, in addition to the overall equilibrium constant. Morrison and Walsh (1987) discriminate between slow and slow, tight-binding inhibitors. These terms are operational, reflecting differences in the analysis of the inhibition. If the interaction can be studied with $I_0 \gg E_0$, so that inhibitor depletion can be ignored, the binding is slow. Slow, tight-binding implies $I_0 \approx E_0$ and the fraction of inhibitor bound to the enzyme must be considered.

Kinetics of Slow Binding

Szedlacsek and Duggleby (1995) emphasise that there is nothing intrinsically different about slow-binding inhibition, merely that the transient-phase kinetics occur over minutes rather than seconds. The analysis here will assume that S remains essentially unchanged throughout the experiment, and that $I_0 \gg E_0$.

As illustrated in Fig. 5 , under these conditions the rate of product formation shows an exponential phase during the approach to equilibrium. Cha (1975) showed that the product concentration P changes with time t as follows

$$P = v_s t + (v_z - v_s)[1 - \exp(-k_{obs}t)]/k_{obs} \tag{12}$$

where v_s is the velocity at equilibrium and v_z is the initial velocity immediately after mixing. The value of v_z will depend on whether the reaction is started by adding enzyme to a mixture of substrate and inhibitor, or by the addition of substrate to preincubated enzyme and inhibitor. In the first case, $v_z = v_0$, the uninhibited velocity (see also Interpretation of k_{obs}).

Procedure

1. Set up the assay as recommended in Subprotocol 1, with substrate and inhibitor in the cuvette. Allow the mixture to reach thermal equilibrium.

2. Add enzyme to start the reaction and follow the release of product until equilibrium is approached. If the consumption of substrate exceeds 10%, adjust the experimental conditions to prevent this.

3. If the rate of inhibition is too fast to allow the collection of sufficient data during the exponential phase, decrease the concentrations of enzyme and inhibitor, while maintaining $I_0 \gg E_0$ if possible. In this context, a 10-fold excess of inhibitor is sufficient. Increasing S will also slow the rate of association.

4. Fit the whole progress curve to eqn.(12) using ENZFITTER. If you have too many data points, eliminate some from the equilibrium phase. Do not be tempted to shorten the experiment, as the final points are necessary to prove that a steady state was reached.

5. Provided the absorbance or fluorescence values are directly proportional to P, that is A or $F = aP + b$ where a and b are constants, these values may be used directly in the calculations.

6. The ENZFITTER editor screen for eqn.(12) is shown in Fig. 6. The displacement term d is added to compensate for any uncertainty in the initial value of P. Note that v_s, v_z, k_{obs} and d are all treated as unknown variables.

7. Make experiments at not less than six values of I_0. Use the values of v_s to calculate $K_{i(app)}$ from the Morrison equation.

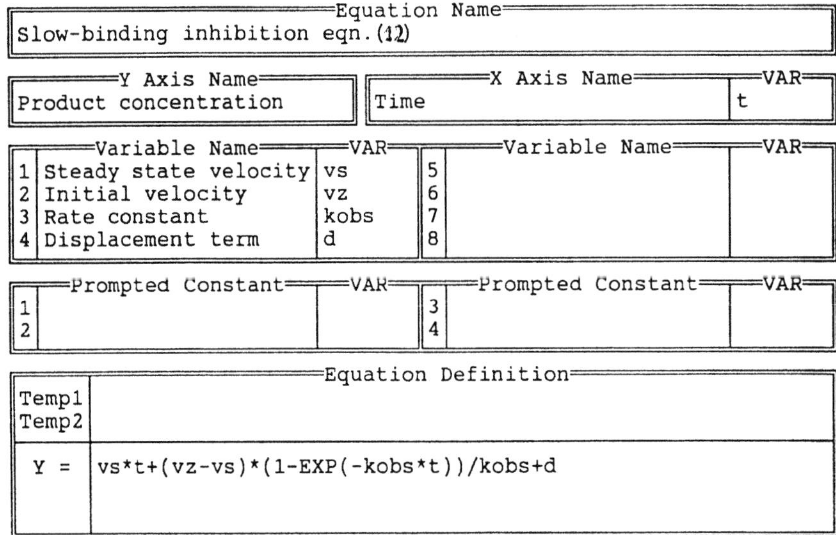

Fig. 6. ENZFITTER equation editor screen for the use of eqn.(12) to analyse slow-binding inhibition.

The simplest treatment of slow binding assumes that it reflects the low concentrations of enzyme and inhibitor necessary to study the interaction. In this case, eqn.(11) takes the form

$$k_{obs} = k_{on(app)}I_0 + k_{off} \tag{13}$$

where $k_{on(app)} = k_{on}/(1+S/K_m)$, because substrate and inhibitor compete for the same binding site. To obtain the kinetic constants, plot k_{obs} versus I_o and measure the slope, $k_{on(app)}$, and the intercept on the ordinate, k_{off}.
Some inhibitors, however, do not conform to this simple mechanism and the slow step is associated with the tightening of an intermediate complex (Cha 1975, Morrison and Walsh 1987)

$$E + I \underset{}{\overset{K_i}{\rightleftharpoons}} EI \underset{k_6}{\overset{k_5}{\rightleftharpoons}} EI^* \tag{14}$$

where k_5 and k_6 are the forward and backward rate constants for the isomerisation, respectively. When $k_6 \ll k_5$, the equilibrium will be in favour of EI^* and the overall dissociation constant $K_i^* \ll K_i$

$$K_i^* = E \cdot I/(EI + EI^*) = K_i k_6/(k_5 + k_6) \tag{15}$$

- When slow binding is due to the tightening of an intermediate complex, the initial velocities v_z calculated from progress curves started with enzyme vary with I_0

$$v_z = v_0/(1 + I_0/K_{i(app)}) \tag{16}$$

where v_0 is the velocity in the absence of inhibitor.

- For the tightened-complex mechanism of inhibition, k_{obs} shows a hyperbolic dependence on I_0 at constant S

$$k_{obs} = k_6 + (k_5 I_0/K_i)/(1 + S/K_m + I_0/K_i) \tag{17}$$

with a lower limit of k_6 and an upper asymptote of $k_5 + k_6$.

- To obtain values of k_5 and k_6, fit the k_{obs} versus I_0 data to eqn.(17) by non-linear regression.

Slow, Tight-Binding Inhibition

Some inhibitors are so potent that they cannot be studied under conditions in which $I_0 \gg E_0$ and allowance must be made for inhibitor depletion (Williams et al. 1979). This implies very dilute solutions, so that intermediate complex formation can be ignored. When depletion of inhibitor occurs, eqn.(12) becomes

$$P = v_s t + \left[(1-\gamma)(v_z - v_s)/\lambda\,\gamma\right] \cdot \ln\left[(1-\gamma e^{-\lambda t})/(1-\gamma)\right] \tag{18}$$

with

$$\gamma = E_0(1 - v_s/v_z)^2/I_0 \tag{19}$$

and

$$\lambda = k_{on(app)}\left[(K_{i(app)} + E_0 + I_0)^2 - 4E_0 I_0\right]^{1/2} \tag{20}$$

where $k_{on(app)} = k_{on}/(1+S/K_m)$. This looks pretty daunting, but the progress curve can be treated by non-linear regression, provided γ is treated as a function of v_z and v_s rather than as an independent variable (Le Bonniec et al. 1995). If the binding is extremely tight, the values of E_0 necessary to demonstrate dissociation may be far below the sensitivity of the assay. In this case the interaction can be studied only under conditions in which $K_{i(app)} \ll E_0$ and I_0. If $S \ll K_m$, eqn.(20) becomes

$$\lambda = k_{on}\left[(E_0 + I_0)^2 - 4E_0 I_0\right]^{1/2} \tag{21}$$

With such tight-binding inhibitors, $K_{i(app)}$ cannot be determined with any precision, but the k_{on} values may be compared. At present, this behaviour has been seen most commonly with protein inhibitors of proteinases (Willenbrock et al. 1993), but with the development of equally potent synthetic inhibitors, similar methods of analysis will be necessary.

Detection of Enzyme Denaturation

The very dilute enzyme solutions employed to study slow, tight-binding interactions may spontaneously lose activity, especially in experiments made over several hours. If this inactivation goes undetected, the kinetic results are severely compromised. A simple test for inactivation was proposed by Selwyn (1965).

1. Record progress curves of P versus t at three enzyme concentrations in the absence of inhibitor. Make the median concentration E_0 that used in the inhibition experiments, and the others $0.5E_0$ and $2E_0$.

2. Follow the reaction for as long as the longest incubation with inhibitor.

3. Plot P versus $E_0 \cdot t$. If the plots can be superimposed, the enzyme is stable throughout the incubations. For an example of the method, see Le Bonniec et al (1995).

Subprotocol 7
Irreversible inhibition

The simplest mechanism of irreversible inhibition assumes the initial formation of a reversible complex which undergoes an intramolecular inactivation

$$E + I \underset{\longleftarrow}{\overset{K_i}{\rightleftharpoons}} E \cdot I \xrightarrow{k_2} E - I \tag{22}$$

The reaction is followed most conveniently in the presence of substrate (Tian and Tsou 1982, Tsou 1987), under pseudo-first-order conditions ($I_0 \gg E_0$). The case where inhibitor and enzyme are of comparable concentration ($I_0 \approx E_0$) has been treated by Wang and Zhao (1997).

Procedure

1. Set up the assay as recommended in Subprotocol 1, with substrate and inhibitor in the cuvette.

2. Start the reaction by adding enzyme and measure product formation with time. The inactivation should be complete before 10% of the substrate is consumed.

3. Under these conditions, the progress curve is given by

$$P = P_\infty[1 - \exp(-k_{\mathrm{obs}} \cdot t)] \tag{23}$$

where P and P_∞ are the concentrations of product at times t and $t = \infty$ (no activity remaining) respectively, and k_{obs} is the apparent first-order inactivation rate constant.

4. To obtain k_{obs}, fit the progress curve to eqn.(23) directly by non-linear regression (Gray and Duggleby 1989). The ENZFITTER equation editor screen for eqn.(23) is shown in Fig. 7. The fluorescence or absorbance values are entered directly (see Subprotocol 6). The displacement term d compensates for any error in the zero time.

5. Make experiments at several values of I_0.

6. When the reaction proceeds via an intermediate complex,

$$k_{obs} = k_2 I_0 / (I_0 + K_{i(app)}) \tag{24}$$

where $K_{i(app)} = K_i(1 + S/K_m)$. If k_{obs}/I_0 is independent of I_0, then $I_0 \ll K_{i(app)}$ and $k_{obs}/I_0 = k_2/K_{i(app)}$. If k_{obs} varies with I_0, fit the values to eqn.(24) by non-linear regression to obtain the individual values of k_2 and $K_{i(app)}$.

7. For examples of the method applied to cysteine proteinases, see Brömme et al. (1989, 1996).

```
============================Equation Name============
 Irreversible inhibition eqn.(23)

=======Y Axis Name======      =====X Axis Name=====    ==VAR==
 Product concentration    |  Time                   |  t

======Variable Name======VAR==   =====Variable Name=====VAR==
1 Final product concn   Pinf   5
2 Rate constant         kobs   6
3 Displacement term     d      7
4                              8

=====Prompted Constant=====VAR==   ====Prompted Constant====VAR==
1                               3
2                               4

===========================Equation Definition===========
Temp1
Temp2

 Y =  | Pinf*(1-EXP(-kobs*t))+d
```

Fig. 7. ENZFITTER equation editor screen for the use of eqn.(23) to analyse irreversible inhibition in the presence of substrate.

Comments

It would be misleading to survey the rich diversity of synthetic inhibitors available. Such a list would be partial at best and the omissions might be those that you would find most useful. The volumes of Methods in Enzymology cited earlier (Barrett 1994, 1995) provide recent information on inhibitors for a wide range of proteolytic enzymes, as does the Handbook of Proteolytic Enzymes (Barrett et al. 1998).

Acknowledgements

I thank Gillian Murphy, Frances Willenbrock and the late Stuart Stone for helpful comments on the manuscript. This chapter is dedicated to the memory of Professor Stone.

References

Barrett AJ (ed) (1994) Proteolytic Enzymes: Serine and Cysteine Peptidases. Methods Enzymol 244:

Barrett AJ (ed) (1995) Proteolytic Enzymes: Aspartic and Metallo Peptidases. Methods Enzymol 248:

Barrett AJ, Rawlings ND, Woessner JF Jr (eds) (1998) Handbook of Proteolytic Enzymes. Academic Press, London

Bieth JG (1995) Theoretical and practical aspects of proteinase inhibition kinetics. Methods Enzymol 248:59-84

Brömme D, Klaus JL, Okamoto K, Rasnick D, Palmer JT (1996) Peptidyl vinyl sulphones: a new class of potent and selective cysteine proteinase inhibitors. Biochem J 315: 85-89

Brömme D, Schierhorn A, Kirshke H, Wiederanders B, Barth A, Fittkau S, Demuth, H-U (1989) Potent and selective inactivation of cysteine proteinases with N-peptidyl-O-acyl hydroxylamines. Biochem J 263:861-866

Cardenas ML, Cornish-Bowden A (1993) Rounding error, an unexpected fault in the output from a recording spectrophotometer - implications for model discrimination. Biochem J 292: 37-40

Cha S (1975) Tight-binding inhibitors-I. Kinetic behaviour. Biochem Pharmacol 24:2177-2185 (see correction Biochem Pharmacol 25:1561)

Chan WW-C (1995) Combination plots as graphical tools in the study of enzyme inhibition. Biochem J 311: 981-985

Cornish-Bowden A (1995a) Fundamentals of Enzyme Kinetics, Revised Edition. Portland Press, London

Cornish-Bowden A (1995b) Analysis of Enzyme Kinetic Data. Oxford University Press, Oxford

Ellis KJ, Duggleby RG (1978) What happens when data are fitted to the wrong equation? Biochem J 171:513-517

Fersht A (1985) Enzyme Structure and Mechanism, 2nd Edition. Freeman, New York, pp 98-120

Gray PJ, Duggleby RG (1989) Analysis of kinetic data for irreversible enzyme inhibition. Biochem J 257:419-424

Groutas WC, Kuang RZ, Venkataraman R, Epp JB, Ruan SM, Prakash O (1997) Structure-based design of a general class of mechanism-based inhibitors of the serine proteinases employing a novel amino acid-derived heterocyclic scaffold. Biochemistry 36: 4739-4750

Knight CG (1995a) Active-site titration of peptidases. Methods Enzymol 248:85-101

Knight CG (1995b) Fluorimetric assays of proteolytic enzymes. Methods Enzymol 248:18-34

Knight CG, Willenbrock F, Murphy G (1992) A novel coumarin-labelled peptide for sensitive continuous assays of the matrix metalloproteinases. FEBS Lett 296:263-266

Le Bonniec BF, Guinto ER, Stone SR (1995) Identification of thrombin residues that modulate its interactions with antithrombin III and α1-antitrypsin. Biochemistry 34:12241-12248

Leatherbarrow RJ (1990) Using linear and non-linear regression to fit biochemical data. Trends Biochem Sci 15:455-458

Morrison JF (1969) Kinetics of the reversible inhibition of enzyme-catalysed reactions by tight-binding inhibitors. Biochim Biophys Acta 185:269-286

Morrison JF, Walsh CT (1987) The behaviour and significance of slow-binding enzyme inhibitors. Adv Enzymol Relat Areas Mol Biol 61:201-301

Sarath G, de la Motte RS, Wagner, FW (1989) Protease assay methods. In: Beynon RJ, Bond, JS (eds) Proteolytic Enzymes, A Practical Approach. IRL Press, Oxford, pp 25-55

Scandura JM, Zhang Y, Van Nostrand WE, Walsh PN (1997) Progress curve analysis of the kinetics with which blood coagulation factor XIa is inhibited by protease nexin-2. Biochemistry 36: 412-420

Selwyn MJ (1965) A simple test for inactivation of an enzyme during assay. Biochim Biophys Acta 105:193-195

Silverman RB (1995) Mechanism-based enzyme inactivators. Methods Enzymol 249:240-283

Sreedharan SK, Verma C, Caves LSD, Brocklehurst SM, Gharbia SE, Shah HN, Brocklehurst K (1996) Demonstration that 1-trans-epoxysuccinyl-L-leucylamido-(4-guanidino)butane (E-64) is one of the most effective low M_r inhibitors of trypsin-catalysed hydrolysis. Characterization by kinetic analysis and by energy minimization and molecular dynamics simulation of the E-64-ß-trypsin complex. Biochem J 316: 777-786

Szedlacsek SE, Duggleby RG (1995) Kinetics of slow and tight-binding inhibitors. Methods Enzymol 249:144-180

Tian W-X, Tsou C-L (1982) Determination of the rate constant of enzyme modification by measuring the substrate reaction in the presence of the modifier. Biochemistry 21:1028-1032

Tsou CL (1987) Kinetics of substrate reaction during irreversible modification of enzyme activity. Adv Enzymol Relat Areas Mol Biol 61: 381-436

Wang M-H, Zhou K-Y (1997) A simple method for determining kinetic constants of complexing inactivation at identical enzyme and inhibitor concentrations. FEBS Lett 412: 425-428

Whiteley CG (1997) Enzyme kinetics: partial and complete competitive inhibition. Biochem Educ 25: 144-146

Willenbrock F, Crabbe T, Slocombe PM, Sutton CW, Docherty AJP, Cockett MI, O'Shea M, Brocklehurst K, Phillips IR, Murphy G (1993) The activity of the tissue inhibitors of metalloproteinases is regulated by C-terminal domain interactions: a kinetic analysis of the inhibition of gelatinase A. Biochemistry 32:4330-4337

Williams JW, Morrison JF (1979) The kinetics of reversible tight-binding inhibition. Methods Enzymol 63:437-467

Williams JW, Morrison JF, Duggleby RG (1979) Methotrexate, a high-affinity pseudo-substrate of dihydrofolate reductase. Biochemistry 18:2567-2573

Yoshino M (1987) A graphical method for determining inhibition parameters for partial and complete inhibitors. Biochem J 248:815-820

Chapter 12

Phosphinic Peptide Libraries for Proteolytic Enzymes

V. DIVE, J. JIRACEK AND A. YIOTAKIS

Introduction

The rapid development of innovative technologies for generating libraries of both variant inhibitors and substrates of these enzymes, in conjunction with other approaches, provides invaluable information, thereby improving description of the factors responsible for the selectivity of these classes of enzymes. Different combinatorial approaches have been described in the application of libraries of peptide substrates to the study of protease specificity (Petithory 1991, Matthews and Wells 1993, McGeehan 1995). Whatever the methods used to generate these libraries, all these approaches rely on their capacity to identify correctly the different products resulting from the action of a protease on these peptide subtrate libraries. As an alternative to these approaches, we elected to develop libraries of transition-state peptide inhibitors and thus used mainly competitive assays to screen these peptide libraries. To this end libraries of phosphinic peptides were developed, since these molecules have proved to be good analogs of the tetrahedral intermediate formed during the hydrolysis of the peptide bond by zinc-metalloproteases or aspactic proteases (Figure 1) (Grams 1996, Abdel-Meguid 1993). Furthermore, in many cases, highly potent inhibitors of these enzymes have been obtained in this class of compounds (Jiracek et al. 1995, 1996, Abdel-Meguid 1993, Yiallouros et al. 1998).

Correspondence to: V. Dive, CEA, Département d'Ingénierie et d'Etudes des Protéines DSV, C.E. de Saclay, Bâtiment 152, Gif sur Yvette Cédex, 91191, France (*phone* (1) 69083585; *fax* (1) 69089071; *e-mail* dive@dsvidf.fr)
J. Jiracek, Academy of Sciences of Czech Republic, Institute of Organic Chemistry and Biochemistry, Flemingovo nám. 2, Praha 6, 16610, Czech Republic
A. Yiotakis, University of Athens, Department of Organic Chemistry, Laboratory of Organic Chemistry, Zografou, Athens, 15771, Greece

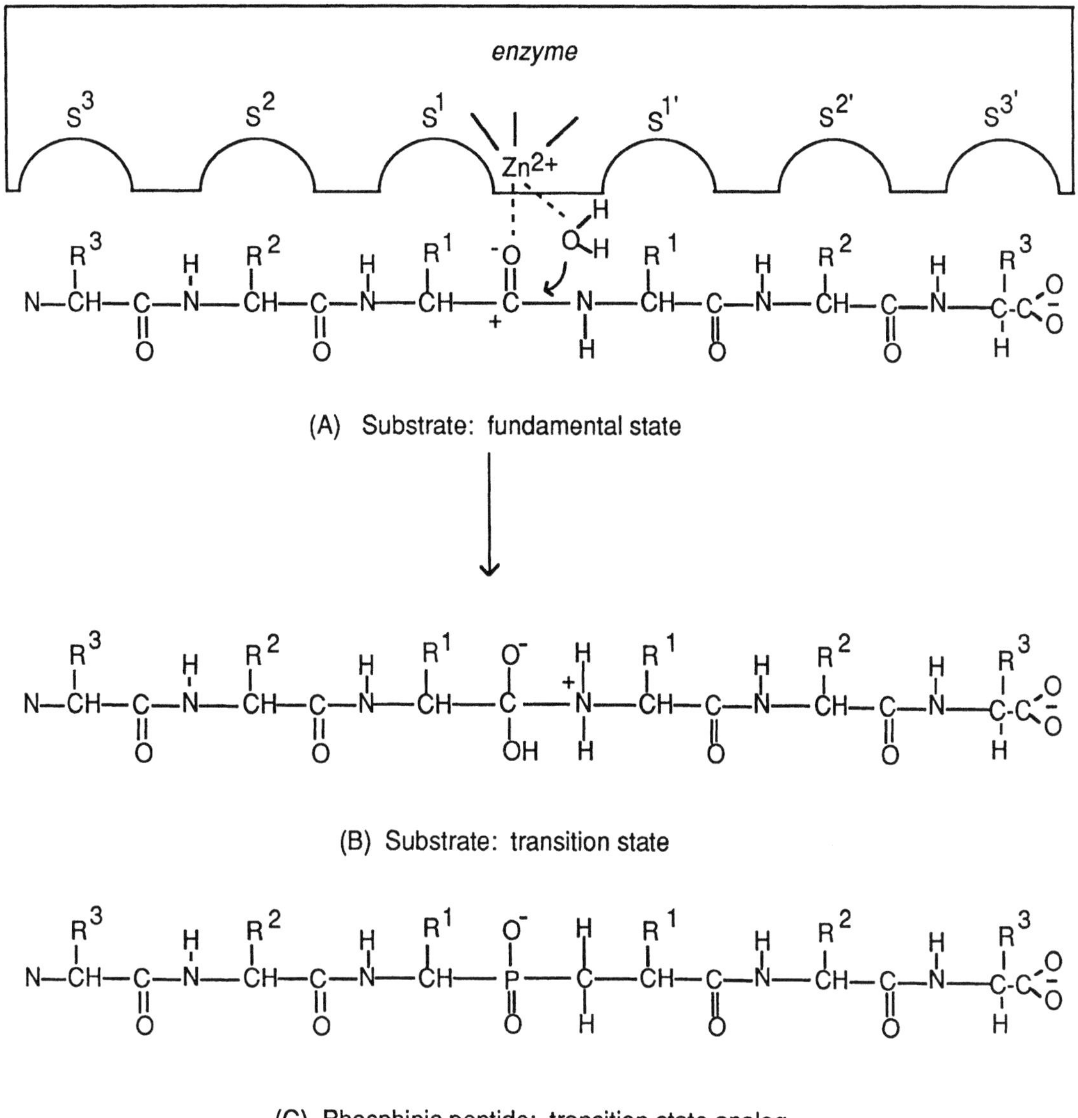

Fig. 1. Schematic diagrams showing the interaction between zinc metalloprotease active site and subtrate or phosphinic inhibitor.

Procedure

Preparation of Phosphinic Peptide Mixtures

The preparation of peptide mixtures implies the use of a solid-phase synthesis strategy. The protocol, developed in our laboratory, is a Fmoc solid-phase strategy, which relies on the introduction of a suitably protected phosphinic dipeptide (Fmoc-XaaY(PO$_{(OAd)}$CH$_2$)Xaa'OH) as a building block during the elongation of the peptide chain (Yiotakis et al, 1996). Given this protocol, it is possible to generate phosphinic peptide mixtures using the method termed mix and split (Furka 1991, Lam 1991). The preparation of 400 phosphinic peptides of general formula Z-$_{(R,S)}$Phe(PO$_2$CH$_2$)Gly-Yaa'-Zaa' by this method is outlined in Figure 2. A first amino acids mixture is prepared by combining twenty different amino acids linked to the resin (Zaa'-R samples). This homogenized mixture is then split into twenty identical samples to which a known amino acid (Yaa') is coupled. This gives rise to twenty different mixtures, each of them containing twenty different dipeptides. The next step is the introduction of the phosphinic dipeptide moiety (Z-$_{(L,D)}$Phe(PO(OAd)-CH$_2$)Gly) into each sample. Side-chain deprotection and cleavage of the peptides from the twenty different resins yielded twenty different phosphinic tetrapeptide mixtures, each containing twenty different tetrapeptides.

Identification of the Most Active Compounds

The mix and split method generates nonoverlapping subsets of mixtures with a unique monomer in the N-terminal position of each subset. The determination of the activity of each subset defines the nature of the residue, which in the N-terminal position optimizes the targeted interaction. Then a second set of peptide mixtures is prepared, but in this case the split position is the N$_{-1}$ position, with each subset containing in the N-terminal position the fixed monomer showing the greatest activity from the previous step. The complexity of the mixture is thus reduced and the process is repeated until a unique peptide is defined. The inhibition potency of each mixture can be determined by adding increasing amounts of inhibitors to a solution of protease, and then measuring the residual enzyme activity, using either a chromogenic or a fluorogenic substrate.

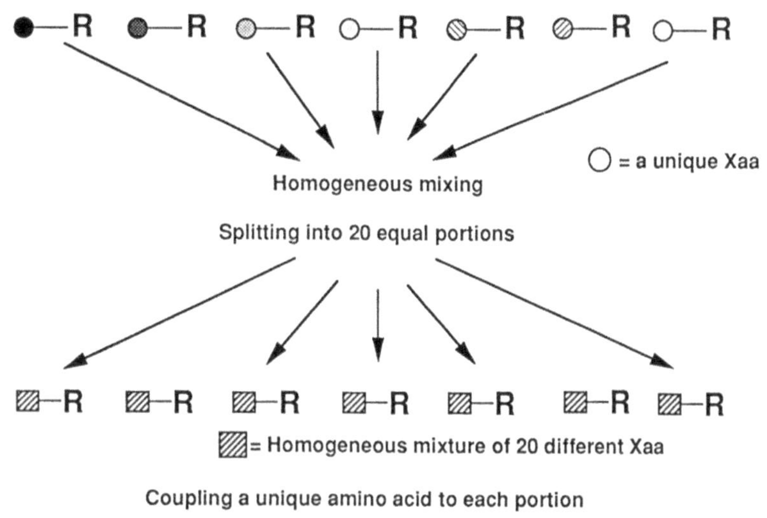

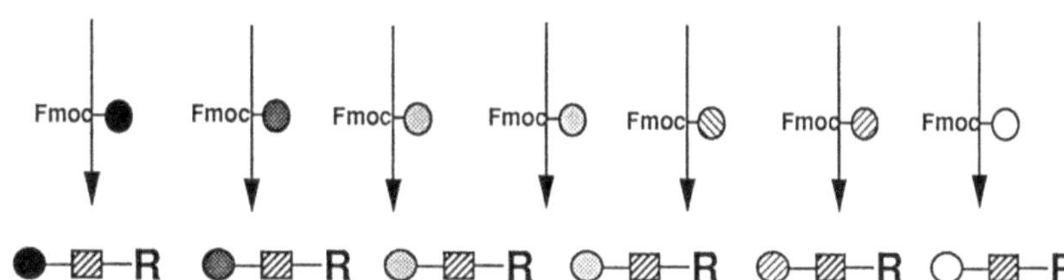

Fig. 2. Preparation of phosphinic peptide mixtures of general formula Z-(L,D)Phe((PO₂ CH₂)Gly-Yaa'-Zaa' by combinatorial chemistry using the plit and mix approach.

Development of a Systematic Approach

Optimization of the inhibitor structure can be assessed by developing a systematic approach, in which the phosphinic peptide mixtures are generated not only by randomization of the peptide sequence, but also by taking into account the size of the peptide, the position of the phosphinic bond in the sequence, as well as the state - free or blocked - of their N- or C-terminal extremities (Figure 3). The screening of these mixtures should define the role of both the peptide size and the position of the phosphinic bond in the peptide sequence on the potency and the selectivity, independently of the sequence of the residues framing the phosphinic block.

Results

Twenty different phosphinic peptide mixtures, of general formula Z-$_{(R,S)}$Phe(PO$_2$CH$_2$)Gly-Yaa'-Zaa', each containing a single amino acid in the Yaa' position, and a mixture of twenty different amino acids in the Zaa', were prepared and screened to develop potent inhibitors of two mammalian zinc-endopeptidases, endopeptidases 24-15 and 24-16. The determination of the relative potency displayed by each mixture makes it possible to establish the preference of both peptidases for the inhibitor Yaa' position. The data reported in Figure 4a indicate that endopeptidase 24-15 prefers inhibitors with a basic residue - an Arg or Lys- in the Yaa' position, while the 24-16 seems to prefer a proline in this position. Based on these results, the synthesis of twenty different phosphinic peptides of general formula Z-$_{(R,S)}$Phe(PO$_2$CH$_2$)Gly-Arg-Zaa', with Zaa' representing twenty different amino acids, was performed in order to obtain specific inhibitor of 24-15. The results reported in Figure 4b show that when an Arg residue was in the Yaa' position, the inhibitors were two orders of magnitude more selective toward 24-15. As compared to the Yaa' position, the identity of the residue in the Zaa' position is less determinent for selectivity. The most potent and selective inhibitor in this series (Z-$_{(R,S)}$Phe(PO$_2$CH$_2$)Gly-Arg-Met) displays Ki values of 0.35nM and 135nM toward endopeptidases 24-15 and 24-16, respectively. Using the same approach, but with phosphinic peptides containing a Z-Phe(PO$_2$CH$_2$)Ala block, the most potent and selective inhibitor of the endopeptidase 24-15 to date was discovered (Z-$_{(R,S)}$Phe(PO$_2$CH$_2$) $_{(R,S)}$Ala-Arg-Met, Ki$_{24-15}$= 70pM and Ki$_{24-16}$= 100nM).

While this approach was successful for the development of potent and selective inhibitors of 24-15, it did not allow us to identify molecules able to discriminate 24-16 from 24-15. Data reported in Figure 4a may explain this

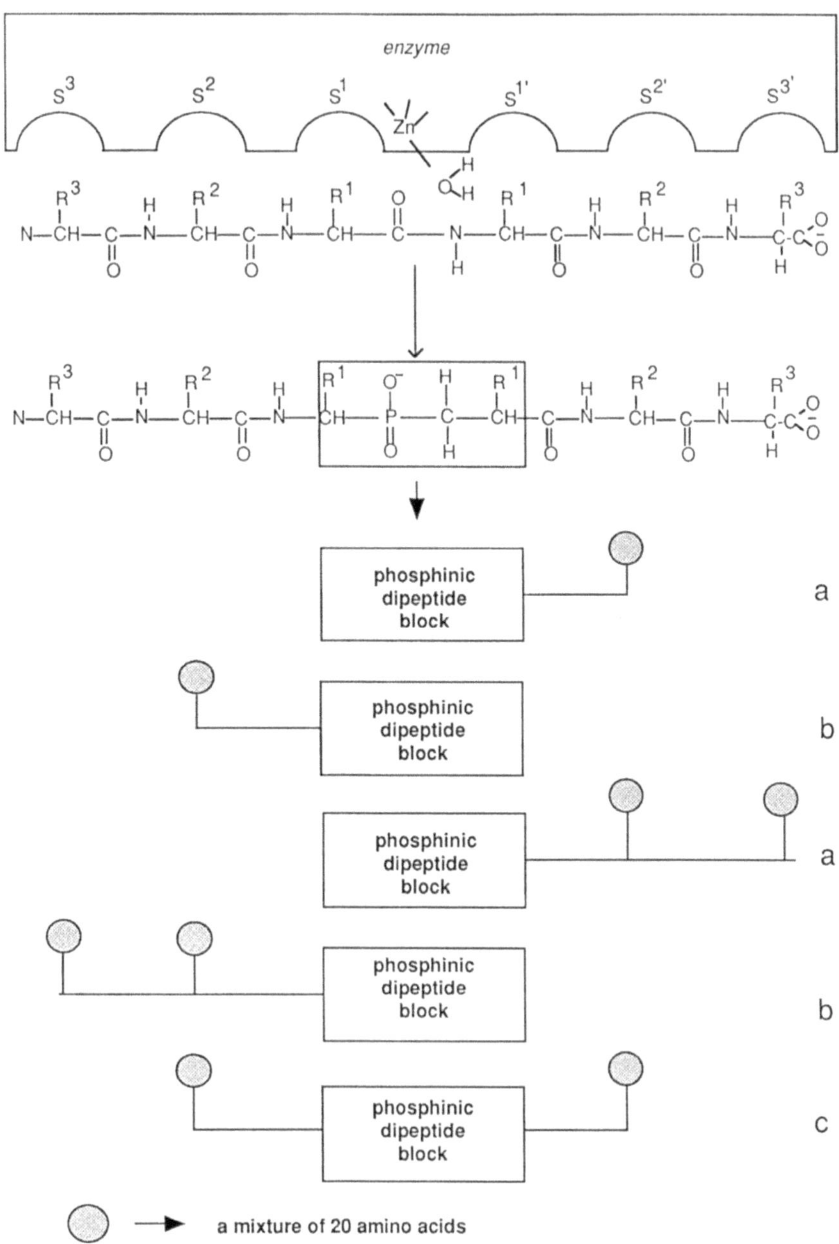

Fig. 3. Schematic representation of a general approach to develop potent and selective inhibitors of zinc-metalloproteases by combinatorial chemistry of phosphinic peptides. For a given peptide size (tripeptide or tetrapeptide in this figure), different peptide mixtures (a,b,c) can be developed according to the position of the phosphinic bond in the sequence.

observation. Among the twenty amino acids, 24-16 prefers inhibitors with a proline in the $P_{2'}$ position. However, as this residue is also well accommodated by 24-15, the presence of the proline in this position cannot provide specific inhibitors. Thus the use of a more systematic approach, as outlined in the Fig. 3, was necessary to discover specific inhibitors of 24-16.

Among different tripeptide and tetrapeptide mixtures which have been developed to obtain inhibitors able to interact specifically with 24-16, the Yaa-$_{(R,S)}$Phe (PO$_2$CH$_2$)Gly-Yaa' mixture proved the most promising in terms of inhibitory selectivity. The deconvolution of this mixture (Fig. 5a) reveals that 24-16 prefers a proline in the Yaa position, while histidine and arginine are better accommodated by 24-15. The potency of the inhibitors, in which the Yaa position is occupied by proline, and the Yaa' position by different residues, again reveals the preference of 24-16 for proline in this position (Fig. 5b). The Pro-$_{(R,S)}$Phe (PO$_2$CH$_2$)Gly-Pro compound, identified from this combinatorial approach, has a Ki value of 4nM toward 24-16 and is three orders of magnitude less potent on 24-15. It thus represents the first potent inhibitor able to discriminate between 24-16 and 24-15 (Jiracek et al. 1996).

The data reported above demonstrate that the devolopment of phosphinic peptide libraries is a powerful strategy for rapid identification of highly potent and selective inhibitors of zinc-metalloproteases. This approach is based on systematic examination, wherever possible, of the effects of different parameters (size, sequence and state of the peptide extremities) on the inhibitor binding specificity. In the present case, the screening of these transition-state analog libraries has provided important and previously unavailable information on the specificity of these peptidases. The selectivity of the $P_{2'}$ position of 24-15 for the arginine is one example. The best specifity constant for 24-15 has been observed for the hydrolysis of dynorphin A_{1-8} (Tyr-Gly-Gly-Phe-Leu-Arg-Arg-Ile) (Orlowski 1989). Cleavage occurs between the leucine and arginine residues, this peptide therefore possesses an arginine in the $P_{2'}$ position. In the same manner, the preference of both the S_2 and $S_{2'}$ subsites of the 24-16 enzyme, unveiled by this study, is in line with several previous observations. These data are in fact in good agreement with the ability of this peptidase to cleave the Pz-peptide

---

Fig. 4. Influence of the Yaa' position for the inhibition of 24-15 (upper part) and 24-16 (lower part) by inhibitor mixtures of general formula -Z-$_{(L,D)}$Phe(PO$_2$CH$_2$)Gly-Yaa'-Zaa' - (Zaa' contains a mixture of 20 different amino acids). The concentration of phosphinic peptides in each mixture was 500nM. **(b)** Selectivity factor (Ki$_{24\text{-}16}$/Ki$_{24\text{-}15}$) displayed by Z-$_{(L,D)}$Phe(-PO$_2$CH$_2$)Gly-Arg-Zaa' peptide mixtures for the inhibition of 24-15 and 24-16.

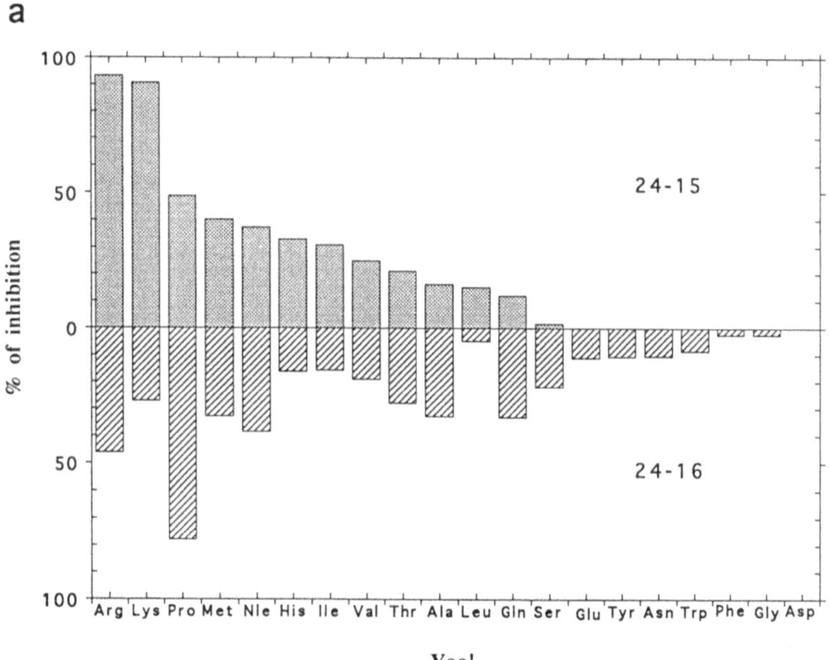
a
100
50
0
50
100
% of inhibition
24-15
24-16
Arg Lys Pro Met Nle His Ile Val Thr Ala Leu Gln Ser Glu Tyr Asn Trp Phe Gly Asp
Yaa'

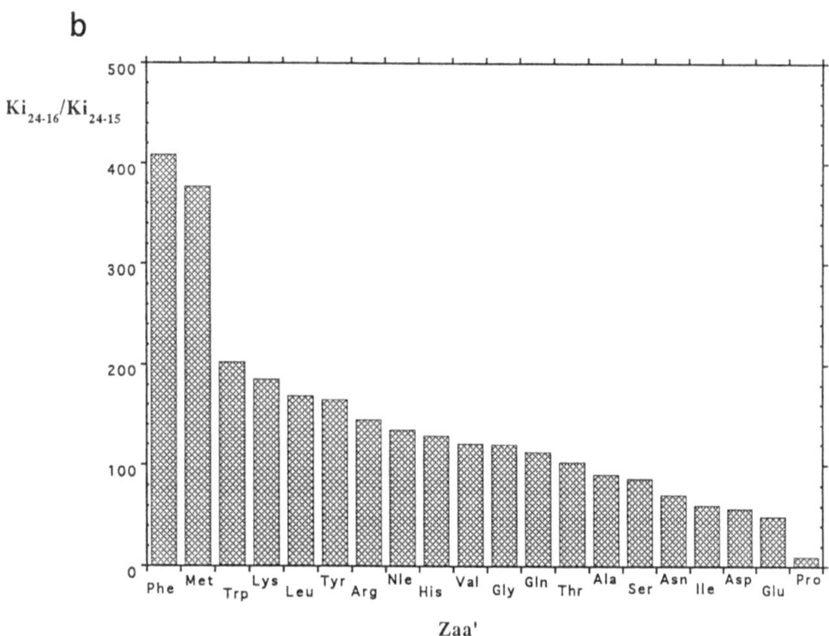
b
Ki_{24-16}/Ki_{24-15}
500
400
300
200
100
0
Phe Met Trp Lys Leu Tyr Arg Nle His Val Gly Gln Thr Ala Ser Asn Ile Asp Glu Pro
Zaa'

(Pz-Pro-Leu-Gly-Pro-$_{(D)}$Arg) between leucine and glycine. Furthermore, our data may explain why the dipeptide Pro-Ile is a specific blocker of the 24-16 (Dauch 1991). We may hypothesize that this peptide interacts with the S_2 and S_1 subsites of 24-16, the specificity of this interaction being insured by the binding of the free proline to the S_2 subsite.

Comments

Prospects An important limitation of the present strategy concerns the molecular diversity of the side chains framing the phosphinic bond. In the coming years, further progress in this area will depend on our ability to develop new synthetic routes to produce the phosphinic blocks more easily. This prospect, in conjunction with the development of the combinatorial chemistry approaches, should provide important insights into the factors controlling the specificity of the proteases and should lead to the discovery of new highly potent and selective inhibitors of zinc-metalloproteases, and of aspartic proteases.

References

Abdel-Meguid SS, Zhao B, Murthy KHM, Winborne E, Choi JK, Desjarlais RL, Minnich, MD, Culp JS, Debouck C, Tomaszek TA, Meek TD, Dreyer GB (1993) Inhibition of human immunodeficiency virus-1 protease by a C_2-symmetric phosphinate. Synthesis and crystallographic analysis. Biochemistry 32, 7972-7980

Dauch P, Vincent JP, Checler F (1991) Specific inhibition of endopeptidase 24-16 by dipeptides. Eur. J. Biochem. 202, 269-267

Furka A, Sebeytyen F, Asgedom M, Dibo G (1991) General method for rapid synthesis of multicomponent peptide mixtures. Int. J. Pept. Prot. Res. 37, 487-493

Grams F, Dive V, Yiotakis A, Yiallouros I, Vassiliou S, Zwilling R, Bode W and Stöcker W. (1996) Structure of astacin with a transition-state analogue inhibitor. Nature Struct. Biol. 3, 671-675

Fig. 5. Influence of the Yaa position for the inhibition of 24-16 (upper part) and 24-15 (lower part) by inhibitor mixtures of general formula Yaa-$_{(L,D)}$ Phe(PO$_2$CH$_2$)Gly-Yaa' (Yaa' contains a mixture of 20 different amino acids). The concentration of phosphinic peptides in each mixture was 1mM for 24-16 and 10 mM for 24-15. **(b)** Influence of the Yaa' position for the inhibition of 24-16 (upper part) and 24-15 (lower part) by inhibitor mixtures of general formula Pro-$_{(L,D)}$ Phe(PO$_2$CH$_2$)Gly-Yaa'. The concentration of phosphinic peptides was 100 nM for 24-16 and 40 mM for 24-15.

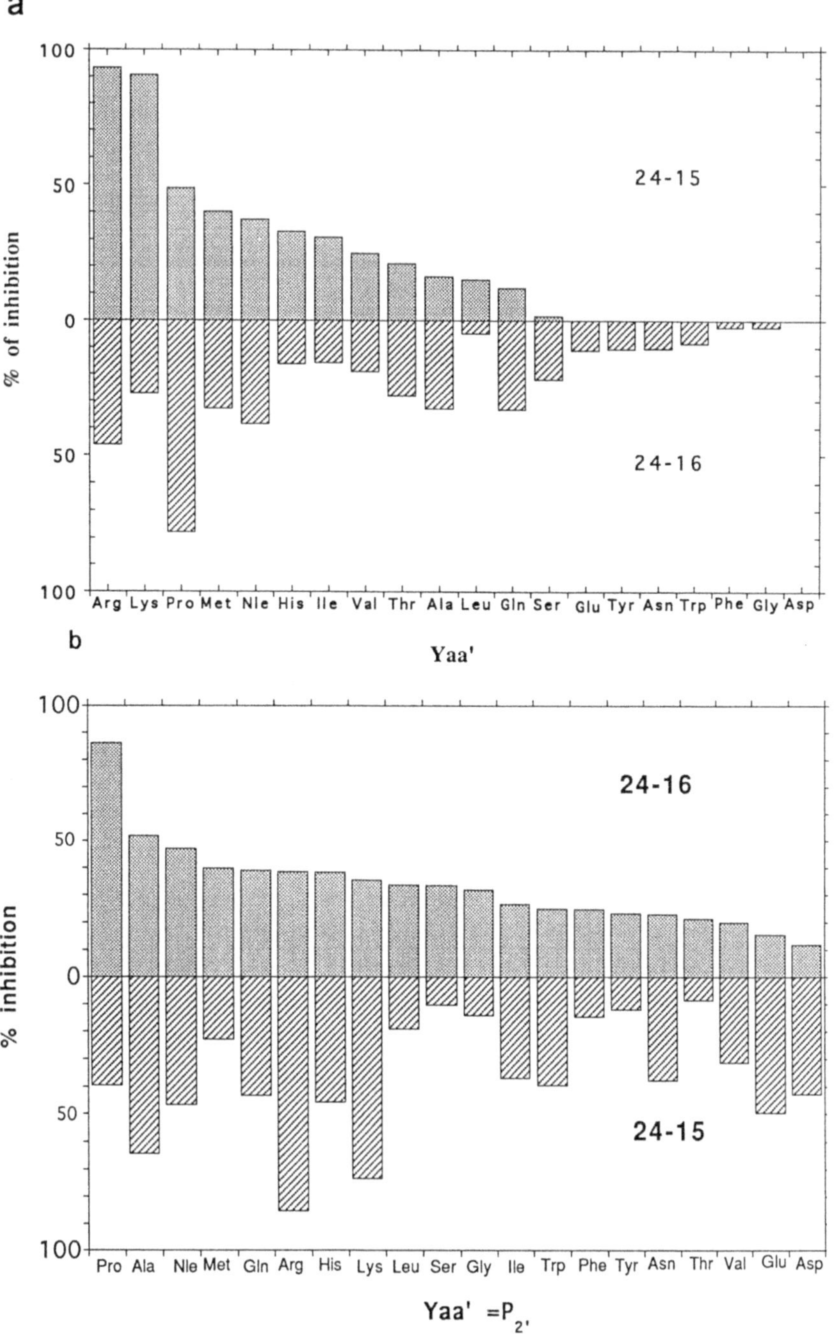
a
100
50
0
50
100
% of inhibition
24-15
24-16
Arg Lys Pro Met Nle His Ile Val Thr Ala Leu Gln Ser Glu Tyr Asn Trp Phe Gly Asp
Yaa'
b
100
50
0
50
100
% inhibition
24-16
24-15
Pro Ala Nle Met Gln Arg His Lys Leu Ser Gly Ile Trp Phe Tyr Asn Thr Val Glu Asp
Yaa' =P_{2'}

Jiracek J, Yiotakis A, Vincent B, Lecoq A, Nicolaou A, Checler F, Dive V (1995) Development of highly potent and selective phosphinic peptide inhibitors of zinc endopeptidase 24-15 using combinatorial chemistry J. Biol. Chem. 270, 21701-21706

Jiracek J, Yiotakis A, Vincent B, Checler F, Dive V (1996) Development of the first potent and selective inhibitor of the zinc endopeptidase 24-1 using a systematic approach based on combinatorial chemistry of phosphinic peptide J. Biol. Chem. 271, 19606-19611

Lam KS, Salmon SE, Hersh EM, Hruby VJ, Kazmierski WM, Knapp RJ (1991) A new type of synthetic peptide library for identifying ligand-binding activity. Nature 354, 82-84

Matthews DJ, Wells JA (1993) Substrate phage: selection of protease substrates by monovalent phage display Science 260 1113-1115

McGeehan GM, Bickett, DM, Wiseman JS, Green M, Berman J (1995) Defined substrate mixtures for mapping of proteinase specificities. In: Barrett AJ (ed) Proteolytic enzymes: aspartic and metallo peptidases. Methods Enzymol. Vol 248. Academic Press, London, pp35-46

Petithory JR, Masiarz FR, Kirsch JF, Santi DV a Malcolm BA (1991) A rapid method for determination of endoproteinase substrate specificity: Specificity of the 3C proteinase from hepatitis A virus. Proc. Natl. Acad. Sci. USA. 88, 11510-11514

Orlowski M, Reznik S, Ayala J, Pierotti AR (1989) Endopeptidase 24-15 from rat testes. Biochem. J. 261, 951-958

Yiallouros I, Vassiliou S, Yiotakis A, Zwilling R, Stöcker W, Dive V (1998) Biochem J 331, 375-379

Yiotakis A, Vassiliou S, Jiracek J, Dive V. (1996) Protection of the hydroxyphosphinyl function of phosphinic dipeptides by adamantyl: Application to the solid-phase synthesis of phosphinic peptides. J. Org. Chem. 61, 6601-6605

Chapter 13

Functional Expression of Recombinant Proteases

DIETER BRÖMME AND BRIAN F. SCHMIDT

Introduction

Proteases play an important role in essentially all physiological processes going from the general protein breakdown for nutrients to the regulation of the programmed cell death. At the same time these proteases are involved in the pathology of almost every imaginable disorder. A growing number of recently discovered proteases are specifically expressed in single tissues, at low expression levels or only at certain developmental stages, and therefore they are difficult to isolate in sufficient quantities with classical biochemical methods. The cloning and expression of these enzymes is frequently the only alternative. Furthermore, recombinant techniques allow directed structural alterations of the enzyme to address mechanistic and functional questions of interest.

Proteases have been expressed in most of the available expression systems (bacterial, viral, yeast, insect cells and mammalian). The choice of which expression system to use depends on the protease and purpose of the follow-up studies, but its out-come is frequently difficult to predict. It will sometimes be necessary to explore more than one expression system for the protease of choice.

In order to produce inactive protease protein for raising antibodies, bacterial expression systems should be favored. Bacterial systems should also be considered for proteases which do not depend on post-translational modifications, in particular on disulfide bond formation. Refolding experiments can be very laborious and time-consuming and often have no guar-

Correspondence to: Dieter Brömme, Mount Sinai School of Medicine, Department of Human Genetics, Box 1498, Fifth Avenue at 100 Street, New YorkNY, 10029, USA (*phone* (212)-659-6753; *fax* (212)-849-2508; *e-mail* brommd01@doc.mssm.edu)
Brian F. Schmidt, FibroGen, 225 Gateway Blvd., South San FranciscoCA, 94080, USA (*phone* (650) 866-7321; *fax* (650) 866-7205; *e-mail* bschmidt@fibrogen.com)

antee of success. If a bacterial system fails to produce functional proteases, and large scale production is not an issue, laboratories are advised to switch to eukaryotic expression systems as soon as possible. Presently, Baculovirus and yeast (*Pichia pastoris*) expression kits are commercially available. Both systems have been shown to produce between 2-50 mg/l of functional mammalian proteases (serine, cysteine and aspartic proteases). Mammalian expression systems are also successful but they are more cost-intensive. They should always be used when mammalian–like posttranslational modifications of the protease are required.

In the following, we will give examples of functional expression of proteases in bacterial, baculovirus, yeast, and mammalian expression sytems and hint to advantages and disadvantages of each expression system.

Subprotocol 1
Expression of Proteases in Bacterial Expression Systems

E. coli Bacteria are generally the first choice for hosts of heterologous protein production since they are easy to manipulate and grow rapidly in inexpensive medium. The gram-negative bacterium, *Escherichia coli*, is the most widely used bacterial host due to the sheer abundance of genetic information and the many different expression systems that are commercially available for **B. subtilis** this organism. Expression in gram-positive bacteria (*e.g. Bacillus subtilis*) can be advantageous since proteins can be directly secreted into the culture medium without utilizing specialized secretory pathways (Wong 1995). Recently, a Bacillus subtilis system has been used to produce two active human serine proteases (Babe 1998, McGrath 1997). However, in general, gram-positive bacteria have not yet found wide use in the production of heterogous proteins due to the presence of extracellular proteases that degrade the proteins of interest, and to a poor understanding of their genetics and secretory machinery.

Major concerns The major concerns for producing active protease in bacteria are insuring high transcription and translation rates, the maintenance of solubility if the enzyme is made in the cytoplasm or periplasm, avoidance of making toxic products, the simplification of the purification, and the avoidance of proteolytic degradation. For *E. coli*, there are a number of different systems available that address these concerns (Olins and Lee 1993, Rosé and Craik 1996). To insure high transcription rates and to reduce toxicity concerns, expression vectors with strong inducible promoters are readily avail-**Expression vectors** able from any major molecular biology vendor or can be easily constructed.

These expression vectors are also usually designed with a strong ribosomal binding site to ensure high translation rates (Gold and Stormo 1990). Translation rates can also be enhanced by using gene fusions (Uhlén and Moks 1990, LaVallie and McCoy 1995) or by translational coupling (Schoner 1990). Solubility, toxicity, and purification problems can in many cases be ameliorated by periplasmic rather than cytoplasmic localization and/ or by the production of fusion or "tagged" proteins. It is important that the fusion protein be designed so that active protease can be released from the fusion protein by proteolytic cleavage (either autocatalytically or by adding an exogenous protease). Inclusion body formation can sometimes be reduced or avoided by growing the bacteria at temperatures higher or lower than the normal 37°C. To avoid proteolytic degradation of the produced protein, *E. coli* strains are available, or can be constructed, that carry protease gene deletions (Gottesman 1990). Changing the location of where the product is produced (from cytoplasm to periplasm or vice versa) can also reduce proteolytic degradation.

Fusion proteins

Unfortunately, no one *E. coli* expression system has yet proven to work well for the production of all proteases or for even one family of proteases. The advantages and disadvantages of different promoters and gene fusions should be studied thoroughly before choosing a given system. It is usually wise to choose an expression system that has been shown to be successful for a protease that is closely related to the one of interest. One should also not give up if the first system fails.

Production of Interleukin-1β-Converting Enzyme (ICE) as a Thioredoxin Fusion Protein in *Escherichia coli*

ICE is a member of a recently discovered but rapidly growing family of cysteine proteases (Kumar 1995, Thornberry and Molineaux 1995). The role ICE-like proteases play in programmed cell death (apoptosis) has increased interest in these proteins. ICE has been produced by a number of groups in *E. coli* as insoluble inclusion bodies that need to be refolded to regain activity. Recently, soluble and active ICE has been made as a thioredoxin fusion protein (Malinowski 1995). Thioredoxin fusion expression systems (LaVallie 1993), suitable for the expression of soluble ICE in the cytoplasm, can be purchased from Invitrogen Corporation.

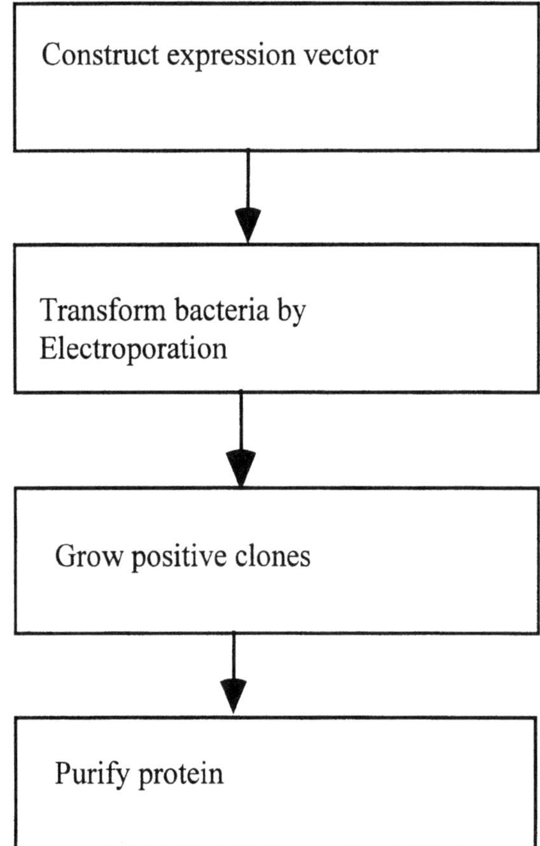

Fig. 1. Scheme for expression in *E. coli*

Materials

- ThioFusion™ Expression System (Cat. No. K350-01, Invitrogen Corporation)
- hICE antibodies (Santa Cruz Biotechnology, Inc.)
- human peripheral blood monocyte cDNA library (Clontech Laboratories, Inc. or self-made)
- Suc-Tyr-Val-Ala-Asp-AMC (Bachem Bioscience, Inc.)
  Additional media and reagents can be purchased from Invitrogen Corp.
- Electroporation apparatus (BioRad Gene Pulser with Pulse Controler or equivalent)
- Luminescence Spectrometer (Perkin-Elmer LS50B or equivalent), or fluorescence plate reader (Titertek Fluoroskan II, ICN or equivalent)

Other reagents and equipment required are listed in the ThioFusion™ Expression System Instruction Manual.

Procedure

1. The human ICE gene can be cloned by PCR or conventional means from commercially available or self-made peripheral blood monocyte cDNA or cDNA library. If using PCR for cloning, the appropriate restriction sites for expression in the pTrxFus vector should be included in the sequence of the PCR cloning primers. Since ICE is self-processing in *E. coli* (Malinowski 1995), the position of the enterokinase cleavage site is not important (as long the ICE gene is cloned in the proper reading frame). Although for some proteases (*e.g.* chymotrypsin-like serine proteases), the correct N-terminal sequence is critical for activity so the cleavage site in the fusion protein must be properly located. Refer to the Instruction Manual for detailed instructions on how to best clone genes in the expression vector, pTrxFus (*Kpn*I and *Bam*HI restriction sites introduced at the 5' and 3' ends, respectively, of the hICE gene can be used for insertion into pTrxFus).

    Expression of ICE

2. Prepare electro-competent GI724 cells as described in the Instruction Manual. Transformation frequencies of up to 10^7/µg DNA can be obtained using 55 µl of cells in a 0.1 cm gap electroporation cuvette with settings of 1800 volts, 25 µF and 200 ohms.

3. Follow the protocol in the Instruction Manual for the expression of the thioredoxin-ICE fusion protein. For soluble expression of ICE, continue to grow the cells at 30°C after induction with tryptophan. Production of ICE can be monitored by Western blots (Burnette 1981) using polyclonal antibodies made against p20 or p10 peptides (the commercially available p10 antibody is more specific in our hands). Active ICE can be assayed using Suc-Tyr-Val-Ala-Asp-AMC by monitoring the fluorescence of the AMC leaving group at 460 nm using a 380 nm excitation wavelength (Thornberry 1994). Antibodies against thioredoxin are also available (Invitrogen Corp.) to analyze the fusion expression products.

Results

About 1 mg of ICE can be produced from 1 liter of culture broth. The insoluble fraction can be refolded and purified to form active enzyme. Purification, refolding and crystallization protocols for ICE can be found in the published literature (*e.g.* Malinowski 1995, Ramage 1995). Detailed protocols for general cloning, expression and protein analysis techniques for production of proteins in *E. coli* can be found in Molecular Cloning, A Laboratory Manual (2nd Edition, Sambrook J, Fritsch EF, Maniatis T, 1989, Cold Spring Harbor Laboratory Press, Cold Spring Harbor, NY) or Current Protocols in Molecular Biology (Ausubel FM, Brent R, Kingston RE, Moore DD, Seidman JG, Smith JA, Struhl K, eds., John Wiley & Sons, Inc., NY). In addition to ICE, examples of heterologous expression of other cysteine protease (e.g. Birch 1995, Burck 1994, Marinez-Abarca 1993, McCall 1994, Taylor 1992, Velasco 1994), metalloprotease (e.g. Freimark 1994, Ho 1994, McKie 1995, Pourmotabbed 1994, Shapiro 1993, Suzuki 1998, Windsor 1991, Ye 1995), serine protease (e.g. Apeler 1997, Collins-Racie 1995, Dergousova 1996, Higaki 1989, Sloan 1997, Shoji 1995, Vishnuvardhan 1997, Wang 1991, Wang 1995) or aspartic protease (e.g. Dame 1994, Ding 1998, Hill 1993, Pfrepper 1997, Rosé 1995) genes in E. coli can be found in the literature.

Troubleshooting

Be sure to perform all the proper controls as outlined in the Instruction Manual. Expression with pTrxFus requires special defined medium and cloning/expression strains which places some constraints on cloning strategies and growth temperatures. Recently, a TrxA fusion system (kit with detailed instruction manual) has become commercially available (pET-32, Novagen, Inc.) based on a T7 promoter driven system (Studier 1990). This system should make cloning easier, gives more flexibility in growth conditions, and allows for the use of *trxB E. coli* strains (Derman 1993). If cost of these kits is an issue, a TrxA–ICE fusion expression vector can be constructed by inserting a cloned *trxA* gene (PCR amplified from *E. coli* genome) in an available expression vector. Alternatively, soluble production of ICE in the periplasm might be possible by using a DsbA fusion which has recently been used successfully for the production of enterokinase (Collins-Racie 1995).

Subprotocol 2
Expression of Proteases in Yeast Expression Systems

Most commonly used yeast species for the functional expression of foreign proteins are *Saccharomyces cerevisae* and *Pichia pastoris*. Both systems allow post-translational modifications such as disulfide bond formation, proteolytic maturation and glycosylation similar to mammalian cells although high mannose glycosylation is sometimes a problem using *Saccharomyces cervisiae*. A specific advantage of the *Pichia* system is the very low level of secretion of native proteins which leaves the foreign gene product as essentially the only protein in the production media (Cregg 1993; Romanos 1995).

Yeast expression systems are recognized to be safe for humans, can be operated in regular laboratories, and are cheap in maintenance. However, commercial users of the *Pichia* system are required to pay significant registration and annual licensing fees.

Both systems utilize the *Saccharomyces cerevisiae* prepro α–mating factor (α–MF) as leader sequence with an ATG initiation codon in their expression vectors. The cDNA of the foreign protein is ligated to the appropriate leader sequence of the transfer vector while maintaining the open reading frame. Whereas the expression vectors for the *S. cerevisiae* system exploit in general ADH or α–MF–promoters and replicate in the yeast cell with the help of the 2μ plasmid origin of replication, the available expression vectors for *P. pastoris* are under the control of the AOX1 promoter and are integrated into the yeast genome. Recently, a methanol-independent *Pichia* expression vectors are commercially available which permit the expression of proteins under the control of the constitutive glyceraldehyde-3-phosphate dehydrogenase (GAP) promoter (Invitrogen, Carlsbad, CA). Several factors that drastically influence the protein expression in *P. pastoris* have been described. These include: copy number of the expression cassette, site and mode of chromosomal integration of the expression cassette, mRNA 5'- and 3'-untranslated blocks, nature of secretion signal, host strain physiology, and endogenous protease activity (Sreekrishna 1997).

Either for *S. cerevisiae* and *P. pastoris*, strains are available which are deficient in endogenous protease. *Saccharomyces,* and probably *Pichia* too, do not express endogenous cysteine endopeptidases. This is of advantage for the expression of cysteine proteases. The only known cysteine protease in yeast is the aminopeptidase bleomycin hydrolase (Enenkel and Wolf 1993).

▦▦ **Materials**

For general protocol
- shaker flask 2l
- incubator (30°C)
- *BJ* 3501 or related yeast strains (Jones 1991)
- SD–Ura media (yeast nitrogen base,w/o amino acids, 6.7g/l; dextrose 20g/l; drop out powder 1.5g/l
- drop out powder: adeneine, 4g; L–Trp, 4g; L–His, 2g; L–Arg, 2g; L–Met, 2g; L–Tyr, 3g; L–Ile, 3g; L–Val, 15g, L–Lys, 3g; L–Phe, 5g; L–Glu, 10g; L–Thr, 20g, L–Leu, 6g; L–Asp, 10g; L–Ser, 37g. Keep as dry powder. Sufficient for 100 l SD–media; autoclave for 30 min.

Saccharomyces cerevisiae

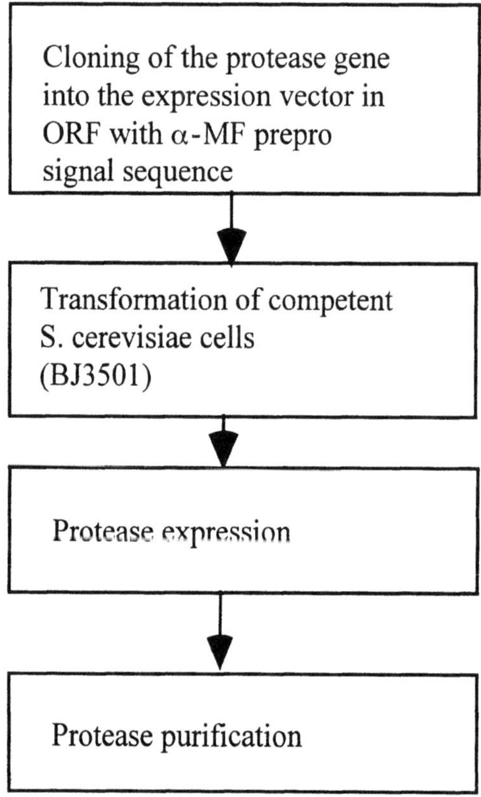

Fig. 2. Scheme for functional protease expression in *Saccharomyces cerevisiae*

- production media: 40g/l casamino acids; 20g/l dextrose; 5g/l ammonium
  sulfate (autoclave for 30 min).
- Processing buffer for recombinant cysteine proteases:
  100 mM sodium acetate buffer, pH 4.5 containing 5 mM dithiothreitol
  and 5 mM EDTA (for cathepsin S; Brömme 1993)

- shaker flasks 100 ml and 2 l
- shaker (30°C)
- *Pichia* Expression Kit (Cat. No. K1710-01, Invitrogen Corporation)
- Media recipes are delivered with the expression kit

For generation of expression vector

Procedure

Clone the coding region of the foreign protease into the open reading frame
at the 3' end of the α–MF prepro leader sequence. If the protease contains its
own signal sequence this sequence must be deleted.

General protocol

Note: The presence of the α–MF–pro region seem not to be essential since
the expression of both fusion proteins, pre–α–MF–procathepsin S and prepro–αMF–procathepsin S resulted in the identical yields.

1. Use 50 µl of competent *BJ* 3501 cells and add 5µl of the expression plasmid.

Transformation of *S. cervisiae*

2. Plate the transformation mix onto SD–ura media containing plates and
   incubate the plates at 30°C for 48 hours.

1. Inoculate 20 ml of SD–ura media with a single colony and shake for 2
   days at 30°C.

Expression culture

2. Then inoculate 1l of production media in a shaker flask and harvest after
   3.5 days.

Note: To optimize the expression, check for protease activity in an initial
culture every 8 to 12 hours with a substrate assay as described for the Baculovirus system.

3. Depending on the location of the protease activity (production media,
   cell pellet or both) process the appropriate fractions:

4. Cellular bound activity

Spin down the cells at 3000 x g, lyse the yeast cells using a French press or glass beater. Resuspend the lysed cells in a processing buffer and incubate at 40°C in a shaker. Follow the increase in activity using a substrate assay and stop at maximum activity.

5. Secreted activity

Concentrate supernatant from 1 l down to 100 ml using a S10Y10 spiral membrane cartridge. Adjust the concentrated solution to appropriate processing conditions, incubate at 40°C and follow the increase in activity.

6. If media and cells contain significant amounts of activity combine and process both fractions.

Note: Although the leader sequence of α–MF is a strong secretory signal the secretion of the protease is not guaranteed. Check always both supernatant and cell pellet for recombinant protein and activity.

Generation of the expression vector Use an expression vector which allows the secretion of your protease (e.g. pPic9 from Invitrogen Corporation). Clone the coding region of the foreign protease into the open reading frame at the 3' end of the α–MF prepro leader sequence. If the protease contains its own signal sequence, this sequence must be deleted. In order to transform his4 *Pichia pastoris* the expression vector has to be linearized by digestion with a restriction endonuclease.

Transformation of *P. pastoris* and screen for recombinant strain

1. *P. pastoris* can be transformed by different methods: a) spheroblast method, b) lithium transformation method and c) by electroporation. The lithium and the electroporation method are more convenient than the spheroblast method but they do not favour a multi-copy integration of the DNA. Detailed protocols for these methods of transformation are delivered with the purchase of the expression kits. After transformation the yeast genome contains your foreign protease and the HIS4 gene.

2. Plate the transformants on a –his agar plate. Only cells which have integrated the HIS4 gene will grow.

3. Patch colonies from step 2 on a –his/+glycerol and a –his/+methanol plate. Select for colonies which grow slowly on the –his/+met plate. In these colonies the AOX1 gene is disrupted (his$^+$/mut– mutants).

Pichia pastoris

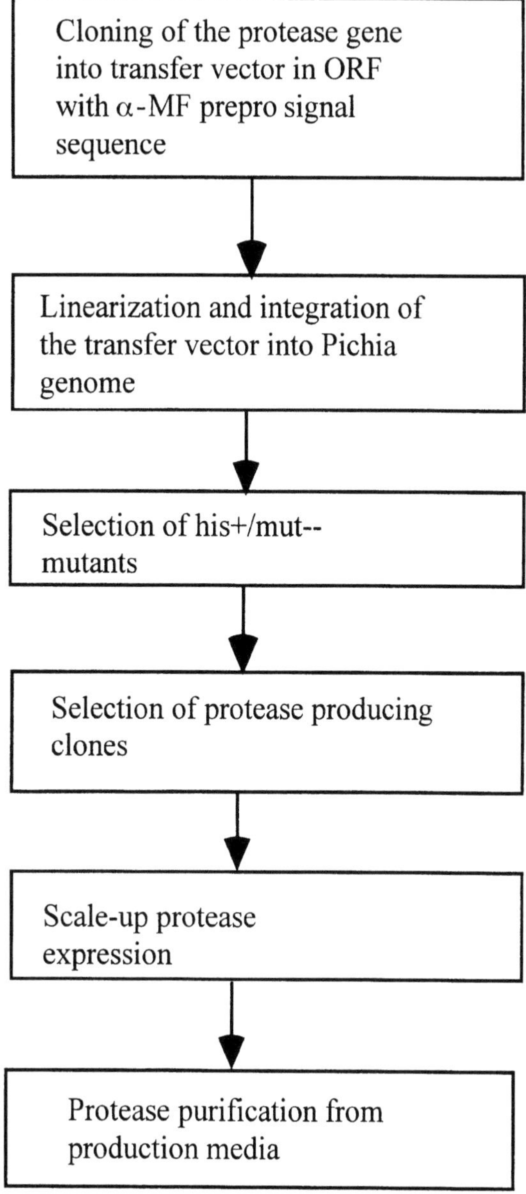

Fig. 3. Scheme for functional protease expression in *Pichia pastoris*

Selection of high producer colonies

1. Select 10-20 his+/mut– colonies and grow for 2 days in media containing glycerol as carbon source at 30°C.

2. Spin out the cells, resuspend the pellet in media containing methanol as a carbon source and grow cells for 4-6 days at 30°C.

3. Harvest the supernatant and analyse the protease expression for activity as described below (baculovirus expression) or by SDS–PAGE and Western blot analysis. Your protease product should be the predominant product in the supernatant. Select for the clone with the highest protein/activity expression.

Scale–up production and purification

1. Grow 50 ml of a selected high producer colony in glycerol containing media overnight at 30°C.

2. Inoculate 1 l of production media containing glycerol with the preculture and let the cell grow up to an OD_{600} of ca.20. At maximal OD_{600} add x mL of methanol. Repeat addition of methanol every 24 h. Monitor increase in activity as described in the baculovirus section and stop at maximum of activity (after 5–6 days).

3. Spin out the cells and concentrate the supernatant using a S10Y10 spiral membrane cartridge. Proceed with a processing step to activate your protease if necessary. Yields are in the range of 10–20 mg/l for cathepsin O2 (D.B. unpublished results).

4. Purify your enzyme with a protease specific protocol. The protease in the concentrated supernatant will already be relatively pure. The use of a strong cationic ion exchange or a hydrophobic interaction chromatography (HIC) resin may reduce the whole purification to one column step to get a protein of 95 % purity (Cregg et al., 1993).

Results

Saccharomyces cervisiae and *Pichia pastoris* have been successfully used for the expression of recombinant cysteine proteases (cathepsins S, L, B, K, V, F, Brömme 1993, Linnervers 1997, Illy 1997, Sun 1997, Roche 1997, Wang 1998, Brömme 1999 papain, Vernet 1993, Ramjee 1996), serine proteases (trypsin, Hedstrom 1992; alkaline protease, Morita 1994; trypsin, Halfon and Craik 1996; tryptase, Niles 1998; enterokinase, Vozza 1996; tissue kallikrein, Chan 1998; granzyme B, Pham 1998; proteinase 3, Harmsen, 1997; carboxypeptidase Y, Ohi, 1996), metalloproteases (TACE, Clarke 1998) and aspartyl proteases (cathepsin E, Yamada 1994).

The *Pichia* expression system seems to be the system of choice for the expression of papain–like cysteine proteases. Cysteine proteases are secreted in their proenzyme form and the yields of protease are generally 6–10mg or higher per liter of culture medium. The *Saccharomyces* system was significantly less efficient in our hands for the expression of the same protease class. Yields of expression for related papain–like cysteine proteases varied between 0.1 mg for unpurified human cathepsin S and papain and 20 mg/L for human cathepsin B (Illy 1997).

In contrast to the *S. cerevisiae* system, *P. pastoris* always secreted sufficient amount of recombinant proteases into the media for our studies. Although the secretory versions of both systems was exploit the same α–MF–secretory signal, secretion in *Saccharomyces*is not reliable. In our hands, procathepsin S was only 40 % in the culture media whereas 60 % was in the cell pellet. Using the same signal sequence for the expression of recombinant papain and cathepsin B, essentially all of the papain was found within the yeast cells whereas cathepsins B was entirely purified from the media supernatant. Thus the presence of a secretory signal in the expression vector is not sufficient to predict the routing of the recombinant protease. In all cases, both the culture supernatant and the cell pellet should be examined for the recombinant protein.

Subprotocol 3
Expression of Protease Genes in Insect Cells

Baculovirus expression vectors have been widely used for the production of recombinant proteins in large quantities. Posttranslational modification of the gene products are similar to mammalian cells offering an advantage to bacterial systems when these modifications are important for follow-up studies. Furthermore, the expression system allows high level production of cytosolic, secreted and membrane-bound proteins.

The inability of the baculovirus (*Autographa californica* nuclear polyhedrosis virus, AcNPV) to infect mammalian cells allows the use of the expression system in regular laboratory facilities. The virus can be easily inactivated with anti microbial disinfectants as used for the disposal of *E. coli* cultures.

The most commonly used host for AcNPV infections are cell lines (*Sf* 9, High Five™) derived from the fall armyworm (*Spodoptera frugiperda*).-Since the baculovirus genome (128Kb) is too large to permit easy manipulations of the DNA, the foreign gene is inserted by homologous recombi-

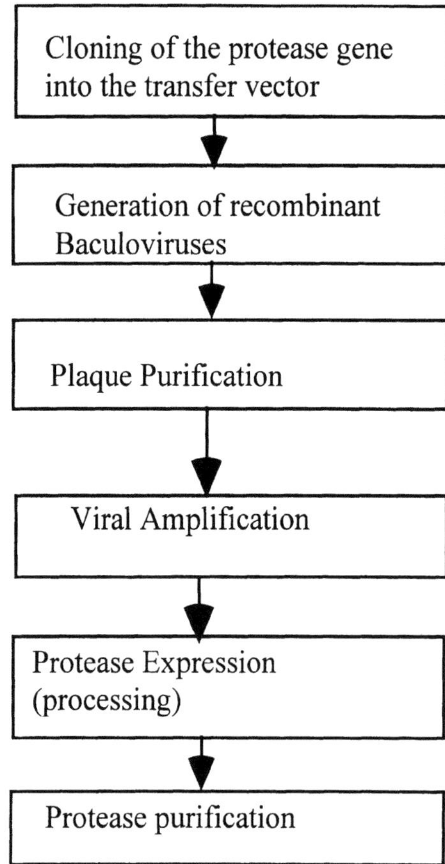

Fig. 4. Scheme for functional protease expression using the baculovirus expression system

nation into the virus via a transfer vector. Routinely used transfer vectors contain the polyhedrin promoter with varying flanking regions and alternative restriction enzyme sites for the insertion of the foreign gene sequence. The cDNA of the selected protease is inserted into the transfer vector as a full length coding sequence starting with an ATG initiation codon and ending with a termination codon. After viral infection, the recombinant protease containing insect cells and/or the medium supernatant are harvested after 48 to 86 hours. Proteases containing a proregion are expressed as inactive precursors and must be activated either autocatalytically or with an appropriate processing enzyme.

Materials

- spinner flask 2 l
- stirring plate
- incubator (28°C)
- Dounce homogenizer
- Baculo Gold™ Starter Kit (Cat.No. 21001K, PharMingen) or MacBac^R 2.0 Kit (K835–01, Invitrogen Corporation), both kits contain linearized baculovirus DNA, transfer vectors, *Sf* 9 cells, Agarose and supplements)
- TMN–FH Insect media (PharMingen)
- Serum-Free Insect media (GibcoBRL)
- Processing buffer for recombinant cysteine proteases:
  100 mM sodium acetate buffer, pH 4.5 containing 5 mM dithiothreitol and 5 mM EDTA (for cathepsin S; Brömme and McGrath 1996)
- 100mM sodium acetate buffer, pH 4.0 containing 2.5 mM dithiothreitol and 2.5 mM EDTA and 0.4 mg/ml pepsin (for cathepsin O2; Brömme 1996)

Procedure

Use a convenient transfer vector (pVL1392/1393; pBlueBac4) and clone the complete coding region of your protease including the ATG initiation codon into the multi-cloning site of the vector using standard methods of molecular biology (2nd Edition, Sambrook J, Fritsch EF, Maniatis T, 1989, Cold Spring Harbor Laboratory Press, Cold Spring Harbor, NY). The distance between the ATG codon and the 3' end of polyhedrin promotor should not be larger than 20 bp.

Generation of the transfer vector

1. Label three T-25 flasks: a) co-transfection; b) positive control, 3) negative control

Generation of recombinant baculoviruses

2. Plate out 2×10^6 *Sf* 9 cells per T-25 flask in 3ml TMN–FH media. Let sit until cells attach (15min). Alternatively High Five™ insect cells can be used which have faster doubling times and adapt easier to serum–free medium and can produce higher yields of secreted proteins.

3. Prepare the transfection mix in a sterile eppendorf tube: combine 0.5 µg linearized BaculoGold™ DNA plus 2 µg recombinant transfer vector, and let sit for 5 min. Then add 1ml of Transfection Buffer B and mix well.

4. For the co-transfection:
   - remove media and add 1ml Transfection Buffer A. Swirl flask a bit to completely cover cells.

- add the 1 ml DNA/Buffer B solution by slowly dropping the mix into the center of the flask and swirl gently

Note: A fine white precipitate should appear after the first addition
For the positive control:

5. remove media and add 3ml fresh TMN–FH media

6. add 50μl wild type *Ac*NPV to the cells
   For negative control
   remove media and add 3 ml fresh TMN–FH media.

7. Incubate all three T-flasks at 28°C for 4 hours.

8. For the co-transfection:

9. remove media from T-flask,

10. add 3 ml fresh TMN–FH media.

11. Incubate all three T–flasks at 28°C for 4 days and then compare for infection. Infected cells are larger and often float.

12. On day 5: remove the supernatant from the co-transfection T-flask, spin out the cells and store the supernatant at 4°C. This is the viral stock.

Plaque Assay to Isolate Single Recombinant Viral clone

1. Label three 10 cm dishes: a) 10-3; b) 10^{-4}; c) 10^{-5}.

2. Plate out 5×10^6 *Sf*9 cells per dish in 12 ml TMN-FH media. Let sit until cells attach (15min).

3. In sterile 6ml tubes, prepare the following serial dilutions of the viral stock in TMN-FH media: a) 10^{-2}; 10^{-3}; 10^{-4}; 10^{-5}.

4. Add 100μl of the viral stock dilutions to the media of the appropriate labelled 10 cm dishes and gently swirl the media.

5. Incubate for 1 hour at 28°C and during incubation prepare Agarplaque™ agarose overlay solution:
   - make up 20 ml of a 1.6 % agarose solution in Serum-Free Insect Media
   - heat to 60°C in the microwave to dissolve agarose(note: do not over-heat as precipitation of media components will result)
   - cool to 45°C in a water bath and then add 20 ml TMN–FH media to a final agarose concentration = 1%.

6. Remove virus inoculum from 10cm dishes and carefully, by slowly pipetting down the side of the dish, add 10ml of the 1% agarose solution to each dish.

7. Place dishes onto level, flat surface until agarose hardens.

8. Return dishes to a humid 28°C incubator for 6 days after which time check for the appearance of plaques.

9. Remove a single plaque by using a sterile glass Pasteur pipette to bore the plaque from the agarose.

10. Transfer the agarose plug containing the plaque to a sterile tube containing 500µl of Gibco SF Media, and rotate overnight at 4°C to elute the viral particles.This material can be used to establish an amplified high-titer viral stock

1. Plate out 7 x10^6 *Sf* 9 cells into a 10cm dish containing 15 ml of SF 900II media and allow cells to attach (15min). **Amplification of viral stock**

2. Add 100 µl of the eluted viral material generated from the agarose plaque plugs and mix gently.

3. Incubate at 28°C for three days.

4. Harvest the media from the dish and spin at 1500 rpm for 10 min to remove any cells from the suspension.

Note: The media should contain approximately 10^7 viral particles /ml

5. Repeat steps 1-4 to obtain the final high titer viral stock of 10^8 viral particles/ml.

6. Store viral stocks at 4°C (good for several weeks).

1. Grow *Sf* 9 cells in 1l SF-900II medium to a density of 2 x 106 cells/ml and infect at a multiplicity of infection of 1.5. **Protease expression**

2. Follow total cell number, cell viability (Trypan Blue) and activity of the recombinant protease in the media supernatant and within the cells every 24 hours and harvest at maximal protease activity.

3. Protease processing assay (cysteine proteases):
Lysosomal cathepsins have to be processed prior to a substrate activity assay either by autocatalytic activation at low pH (50 mM acetate buffer

4–4.5) or with pepsin (0.1mg/ml) at pH 4.0 for 30 min. For the activity assay use 1 ml of the cell culture, spin out the cells, lyse the in the assay buffer resuspended cells by a 5–times passage through a 22G1 1 1/2 needle and adjust the pH of the clear supernatant with 1 M acetic acid to the required value (4.0 to 4.5). Add dithiotreitol and EDTA to a final concentration of 2.5 mM and incubate for 30 min at 40 °C. Use an aliquot to measure the activity against a convenient substrate (e.g. Z-Phe-Arg-MCA for papain-like cysteine proteases). Proteases expressed as mature enzymes may be directly analyzed at their pH optimum.

4. Depending where the activity is found (in the media supernatant, the cell pellet or in both) process the appropriate fractions.
   - For cell activity:
     Spin out the cells (3000 x g), lyse them with a Dounce homogenizer, resuspend the lysate in a processing buffer and incubate at 40°C in a shaker. Monitor the increase of activity with a substrate assay and stop at maximal activity.
   - For media supernatant activity:
     Concentrate the supernatant from 1L down to 100ml using a S10Y10 spiral membrane cartridge. Adjust the concentrate to the conditions for processing, incubate at 40°C and follow the increase of activity.
   - If media and cell pellet contain significant amounts of activity process both fractions.

Note: If cysteine proteases are expressed in the baculovirus system an endogenous viral and host cell cysteine protease may interfere with the recombinant protease. The host cell protease can be easily inactivated by a short time exposure of the activity mixture to pH 7.0 for 5 min if the recombinant protease is stable at this pH. The viral cysteine protease can be detected by using Z-Arg-Arg-MCA as substrate. The dibasic substrate is 3 times more efficiently hydrolyzed than Z-Phe-Arg-MCA. Besides cathepsin B most known cysteine proteases are unable to hydrolyze the Z-Arg-Arg-MCA substrate (Brömme and Okamoto 1995).

- The processed mature enzyme can be purified using conventional column fractionation steps.

Results

Proteases of different classes (serine proteases: pro-kallikrein, Fertig et al. 1993; furin, Bravo 1994; Kex2p, Germain 1992; plasmin, Mhashilkar 1993; Thomson 1995; pancreatic necrosis virus protease, Magyar and Dobos 1994; Kex-1, Latchinian-Sadeck and Thomas 1994; tryptase and chymase, Wang 1998; human prostate specific antigen , Kurkela 1995; neuropsin, Shimizu 1998, caldecrin, Yoshino-Yasuda 1998; granzyme B, Xia 1998; hepatitis C protease, Sali 1998; cysteine proteases: ICE, Wang 1994; falsipain, Salis 1995; papain, Vernet 1990; cathepsin S, Brömme and McGrath 1996; cathepsin O2/K, Brömme 1996, Bossard 1996; cathepsin B, Steed 1998, caspases, Fassy 1998; aspartyl proteases: thermopsin, Lin 1992, and metallo-proteinases: TACE, Patel, 1998; MMP9, George 1997) have been successfully expressed in the Baculovirus system.

Yields of the protease expression vary from 1 to 60 mg/l cell culture and are difficult to predict. Caution is to be taken with contamination of your recombinant protease with endogenous host cell or viral proteases. Our own result revealed the expression of a baculovirus specific cysteine protease (Brömme and Okamoto 1995) and a host cell thiol-dependent protease. The latter one is highly antigenic (unpublishesd results, DB). Probably this protease and two other proteolytic activities have been described by Naggie and Bently 1998. Trials to produce polyclonal antibodies against purified recombinant cathepsins S and O2 resulted in antibodies against the host cell protease although N-terminal sequencing displayed only the recombinant enzymes. Minuscule amounts of the *Sf* 9 protease were obviously sufficient to induce the antibody production.

Comments

Detailed protocols for using the Baculovirus Expression system are available with the expression kits from PharMingen and Invitrogen Corporation.

The interested readers may refer to comprehensive compilations of the Baculovirus expression system in recently published books: O'Reilly DR, Miller LK, Luckow VA (1992) Baculovirus Expression Vectors, Freeman WE, New York; King LA, Possee RD (1992) The Baculovirus Expression System: A Laboratory Guide, Chapman and Hall, London; Richardson CD (1995) Baculovirus Expression Protocols, Methods in Molecular Biology, Vol. 39, Humana Press, Totowa, New Jersey.

Subprotocol 4
Expression of Proteases in Mammalian Expression Systems

Production of human proteases in mammalian cells may be the best choice if proper posttranslational processing (glycosylation, phosphorylation, membrane anchoring *etc.*) or special cofactors are critical for enzyme activity or stability. In general, mammalian cells can produce large complex multimeric proteases from higher eukaryotes much better than other hosts. For functional assays, expression in mammalian cells may be the only option. The negative aspects of mammalian cells are their fastidious growth and handling requirements, and the expensive media and equipment needed for their production. Typically, one would use an Adeno, vaccinia, or Sindbis virus based expression system for rapid production of a protease (especially useful for the production of toxic proteases). For testing the utility of a given gene construction (promoters, enhancers, *etc.*) or for the production of relatively small amounts of a number of different proteases (*e.g.* mutants), a transient transfection system is usually preferred. However, for the repetitive production of large amounts of a protease, expression from a stable transformant, usually in Chinese hamster ovary cells (CHO), is the preferred production system. There are many choices for vectors (Kaufman 1990a), promoters (constitutive or inducible), enhancers, splice signals (Kriegler 1990), selection/amplification methods (Kaufman 1990b), and gene transfer techniques (Keown 1990). Many different types of mammalian expression systems are now commercially available. In general, it is best to start with the mammalian system that has already been successful for the protease of interest or the closest related protease.

Production of Human Neutral Endopeptidase (NEP) in CHO Cells

NEP is a zinc metalloprotease that can cleave small peptides like enkephalins and tachykinins. Thus, active NEP may have a therapeutic potential in being able to suppress acute inflammation and hyperimmune responses. NEP is glycosylated, membrane-anchored, and has a number of intramolecular disulfide bridges making it a good choice for expression in mammalian cells.

Materials

- pCIShENK (Malfroy-Camine and Schofield 1990), or a human placenta cDNA library (Clontech Laboratories, Inc. or self made) and an appropriate mammalian expression vector (see below)
- Chinese hamster ovary/dihydrofolate reductase minus cell (CHO/ **Cells** DHFR–, *e.g.* American Type Culture Collection ATCC#9096)

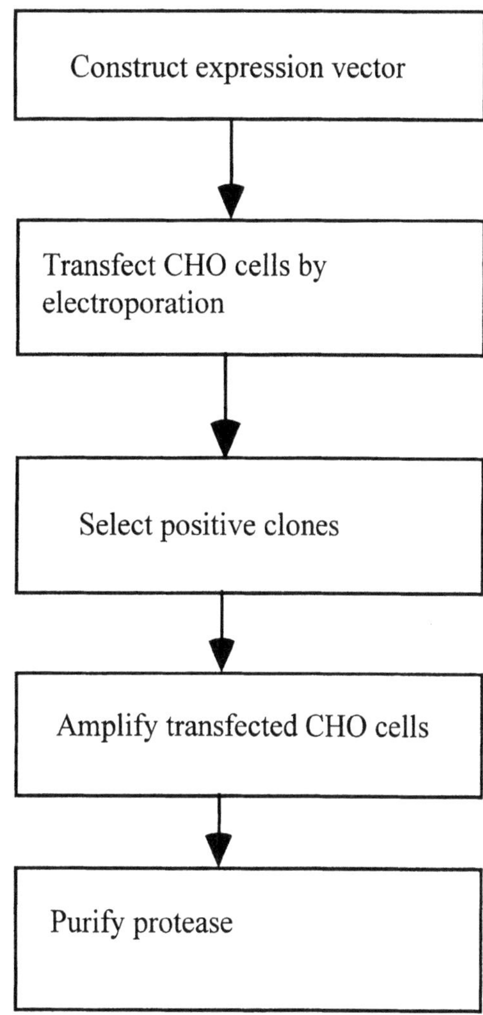

Fig. 5. Scheme for expression in CHO cells

- F-12 growth medium, F-12 nutrient mixture (Ham) with 10% dialyzed fetal bovine serum, 100 µg/ml streptomycin, and 100 U/ml penicillin G (Life Technologies, Inc.)
- F-12 HT growth medium, F12 growth medium with 0.1 mM hypoxanthine, 0.01 mM thymidine
- Phosphate Buffered Saline (PBS), calcium and magnesium free (Life Technologies, Inc.)
- [+] Amethopterin (methotrexate), cell culture grade (Sigma Chemical Company)
- Trypsin-EDTA (0.05% trypsin, 0.53 mM EDTA in Hank's balanced salts, (Life Technologies, Inc.)
- Inverted microscope and hemocytometer (for cell counting)
- Electroporation apparatus (BioRad Gene Pulser or equivalent)
- Refrigerated tabletop centrifuge
- Humidified 5% CO_2/95% air incubator

Procedure

Vector construction

The expression plasmid, pCIShENK (Malfroy-Camine and Schofield 1990), should be obtained or an equivalent vector constructed. The human NEP gene can be cloned from a human placenta cDNA library. An expression vector, equivalent to pCIShENK, can then be constructed by inserting the cloned NEP gene into an appropriate mammalian expression vector (*e.g.* pCI-neo, Promega Corp.; or pcDNA3, Invitrogen Corp.; the DHFR gene, for amplification, could be cloned into these vectors or cotransfected with the expression plasmid in the CHO cells). Vectors with a viral origin of replication should be used for transient expression in order to produce enough protease to assay (see Troubleshooting below).

Transfection

1. Grow CHO cells in F-12 HT growth medium.

2. Trypsinize about 10^8 cells.

3. Collect cells by centrifugation at 1500 rpm for 5 min. Resuspend cell pellet in 8 ml ice cold PBS. Spin cells down and resuspend in fresh buffer again. Repeat spins and resuspensions two more times (final resuspension should be about 6 x 10^6 cells/ml).

4. Add 0.8 ml of suspended cells to an electroporation cuvette (0.4 cm electrode gap). Keep on ice. Add 10 µg of linearized plasmid DNA (cut at a restriction site outside of the expression/selection genes) and mix by

gentle pipetting. Let the cuvette sit on ice for 5 min. Electroporate the cells at 300 volts with a capacitance setting of 960 µF. After electroporation, place the cuvette on ice for 10 minutes.

5. Transfer the cuvette contents to 15 ml of F–12 growth medium in a 10 cm dish (this medium selects for DHFR positive clones). Use 1 ml of the growth medium to rinse the cuvette and add this to the dish. Place dish in incubator at 37°C. Next day, trypsinize plates, count cells, and plate out on 20–30 10 cm dishes at 500 cells per dish in F–12 growth medium.

1. Feed dishes every 2–3 days with fresh F–12 growth medium. After 12–14 days, pick 3–5 colonies and transfer to 24-well cluster plates. Grow to confluence (6–10 days) and transfer cells to 6–well cluster plates and grow to confluence.

Selection/ amplification

2. Remove medium and harvest trypsinized cells for NEP activity assays (Malfroy–Camine and Schofield 1990) and Western blots (Burnette 1981). A portion of the cells should be frozen and stored on liquid nitrogen for future use.

3. Pick best NEP producing colonies for methotrexate amplification. Start cells on F–12 growth medium with 10 nM methotrexate. (Caution: methotrexate is a potent carcinogen and should be handled with care. Always wear gloves.)

4. Amplify using standard protocols (*e.g.* Kaufman 1990b). The best NEP producers after amplification can be used for larger scale production (Mather 1990).

Results

Production of NEP in methotrexate amplified CHO cells can yield about 1 mg per liter of culture medium or more. The cell culture conditions can be optimized and scaled up to produce commercial levels of NEP using standard techniques (Mather 1990). Additional protocols for general cloning, electroporation, expression/amplification, protein analysis and cell culture techniques can be found in Molecular Cloning, A Laboratory Manual (2nd Edition, Sambrook J, Fritsch EF, Maniatis T, 1989, Cold Spring Harbor Laboratory Press, Cold Spring Harbor, NY) or Current Protocols in Molecular Biology (Ausubel FM, Brent R, Kingston RE, Moore DD, Seidman JG, Smith JA, Struhl K, eds., John Wiley & Sons, Inc., NY). In addition to NEP, other metalloprotease genes (e.g. Benbow 1996, Freije 1994, Freimark 1994,

Kelly, 1996, Pei 1994, Russo 1997, Schmidt 1994, Smyth 1995), as well as serine protease (*e.g.* Caputo 1993, Cheng 1993, Molloy 1992, Sinha 1992, Urata 1993) and aspartic protease (*e.g.* Richo and Conner 1994, Tsukuba 1993) genes, have been expressed in mammalian cells.

Troubleshooting

It is recommended that a number of controlled experiments be performed to characterize the CHO (DHFR deficient) cell line before attempting to express the protease of interest.

- Confirm that CHO cells are DHFR deficient by comparing growth and cell phenotype in F–12 HT growth medium versus F–12 growth medium (can use a six-well cluster plate and inoculate each well with 50,000 cells). Cells grow slowly, have a flattened morphology, and die when monitored for DHFR deficiency in F–12 growth medium.

- Optimize conditions for CHO cells. One can vary buffer composition, DNA amounts (*e.g.* 1, 5, or 10 µg), and capacitance/voltage settings (*e.g.* 750, 800, 850 or 900 volts at 25 µF; 200, 250, 300, or 350 volts at 960 µF) to determine the best transfection conditions for a particular CHO cell line. One can use a reporter vector (*e.g.* pSV-β-galactosidase with a β-galactosidase enzyme assay system, Promega Corp.) to determine transfection efficiencies.

- Verify that the constructed expression plasmid functions to produce the desired protease. This can be done using a transient transfection protocol using COS cells (the vector should have a viral replication origin). Protocols for transient expression of proteins are available (*e.g.* in Current Protocols in Molecular Biology, see above) and have been used to produce NEP (Devault 1988, Gorman 1989). In addition to DHFR, it is sometimes useful to have an additional selectable marker on the expression plasmid flanking the NEP gene (*e.g.* neomycin resistance) to help select the integration of the NEP gene. One can select for drug resistance and then verify the DHFR phenotype of these cells (by growth in F–12 growth medium) before assaying for NEP production.

References

Apeler H, Gottschalk U, Guntermann D, Hansen J, Massen J, Schmidt E, Schneider KH, Schneidereit M, Rubsamen-Waigmann H (1997) Expression of natural and synthetic genes encoding herpes simplex virus 1 protease in *Escherichia coli* and purification of the protein. Eur J Biochem 247:890-895

Babé LM, Yoast, S, Dreyer, M, Schmidt, BF (1998) Heterologous expression of human granzyme K in *Bacillus subtilis* and characterization of its hydrolytic activity *in vivo*. Biotechnol Appl Biochem 27:117-124

Benbow U, Buttice G, Nagase H, Kurkinen M (1996) Characterization of the 46-kDa intermediates of matrix metalloproteinase 3 (stromelysin 1) obtained by site-directed mutation of phenylalanine 83. J Biol Chem 271:10715-10722

Birch GM, Black T, Malcolm SK, Lai MT, Zimmerman RE, Jaskunas SR (1995) Purification of recombinant human rhinovirus 14 3C protease expressed in Escherichia coli. Protein Expr Purif 6:609-618

Bossard MJ, Tomaszek TA, Thompson SK, Amegadzie BY, Hanning CR, Jones C, Kurdyla JT, McNulty DE, Drake FH, Gowen M, Levy MA (1996) Proteolytic activity of human osteoclast cathepsin K. Expression, purification, activation, and substrate identification. J Biol Chem 271:12517-24

Bravo DA, Gleaso JB, Sanchez RI, Roth RA, Fuller RS (1994) Accurate and efficient cleavage of the human insulin proreceptor by the human proprotein-processing protease furin. Characterization and kinetic parameters using the purified, secreted soluble protease expressed by recombinant baculovirus. J Biol Chem 269: 25830-25837

Brömme D, Bonneau PR, Lachance P, Storer AC (1994) Engineering the S_2 subsite specificity of human cathepsin S. J Biol Chem 269: 30238-30242

Brömme D, Bonneau PR, Lachance P, Wiederanders B, Kirschke H, Peters C, Thomas DY, Storer AC, T. Vernet (1993) Functional expression of human cathepsin S in *Saccharomyces cerevisiae*: Purification and characterization of the recombinant enzyme". J Biol Chem 268: 4832-4838

Brömme D, McGrath ME (1996) High level expression and crystallization of recombinant human cathepsin S. Protein Science (in press)

Brömme D, Okamoto K (1995) The baculovirus cysteine protease has a cathepsin B-like S2 subsite specificity. Biological Chemistry Hoppe-Seyler 376: 611-615

Brömme D, Okamoto K, Wang B, Biroc S (1996) Human cathepsin O2, a matrix protein degrading cysteine protease expressed in osteoclasts. Functional expression of human cathepsin O2 in *Spodoptera frugiperda* and characterization of the enzyme. J Biol Chem 271: 2126-2132

Brömme D, Li Z, Barnes M, Mehler E (1999) Human Cathepsin V: Functional expression, tissue distribution, electrostatic surface potential, enzymatic characterization, and chromosomal localization. Biochemistry 38:2377-2385

Burck PJ, Berg DH, Luk TP, Sassmannshausen LM, Wakulchik M, Smith DP, Hsiung HM, Becker GW, Gibson W, Villarreal EC (1994) Human cytomegalovirus maturational proteinase: expression in *Escherichia coli*, purification, and enzymatic characterization by using peptide substrate mimics of natural cleavage sites. J Virol 68:2937-2946

Burnette WN (1981) Western blotting: electrophoretic transfer of proteins from sodium dodecyl sulfate-polyacrylamide gels to unmodified nitrocellulose and radiographic detection with antibody and radioiodinated protein A. Anal Biochem 112:195-203

Caputo A, Garner RS, Winkler U, Hudig D (1993) Activation of recombinant murie cytotoxic cel proteinase-1 requires deletio of an amino-terminal dipeptide. J Biol Chem 268:17672-17675

Chan H, Springman EB, Clark JM (1998) Expression and characterization of human tissue kallikrein variants. Protein Expr Purif 12:361-70

Cheng D, Yu W, Han S, Li X, Li F, Hu B, Fang J, Huang C (1993) High level expression of human prourokinase in Chinese hamster ovary cells. Chin J Biotechnol 9:151-159

Clarke HR, Wolfson MF, Rauch CT, Castner BJ, Huang CP, Gerhart MJ, Johnson RS, Cerretti DP, Paxton RJ, Price VL, Black RA (1998) Expression and purification of correctly processed, active human TACE catalytic domain in *Saccharomyces cerevisiae*. Protein Expr Purif 13:104-10

Collins-Racie LA, McColgan JM, Grant KL, DiBlasio-Smith EA, McCoy JM, LaVallie ER (1995) Production of recombinant bovine enterokinase catalytic subunit in *Escherichia coli* using the novel seretory fusion partner DsbA. Biotechnology 13:982-987

Cregg JM, Vedvick TS, Raschke, WC (1993) Recent advances in the expression of foreign genes in *Pichia pastoris*. Bio/Technology 11: 905-910

Dame JB, Reddy GR, Yowell CA, Dunn BM, Kay J, Berry, C (1994) Sequence, expression and modeled structure of an aspartic proteinase from the human malaria parasite *Plasmodium falciparum*. Mol Biochem Parasitol 64:177-190

Dergousova NI, Amerik AYu, Volynskaya AM, Rumsh LD (1996) HIV-I protease. Cloning, expression, and purification. Appl Biochem Biotechnol 61:97-107

Derman AI, Prinz WA, Belin D, Beckwith, J (1993) Mutations that allow disulfide bond formation in the cytoplasm of *Escherichia coli*. Science 262:1744-1747

Devault A, Nault C, Zollinger M, Fournie-Zaluski M-C, Roques, BP, Crine P, Boileau G (1988) Expression of neutral endopeptidase (enkephalinase) in heterologous COS-1 cells. J Biol Chem 263:4033-4040

Ding YS, Owen SM, Lal RB, Ikeda RA (1998) Efficient expression and rapid purification of human T-cell leukemia virus type 1 protease. J Virol 72:3383-3386

Enenkel C, Wolf DH (1993) BLH1 codes for a yeast thiol aminopeptidase, the equivalent of mammalian bleomycin hydrolase. J Biol Chem 268: 7036-7043

Fassy F, Krebs O, Rey H, Komara B, Gillard C, Capdevila C, Yea C, Faucheu C, Blanchet AM, Miossec C, Diu-Hercend A (1998) Enzymatic activity of two caspases related to interleukin-1beta-converting enzyme. Eur J Biochem 253:76-83

Fertig G, Rahn HP, Angermann A, Kloppinger M, Miltenburger HG (1993) Biotechnological aspects of the production of human pro-kallikrein using the AcNPV-baculovirus-expression system. Cytotechnology 11: 67-75

Freije JMP, Díaz-Itza I, Balbín M Sánchez LM, Blasco R, Tolivia J, López-Otín C (1994) Molecular cloning and expression of colagenase-3, a novel human matrix metalloproteinase produced by breast carcinomas. J Biol Chem 269:16766-16773

Freimark BD, Feeser WS, Rosenfeld SA (1994) Multiple sites of the propeptide region of human stromelysin-1 are required for maintaining a latent form of the enzyme. J Biol Chem 269:26982-26987

George HJ, Marchand P, Murphy K, Wiswall BH, Dowling R, Giannaras J, Hollis GF, Trzaskos JM, Copeland RA (1997) Recombinant human 92-kDa type IV collagenase/gelatinase from baculovirus-infected insect cells: expression, purification, and characterization. Protein Expr Purif 10:154-61

Germain D, Vernet T, Boileau G, Thomas DY (1992) Expression of the *Saccharomyces cerevisiae* Kex2p endoproteases in insect cells. Eur J Biochem 204: 121-126

Gold L, Stormo GD (1990) High-level translation initiation. Meth Enzymol 185:89-93
Gorman CM, Gies D, Schofield PR, Dado-Fong H, Malfroy B (1989) Expression of enzymatically active enkephalinase (neutral endopeptidase) in mammalian cells. J Cellular Biochem 39:277-284
Gottesman S (1990) Minimizing proteolysis in *Escherichia coli*: genetic solutions. Meth Enzymol 185:119-129
Halfon S, Craik CS (1996) Regulation of proteolytic activity by engineered tridentate metal binding loop. J Am Chem Soc 118: 1227-1228
Harmsen MC, Heeringa P, van der Geld YM, Huitema MG, Klimp A, Tiran A, Kallenberg CG (1997) Recombinant proteinase 3 (Wegener's antigen) expressed in Pichia pastoris is functionally active and is recognized by patient sera. Clin Exp Immunol 110:257-64
Hedstrom L, Szilagyi L, Rutter WJ (1992) Converting trypsin to chymotrypsin: The role of surface loops. Science 255: 1249-1253
Higaki JN, Evnin LB, Craik CS (1989) Introduction of a cysteine protease active site into trypsin. Biochemistry 28:9256-9263
Hill J, Montgomery DS, Kay J (1993) Human cathepsin E produced in *E. coli*. FEBS Lett. 326:101-104
Ho TF, Qoronfleh MW, Wahl RC, Pulvino TA, Vavra KJ, Falvo J, Banks TM, Brake PG, Ciccarelli RB (1994) Gene expression, purification and characterization of recombinant human neutrophil collagenase. Gene 146:297-301
Illy C, Quraishi O, Wang J, Purisima E, Vernet T, Mort JS (1997) Role of the occluding loop in cathepsin B activity. J Biol Chem 272:1197-202
Jones EW (1991) Tackling the protease problem in *Saccharomyces cerevisiae*. Methods Enzymol 194: 428-453
Kaufman RJ (1990a) Vectors used for expression in mammalian cells. Meth Enzymol 185:487-511
Kaufman RJ (1990b) Selection and coamplification of heterologous genes in mammalian cells. Meth Enzymol 185:537-566
Kelly JM, O'Connor MD, Hulett MD, Thia KYT, Smyth MJ (1996) Cloning and expression of the recombinant mouse natural killer cell granzyme Met-ase-1. Immunogenetics 44:340-350
Keown WA, Campbell CR, Kucherlapati RS (1990) Methods for introducing DNA into mammalian cells. Meth Enzymol 185:527-537
Kriegler M (1990) Asembly of enhancers, promoters, and splice signals to control expression of transferred genes. Meth Enzymol 185:512-527
Kumar S (1995) ICE-like proteases in apoptosis. Trends Biochem Sci 20:198-202
Kurkela R, Herrala A, Henttu P, Nai H, Vihko P (1995) Expression of active, secreted human prostate-specific antigen by recombinant baculovirus-infected insect cells on a pilot-scale. Biotechnology (N Y) 13:1230-4
Latchinian-Sadeck L, Thomas DY (1994) Secretion, purification and characterization of a soluble form of the yeast Kex1-encoded protein from insect cell cultures. Eur J Biochem 219: 647-652
LaVallie ER, DiBlasio EA, Kovacic S, Grant KL, Schendel PF, McCoy JM (1993) A thioredoxin gene fusion system that circumvents inclusion body formation in the *E. coli* cytoplasm. Biotechnology 11:187-193
LaVallie ER, McCoy JM (1995) Gene fusion expression systems in *Escherichia coli*. Curr Opin Biotechnol 6:501-506
Lin X, Liu M, Tang J (1992) Heterologous expression of thermopsin, a heat stable acid proteinase. Enzyme Microb. Technol 14: 696-701

Linnevers CJ, McGrath ME, Armstrong R, Mistry FR, Barnes MG, Klaus JL, Palmer JT, Katz BA, Bromme D (1997) Expression of human cathepsin K in *Pichia pastoris* and preliminary crystallographic studies of an inhibitor complex. Protein Sci 6:919-21

Magyar G, Dobos P. (1994) Evidence for the detection of the infectious pancreatic necrosis virus polyprotein and the 17-kDa polypeptide in infected cells and of the NS protease in purified virus. Virology 204: 580-589

Malfroy-Camine B, Schofield PR (1990) Compositions and methods for the synthesis and assay of a mammalian enkephalinase. U.S. Patent 4,960,700

Malinowski JJ, Grasberger BL, Trakshel G, Huston EE, Helaszek CT, Smallwood AM, Ator MA, Banks TM, Brake PG, Ciccarelli RB, Jones BN, Koehn JA, Kratz D, Lundberg N, Stams T, Rubin B, Alexander RS, Stevis PE (1995) Production, purification, and crystallization of human interleukin-1β converting enzyme derived from an *Escherichia coli* expression system. Protein Sci 4:2149-2155

Martinez-Abarca F, Alonso MA, Carrasco L (1993) High level expression in *Escherichia coli* cells and purification of poliovirus protein 2Apro. J Gen Virol 47:2645-2652.

Mather JP (1990) Optimizing cell and culture environment for production of recominant proteins. Meth Enzymol 185:567-577

McCall JO, Kadam S, Katz L (1994) A high capacity microbial screen for inhibitors of human rhinovirus protease 3C. Biotechnology 12:1012-1016

McGrath, ME, Osawa, AE, Barnes, MG, Clark, JM, Mortara, KD, Schmidt, BF (1997) Production of crystallizable human chymase from a *Bacillus subtilis* system. FEBS Lett 413:486-488

McKie N, Dando PM, Brown MA, Barrett AJ (1995) Rat thimet oligopeptidase: large-scale expression in *Escherichia coli* and characterization of the recombinant enzyme. Biochem J 309:203-207

Mhashilkar AM, Viswanatha T, Chibber BA, Castellino FJ (1993) Breaching the conformational integrity of the catalytic triad of the serine prorease plasmin: localized disruption of a side chain of His-603 strongly inhibitis the amidolytic activity of human plasmin. Proc Natl Acad Sci. USA 90: 5374-5377

Molloy S, Bresnahan P, Leppla SH, Klimpe KR, Thomas G (1992) Human furin is a calcium dependent serine endoprotease that recognizes the sequence Arg-X-X-Arg and efficiently cleaves the anthrax toxin protective antigen. J Biol Chem 267:16396-16402

Morita S, Kuriyama M, Maejima K, Kitano K (1994) Cloning and nucleotide sequence of the alkaline protease gene from Fusarium sp. S–19–5 and expression in *Saccharomyces cerevisiae*. Biosci Biotechnol Biochem 58: 621-626

Naggie S, Bentley WE (1998) Appearance of protease activities coincides with p10 and polyhedrin-driven protein production in the baculovirus expression system: effects on yield. Biotechnol Prog 14:227-32

Niles AL, Maffitt M, Haak-Frendscho M, Wheeless CJ, Johnson DA (1998) Recombinant human mast cell tryptase beta: stable expression in *Pichia pastoris* and purification of fully active enzyme. Biotechnol Appl Biochem 28(Pt 2):125-131

Ohi H, Ohtani W, Okazaki N, Furuhata N, Ohmura T (1996) Cloning and characterization of the *Pichia pastoris* PRC1 gene encoding carboxypeptidase Y. Yeast 12:31-40

Olins PO, Lee SC (1993) Recent advances in heterologous gene expression in *Escherichia coli*. Curr Opin Biotechnol 4:520-525

Patel IR, Attur MG, Patel RN, Stuchin SA, Abagyan RA, Abramson SB, Amin AR (1998) TNF-alpha convertase enzyme from human arthritis-affected cartilage: isolation of cDNA by differential display, expression of the active enzyme, and regulation of TNF-alpha. J Immunol 160:4570-9

Pei D, Majmudar G, Weiss SJ (1994) Hydrolytic inactivation of a breast carcinoma cell-derived serpin by human stromelysin-3. J Biol Chem 269:25849-25855.

Pfrepper KI, Lochelt M, Schnolzer M, Flugel RM (1997) Expression and molecular characterization of an enzymatically active recombinant human spumaretrovirus protease. Biochem Biophys Res Commun 237:548-553

Pham CT, Thomas DA, Mercer JD, Ley TJ (1998) Production of fully active recombinant murine granzyme B in yeast. Biol Chem 273:1629-33

Pourmotabbed T, Solomon TL, Hasty KA, Mainardi CL (1994) Characteristics of 92 kDa type IV collagenase/gelatinase produced by granulocytic leukemia cells: structure, expression of cDNA in *E. coli* and enzymic properties. Biochim Biophys Acta 1204:97-107

Ramage P, Cheneval D, Chvei M, Graff P, Hemmig R, Heng R, Kocher HP, Mackenzie A, Memmert K, Revesz L, Wishart W (1995) Expression, refolding, and autocatalytic proteolytic processing of the interleukin-1 beta-converting enzyme precursor. J Biol Chem 270:9378-9383

Ramjee MK, Petithory JR, McElver J, Weber SC, Kirsch JF (1996) A novel yeast expression/secretion system for the recombinant plant thiol endoprotease propapain. Protein Eng 9:1055-61

Richo GR, Conner GE (1994) Structural requirements of procathepsin D activation and maturation. J Biol Chem 269:14806-14812

Roche L, Dowd AJ, Tort J, McGonigle S, McSweeney A, Curley GP, Ryan T, Dalton JP (1997) Functional expression of Fasciola hepatica cathepsin L1 in Saccharomyces cerevisiae. Eur J Biochem 245:373-80

Romanos M (1995) Advances in the use of Pichia pastoris for high level gene expression. Current Opinion in Biotechnol 6: 527-533

Rosé JR, Babé LM, Craik, CS (1995) Defining the level of human immunodeficiency virus type 1 (HIV-1) protease activity required for HIV-1 particle maturation and infectivity. J Virology 69:2751-2758

Rosé JR, Craik CS (1996) The art of expression: sites and strategies for heterologous expression. In: Protein Engineering and Design, Academic Press, Inc., San Diego, CA, pp 75-104

Russo G, Gast A, Schlaeger EJ, Angiolillo A, Pietropaolo C (1997) Stable expression and purification of a secreted human recombinant prethrombin-2 and its activation to thrombin. Protein Expr Purif 10:214-225

Salas F, Fichman J, Lee GK, Scott MD, Rosenthal PJ (1995) Functional expression of falcipain, a Plasmodium falciparum cysteine proteinase, supports its role as a malarial hemoglobinase. Infect Immun 63: 2120-2125

Sali DL, Ingram R, Wendel M, Gupta D, McNemar C, Tsarbopoulos A, Chen JW, Hong Z, Chase R, Risano C, Zhang R, Yao N, Kwong AD, Ramanathan L, Le HV, Weber PC (1998) Serine protease of hepatitis C virus expressed in insect cells as the NS3/4A complex. Biochemistry 37:3392-401

Schmidt M, Kröger B, Jacob E, Seulberger H, Subkowski T, Otter R, Meyer T, Schmalzing G, Hillen H (1994) Molecular characterization of human and bovine endothelin converting enzyme (ECE-1) FEBS Lett 356:238-243

Schoner BE, Belagaje, RM, Schoner RG (1990) Enhanced translational efficiency with two-cistron expression system. Meth Enzymol 185:94-103

Shapiro SD, Kobayashi DK, Ley TJ (1993) Cloning and characterization of a unique elastolytic metalloproteinase produced by human alveolar macrophages. J Biol Chem 268:23824-23829

Shimizu C, Yoshida S, Shibata M, Kato K, Momota Y, Matsumoto K, Shiosaka T, Midorikawa R, Kamachi T, Kawabe A, Shiosaka S (1998) Characterization of recombinant and brain neuropsin, a plasticity-related serine protease. J Biol Chem 273:11189-96

Shoji I, Suzuki T, Chieda S, Sato M, Harada T, Chiba T, Matsuura Y, Miyamura T (1995) Proteolytic activity of NS3 serine proteinase of hepatitis C virus efficiently expressed in *Escherichia coli*. Hepatology 22:1648-1655

Sinha U, Hancock TE, Lin PH, Hollenbach S, Wolf DL (1992) Expression, purification, and characterization of inactive human coagulation factor Xa (Asn^{322}Ala^{419}) Protein Express Purification 3:518-524

Sloan JH, Loutsch JM, Boyce SY, Holwerda BC (1997) Expression and characterization of recombinant murine cytomegalovirus protease. J Virol 71:7114-7118

Smyth MJ, McGuire MJ, Thia KY (1995) Expression of recombinant human granzyme B. A processing and activation role for dipeptidyl peptidase I. Immunol 154:6299-6305

Sreekrishna K, Brankamp RG, Kropp KE, Blankenship DT, Tsay JT, Smith PL, Wierschke JD, Subramaniam A, Birkenberger,LA (1997) Strategies for optimal synthesis and secretion of heterologous proteins in the methylotrophic yeast *Pichia pastoris*. Gene 190: 55-62

Steed PM, Lasala D, Liebman J, Wigg A, Clark K, Knap AK (1998) Characterization of recombinant human cathepsin B expressed at high levels in baculovirus. Protein Sci 7:2033-7

Studier FM, Rosenberg AH, Dunn JJ, Dubendorff JW (1990) Use of T7 RNA polymerase to direct expression of cloned genes. Meth Enzymol 185:60-89

Sun J, Bottomley SP, Kumar S, Bird PI (1997) Recombinant caspase-3 expressed in Pichia pastoris is fully activated and kinetically indistinguishable from the native enzyme. Biochem Biophys Res Commun 238:920-4

Suzuki K, Kan CC, Hung W, Gehring MR, Brew K, Nagase H (1998) Expression of human pro-matrix metalloproteinase 3 that lacks the N-terminal 34 residues in *Escherichia coli*: autoactivation and interaction with tissue inhibitor of metalloproteinase 1 (TIMP-1). Biol Chem 379:185-191

Taylor MA, Pratt KA, Revell DF, Baker KC, Sumner IG, Goodenough PW (1992) Active papain renatured and processed from insoluble recombinant propapain expressed in *Escherichia coli*. Protein Eng 5:455-459

Thomson DR, Newcomb WW, Brown JC, Homa FL (1995) Assembly of the herpes simplex virus capsid: reqiuirement for the carboxyl-terminal twenty-five amino acids of the proteins encoded by the UL26 and UL26.5 genes. J Virol 69: 3690-3703

Thornberry NA (1994) Interleukin-1β converting enzyme. Meth Enzymol 244:615-631

Thornberry NA, Molineaux SM (1995) Interleukin-1β convering enzyme: a novel cysteine protease requicd for IL-1β production and implicated in programmed cell death. Protein Sci 4:3-12

Tsukuba T, Hori H, Azuma T, Takahashi T, Taggart RT, Akamine A, Ezaki M, Nakanishi H, Sakai H, Yammamoto K (1993) Isolation and characterization of recombinant human cathepsin E expressed in Chinese hamster ovary cells. J Biol Chem 268:7276-7282

Uhlén M, Moks T (1990) Gene fusions for enhanced translation of foreign genes in *Escherichia coli*. Meth Enzymol 185:129-143

Urata H, Karnik SS Graham RM, Hussain A (1993) Dipeptide processing activates recombinant human prochymase. J Biol Chem 268:24318-24321

Velasco G, Ferrando AA, Puente XS, Sanchez LM, Lopez-Otin C (1994) Human cathepsin O. Molecular cloning from a breast carcinoma, production of the active enzyme in *Escherichia coli*, and expression analysis in human tissues. J Biol Chem 269:27136-27142

Vernet T, Chatellier J, Tessier DC, Thomas, DY (1993) Expression of functional papain precursor in *Saccharomyces cerevisiae*: rapid screening of mutants. Protein Engineering 6: 213-219

Vernet T, Tessier DC, Richardson C, Laliberte F, Khouri HE, Bell AW, Storer AC, Thomas DY (1990) Secretion of functional papain precursor from insect cells. J Biol Chem 265: 16661-16666

Vishnuvardhan D, Kakiuchi N, Urvil PT, Shimotohno K, Kumar PK, Nishikawa S (1997) Expression of highly active recombinant NS3 protease domain of hepatitis C virus in *E. coli*. FEBS Lett 402:209-212

Vozza LA, Wittwer L, Higgins DR, Purcell TJ, Bergseid M, Collins-Racie LA, LaVallie ER, Hoeffler JP (1996) Production of a recombinant bovine enterokinase catalytic subunit in the methylotrophic yeast *Pichia pastoris*. Biotechnology 14, 77-81

Wang B, Shi GP, Yao PM, Li Z, Chapman HA, Brömme D (1998) Human Cathepsin F. J Biol Chem 273:32000-32008

Wang J, Chao J, Chao L (1991) Purification and characterization of recombinant tissue kallikrein from *Escherichia coli* and yeast. Biochem J 276:63-71

Wang XM, Helaszek CT, Winter LA, Lirette RP, Dixon DC, Ciccarelli RP, Kelley MM, Malinowski JJ, Simmons SJ, Huston EE et al. (1994) Production of active human interleukin-1 beta-converting enzyme in a baculo expression system. Gene 145: 237-237

Wang Z, Walter M, Selwood T, Rubin H, Schechter NM (1998) Recombinant expression of human mast cell proteases chymase and tryptase. J Biol Chem 379:167-74

Wang ZM, Rubin H, Schechter NM (1995) Production of active recombinant human chymase from a construct containing the enterokinase cleavage site of trypsinogen in place of the native propeptide sequence. Biol Chem Hoppe-Seyler 376:681-684

Windsor LJ, Birkedal-Hansen H, Birkedal-Hansen B, Engler JA (1991) An internal cysteine plays a role in the maintenance of the latency of human fibroblast collagenase. Biochemistry 30:641-647

Wong S-L (1995) Advances in the use of *Bacillus subtilis* for the expression and secretion of heterologous proteins. Curr Opin Biotechnol 6:517-522

Xia Z, Kam CM, Huang C, Powers JC, Mandle RJ, Stevens RL, Lieberman J (1998) Expression and purification of enzymatically active recombinant granzyme B in a baculovirus system. Biochem Biophys Res Commun 243:384-9

Yamada M, Azuma T, Matsuba T, Lida H, Suzuki H, Yamamoto K, Kohli Y, Hori H (1994) Secretion of human intracellular aspartic proteinase cathepsin E expressed in the methylotrophic yeast, *Pichia pastoris* and characterization of produced recombinant cathepsin E. Biochim Biophys Acta 1206: 279-285

Ye Q–Z, Johnson LL, Yu AE, Hupe D (1995) Reconstructed 19 kDa catalytic domain of gelatinase A is an active proteinase. Biochemistry 34:4702-4708

Yoshino-Yasuda I, Kobayashi K, Akiyama M, Itoh H, Tomomura A, Saheki T (1998) Caldecrin is a novel-type serine protease expressed in pancreas, but its homologue elastase IV, is an artifact during cloning derived from caldecrin gene. J Biochem (Tokyo)123:546-54

Part III

The Use of Proteolytic Enzymes as Biochemical Tools

Proteases in Peptide Mapping and Sequencing

JOSEF KELLERMANN

Introduction

The ability of some proteases to cleave polypeptide chains at restricted cleavage sites makes them important tools for the elucidation of the primary and even higher-order structure of proteins. This chapter provides an introduction into the methodology of enzymatic protein fragmentation. In addition, a series of complementary chemical methods are included which are indispensable for protein sequence analysis and peptide mapping. The specificity of most endopeptidases show a strong preference for particular amino acid residues or short sequence recognition sites. The chemical neighborhood of these residues, i.e. the character of their side chains, like hydrophobicity or size, may influence the rate of hydrolysis.

Based on the extent of the reaction, two different proteolytic approaches can be distinguished, complete and incomplete (limited) proteolysis. A proteolytic reaction, which is allowed to reach completion, creates an equimolar set of peptides. Further addition of the same protease will not influence this pattern.

Complete cleavage of peptide bonds at selected amino acid residues may also be achieved by some chemical reagents i.e. cyanogen bromide at methionine or hydroxylamine between asparagine and glycine. A variety of chemical methods for cleavage have been reviewed (Kaspar 1975; Fontana 1986).

If proteolytic fragmentation is not allowed to reach completion (limited proteolysis), different sets of peptides can be achieved. Further proteolysis can be inhibited by removal of the protease, addition of an inhibitor or by changing reaction conditions (pH, temperature etc.) in a way that prote-

Josef Kellermann, Max-Planck-Institut für Biochemie, Arbeitsgruppe Proteinanalytik, Am Klopferspitz 18a, Martinsried, 82152, Germany (*phone* (089) 8578 2484; *fax* (089) 85782802; *e-mail* kellerma@biochem.mpg.de)

olysis no longer proceeds. This may be necessary to study initial cleavage products, to monitor the time course of a reaction or to generate large peptide fragments. The proteolytic digestion of a native protein may also be limited by the resistance of the substrate against complete proteolysis, caused by its compact higher-order structure. Examples for this kind of limited proteolysis in vivo are zymogen activation, prohormone processing or cleaving out peptides of a polypeptide chain.

In most cases both peptide mapping and sequencing of proteins require a complete fragmentation of the protein, resulting in a strongly defined and reproducible peptide pattern.

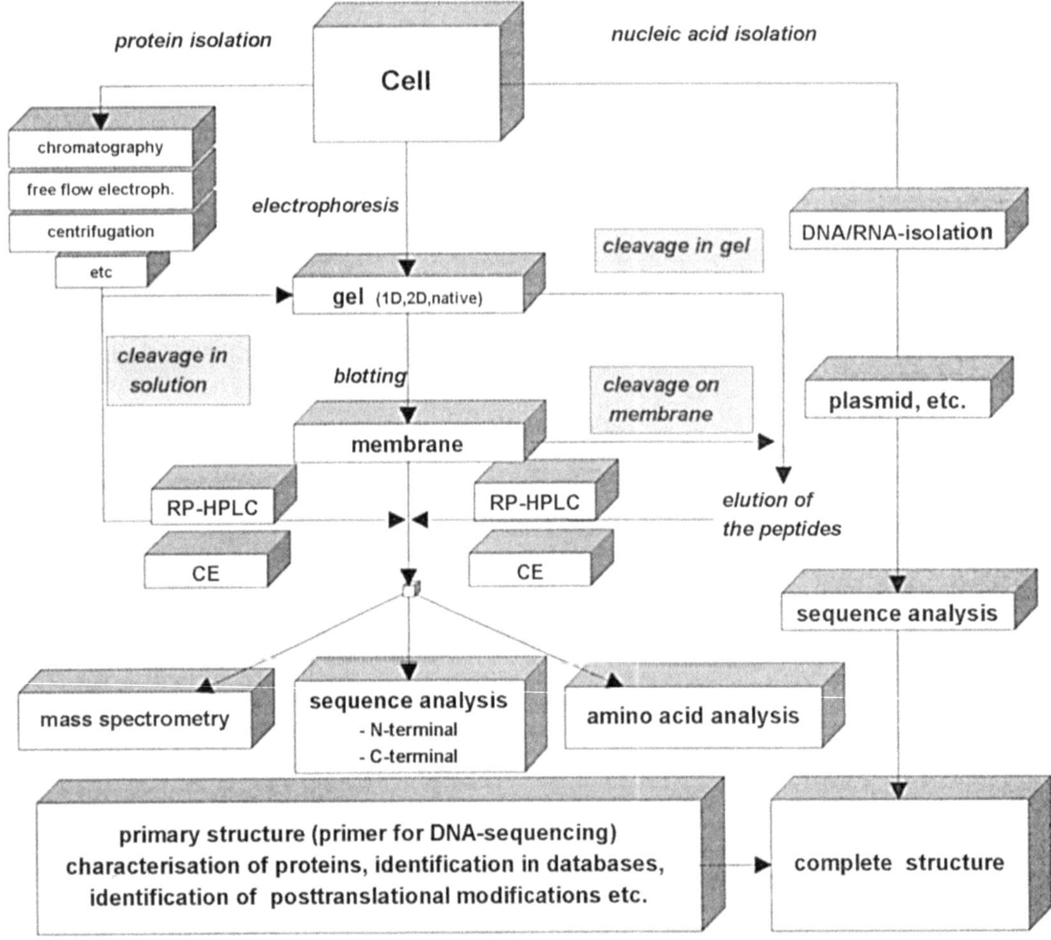

Fig. 1. Strategy for the characterization of proteins depending on the method of purification

Strategy

For peptide mapping or sequencing a protein must be purified to near homogeneity. Starting from a complex mixture of proteins as present in a complete cell, a variety of chromatographic or electrophoretic steps may be necessary until a protein is pure enough for characterization. Depending on the way of purification there are three possibilities a protein can be applied for a cleavage reaction:

- in solution,

- bound onto a membrane and

- in a polyacrylamide matrix (Fig. 1).

Proteins in solution are supposed to be the simplest starting material to perform a proteolytic degradation. The problem often is their solubility. The solubilization buffer often contains chemicals like detergents or salts, which may affect either protease activity or the subsequent chromatographic separation.

Electrophoretic separation by 1-dimensional or 2-dimensional gel electrophoresis and subsequent electroblotting onto an inert membrane are the most powerful and effective strategies for protein microisolation. Suitable membranes are either polyvinylidene difluoride (PVDF) (Pluskal et al. 1986; Matsudeira 1987) or siliconized glass fibres (Eckerskorn et al. 1988). The protein bands can be visualized after staining them with, e.g. Coomassie Blue, Amido Black or Ponceau S. Bands of interest can be excised and applied for N-terminal sequencing or fragmentation.

Proteins can also be cleaved directly within the polyacrylamide matrix. After the separation by 1D- or 2D-SDS PAGE the proteins are visualized by Coomassie or silver staining and excised. The gel pieces are washed, shrunk with acetonitrile and then buffer and protease is added. During reswelling of the matrix the enzyme is sucked into the gel and digestion starts. The recovery of the peptides can be achieved by application of the gel pieces onto a stacking gel and subsequent separation (Cleveland et al. 1977) followed by electroblotting (Kennedy et al. 1988) or by elution of the peptides from the gel matrix (Eckerskorn and Lottspeich, 1989). The direct digestion in the gel matrix reduces the number of handling steps and is therefore advantageous for small amounts.

Denaturation and Reduction

As mentioned already, secondary and tertiary structures of a native protein make cleavage sites not always accessible to proteases. Denaturation destroys the higher-order structure of a protein, bringing the conformation nearly to a random coil. For denaturation steps, urea, guanidinium hydrochloride and detergents are used. The latter ones are mainly used for solubilization of membrane proteins.

Denaturation by urea often causes carbamylation of the protein, blocking the N-terminus by contamination of the urea with its degradation products (e.g. cyanate ions). Therefore, a solution of 6M guanidine hydrochloride is often preferred as denaturing agent. Diluting this solution to a 1M concentration often makes it compatible with the activity of many proteases

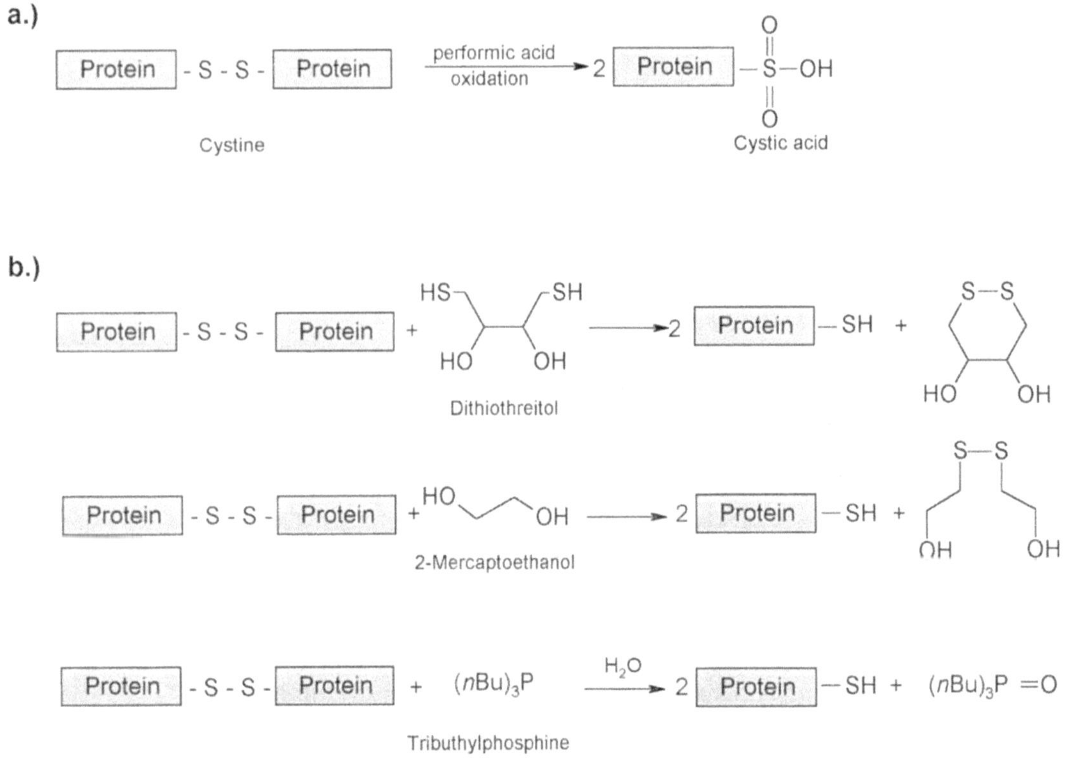

Fig. 2. Cleavage of disulfide bonds. (a) Oxidation of disulfide bonds with performic acid to cysteic acid; (b) Reduction of disulfide bonds with dithiothreithol, ß-mercaptoethanol and tributylphosphine

(Riviere et al. 1992). Disulfide bonds also stabilize the compact structure of a protein thereby hindering proteolysis. Furthermore, cleavage fragments are difficult to characterize, if still linked together by one or more disulfide bonds.

Therefore, a reduction of the S-S bonds, followed by a protection of the cysteines by alkylation helps to make proteines more accessible to proteases.

The cleavage of the S-S bonds by reduction, yielding two cysteines, is achieved by dithiothreithol (DTT), 2-mercaptoethanol or tributylphosphine (Fig. 2). DTT (also Clelands reagent) (Cleland 1964) often is preferred, for it has a low redox potential and the reaction needs only a few minutes. For vapor phase reaction tributylphosphine is used (Amons 1987). A further possibility to cleave S-S bonds is given by oxidation with performic acid (Hirs 1967a).

Alkylation

The second step, after cleaving the S-S bonds, is the modification of the thiol groups (Fig. 3), to stabilize them. Cysteine may be destroyed during Edman degradation and can not be identified without derivatization after reduction. Most frequently used reagents for alkylation are 4-vinylpyridine (Raftery 1966) with its reaction product 4-(pyridylethyl)cysteine, iodacetic acid, yielding S-(carboxymethyl)cysteine (Crestfield 1963) and iodoacetamide with its product S-(carboxyamidomethyl)cysteine (Allen 1989). The incubation time for pyridylethylation should be followed carefully, for extended exposure of a protein causes several side reactions of 4-vinylpyridine with other amino acid residues (i.e. histidine, tryptophan and methionine may be modified). A separation step should therefore immediately follow the pyridylethylation. 4-(pyridylethyl)cysteine has the advantage of exhibiting an additional absorption maximum at 256 nm, which allows to identify peptides containing cysteine residues.

Incompletely reacted acrylamide monomers in polyacrylamide gels are known to attack thiol groups. This reaction is also often used in S-alkylation especially as a one step reaction prior to SDS-gelelectrophoresis (Brune 1992).

A large variety of other protein modifications have been described (Glazer et al. 1975, Darbre 1986) without playing a significant role in protein characterization on microscale. The need of large amounts and the formation of multiple reaction products often make them unsuitable for microscale application.

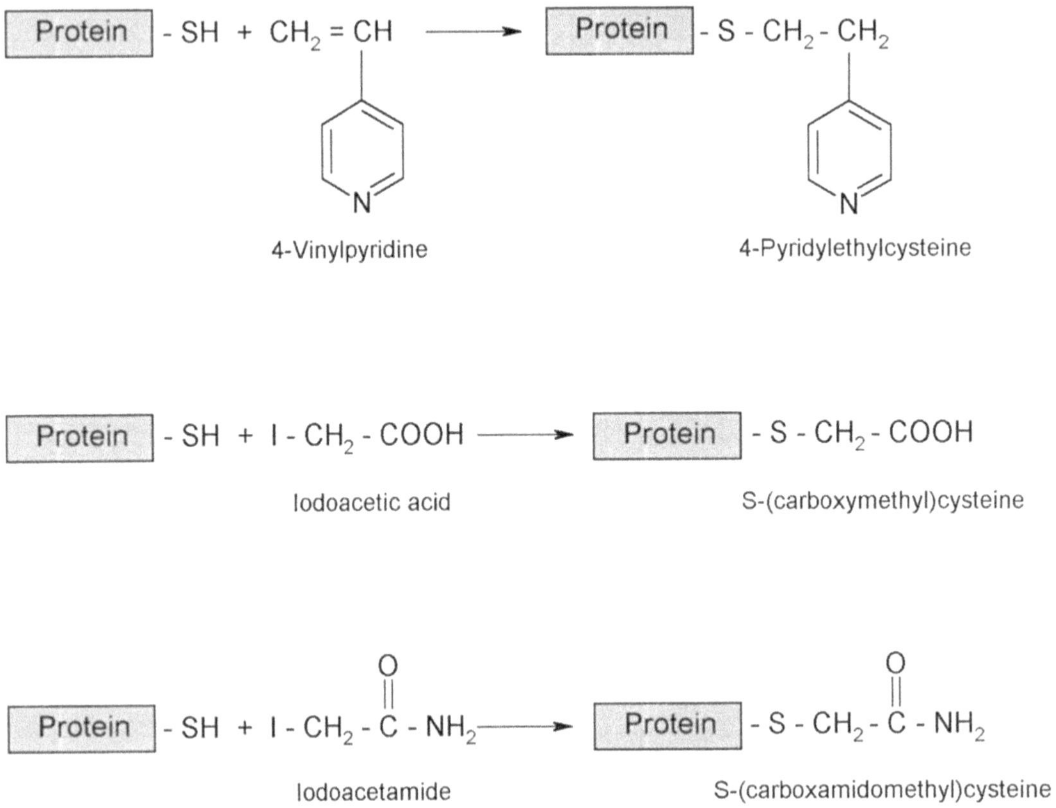

Protein – SH + CH$_2$ = CH ⟶ Protein – S – CH$_2$ – CH$_2$

4-Vinylpyridine   4-Pyridylethylcysteine

Protein – SH + I – CH$_2$ – COOH ⟶ Protein – S – CH$_2$ – COOH

Iodoacetic acid   S-(carboxymethyl)cysteine

Protein – SH + I – CH$_2$ – C(=O) – NH$_2$ ⟶ Protein – S – CH$_2$ – C(=O) – NH$_2$

Iodoacetamide   S-(carboxamidomethyl)cysteine

Fig. 3. Alkylation of cysteine residues. Alkylation of cysteine with 4-vinylpyridine to 4-pyridylethylcysteine, with iodoacetic acid to S-(carboxymethyl)cysteine and with iodoacetamide to S-(carboxamidomethyl)cysteine

Table 1. Chemical modification of cysteine residues

Reagent	Product	Reference
performic acid	cysteic acid	Hirs (1967)
sulfite	S-sulfocysteine	Bayley and Cole (1959)
iodoacetic acid	S-carboxymethylcysteine	Crestfield (1963)
iodoacetamide	S-carboxamidomethylcysteine	Allen (1989)
ethyleneimine	S-(2-aminoethyl)cysteine	Lindley (1956)
4-vinylpyridine	4-(pyridyl) ethylcysteine	Friedman (1970)
acrylamide	Cys-S-ß-propionamide	Brune (1992)

Fragmentation

The choice of the way of fragmentation of a protein depends on the purpose of the fragmentation and the already known characteristics of the protein. If the amino acid composition or even the primary structure is already known, a specific strategy might be chosen, to get some large or many small fragments. A hypothetical protein of 300 amino acid residues can be created from a protein database (NBRF-PIR) taking the total number of amino acid residues versus the number of database entries (Kellner 1994). This shows the average number of specific fragments and their average length, when cleaved with a specific proteinase or by a chemical method (Table 2).

The fragmentation pattern for a given protein under constant conditions is highly reproducible and very specific for that protein. The separation of the peptide mixture by electrophoresis (SDS-PAGE or capillary electrophoresis) or by chromatographic methods (HPLC) results in a characteristic fingerprint or peptide map. These peptides can be characterized further and identified by mass spectrometry or Edman-sequencing.

Classification of Proteinases

Proteases are classified into endo- and exopeptidases. **Endopeptidases** cleave the protein backbone, generating a highly specific peptide pattern for each protein. Many of the proteases cleave C- or N-terminal of charged residues (endopeptidase Lys-C, trypsin, endopeptidase Asp-N). Proteases of

Table 2. Theoretical number and average length of peptide fragments of a hypothetical 300 residue protein calculated from a protein database (NBRF-PIR)

Specific enzyme or chemical cleavage	Amino Acid Residue	Average Length (residues)	Number of fragments
chymotrypsin	Leu,Phe,Trp,Tyr	6	54
trypsin	Lys, Arg	9	35
endopeptidase Glu-C	Glu	15	20
endopeptidase Lys-C	Lys	16	19
endopeptidase Arg-C	Arg	18	17
endopeptidase Asp-N	Asp	18	17
cyanogen bromide	Met	38	8
BNPS-scatole	Trp	60	5

Table 3. Enzymes used for protein structure characterization

Enzyme	EC-Number	Type	Specificity	optimal pH-region	Inhibitors	Reference
Endopeptidases					general inhibitor: α_2M	
chymotrypsin	3.4.21.1	Ser	Tyr/Phe/Trp/Leu*Xaa	7.5-8.5	aprotinin, DFP, PMSF	Kaspar (1970)
trypsin)	3.4.21.4	Ser	Arg/Lys*Xaa	8.0-9.0	TLCK, DFP, PMSF	Allen (1989)
endoproteinase Glu-C(glutamyl endopeptidase)	3.4.21.19	Ser	Glu/Asp*Xaa	8.0	DFP, 3,4-dichloro-isocumarin	Drapeau et al. (1972)
endoproteinase Lys-C (lysyl endopeptidase)	3.4.21.50	Ser	Lys*Xaa	7.5-8.5	DFP, TLCK aprotinin, leupeptin	Allen (1989)
endoproteinase Arg-C(clostripain)	3.4.22.8	Cys	Arg*Xaa	8.0-8.5	oxidants, EDTA, Co2+, Cu^{2+}, citrate, borate	Schenkein et al. (1977)
endoproteinase Asp-N (peptidyl-Asp metallo endopeptidase)	3.4.24.33	Zn	Xaa*Asp/Glu/cysteic acid	6.0-8.0	EDTA, o-phenanthroline	Drapeau (1980)
pancreatic elastase	3.4.21.36	Ser	Ala/Val/Ile/Leu/Gly*Xaa	8.9	DFP, α_1-antitrypsin, PMSF, elastinal	Barrett (1981)
pepsin A	3.4.23.1	Asp	Xaa*Uaa*Xaa cleaves C- or N-terminally of hydrophobic (aromatic) residues	2.0-4.0	pepstatin, 4-bromo-phenacylbromide	Fruton (1977)
subtilisin	3.4.21.62	Ser	Uaa*Xaa , unspecificprefers large hydrophobic, residues in P_1	7.0-11.0	DFP, PMSF indole, phenole	Serano et al. (1984)
thermolysin	3.4.24.27	Zn	Xaa*UaaUaa preferentially Leu, Phe	7.0-9.0	EDTA, o-phenanthroline, phosphoramidon	Heinrickson (1977)
papain	3.4.22.2	Cys	Uaa-Xaa*Xbb Xbb $\neq$ Val	7.0-9.0	iodoacetate, iodoacetamide, TLCK, TPCK	Cleveland et al. (1977)

Table 3. Continuous

Enzyme	EC-Number	Type	Specificity	optimal pH-region	Inhibitors	Reference
pronase			unspecific mixture of enzymes	7.5	no specific inhibitor	Sambrook et al. (1989)
proteinase K (endopeptidase K)	3.4.21.64	Ser	Uaa*Xaa (similar to subtilisin)	7.0	SH-blocking reagents (iodoacetamide etc.)	Ebeling et al. (1974)
thrombin	3.4.21.5	Ser	Uaa-Uaa-Pro-Arg/Lys* Xaa Xaa = Gly in fibrinogen	7.5	DFP, TLCK, PMSF, leupeptin, STI, hirudin	Baugham (1970)
coagulation factor Xa (Stuart factor	3.4.21.61	Ser	Ile-Glu-Gly-Arg*Xaa Xaa = Thr or Ile thrombokinase) in prothrombin	8.3	DFP, PMSF, STI	Owen et al. (1974)
enteropeptidase (enterokinase)	3.4.21.9	Ser	(Asp)4-Lys*Xaa Xaa = Ile in trypsinogen	8.0	DFP, TLCK	Light & Janska (1989)
Exopeptidases						
Carboxypeptidases						
carboxypeptidase P (lysosomal Pro-X carboxypeptidase)	3.4.16.2	Ser	Pro*Xaa-COOH releases C-terminal residues adjacent to Pro	4.0-5.5	DFP, iodoacetate	Yokoyama et al. (1981)
carboxypeptidase C (cathepsin A carboxypeptidase Y)	3.4.16.5	Ser	Xaa*Xaa-COOH releases C-terminal residues, unspecific	4.0-6.5	DFP, SH-blocking reagents, PMSF, ZPCK, aprotinin	Tschesche & Kupfer (1972) Hayashi et al. (1973)
carboxypeptidase A	3.4.17.1	Zn	Xaa*Xbb-COOH Xbb≠Asp,Glu,Arg, Lys,Pro	7.0-8.0	chelators, pyrophosphate, oxalate	Petra (1970)
carboxypeptidase B	3.4.17.2	Zn	Xaa*Arg/Lys-COOH (preferentially)	7.0-9.0	Chelators, basic amino acids	Folk (1970)

Table 3. Continuous

Enzyme	EC-Number	Type	Specificity	optimal pH-region	Inhibitors	Reference
Aminopeptidases						
cathepsin C (dipeptidyl dipeptidase I)	3.4.14.1	Cys	Xaa-Xbb*Xcc (Xaa≠Lys/Arg; Xbb≠Pro) releases N-terminal dipeptides	5.5	Iodoacetate, formaldehyde	Lynn & labow (1985)
Acylaminoacyl peptidase	3.4.19.1	Ser	acetyl/formyl-Xaa*Xbb-Xcc preference Xaa=Ser/Ala/Met poor Xaa=Gly/Asp/Asn/Pro)	7.5-9.0	DFP	Farries et al. (1991)
pyroglutammyl peptidase I	3.4.19.3	Cys	n-pGlu*Xbb- (Xbb≠Pro)	7.0-9.0	SH-blocking reagents (iodoacetamide)	Armentrout & Doolittle (1969)
Glucosidases						
N-glycosidase A or F (peptide-N4-(N-acetyl-ß-glucosaminyl) asparagine amidase)	3.5.1.52		releases N-acetyl-ß-D-glucosamine and a peptide containing Asp	4.5-7.0		Maley (1989)
O-glycosidase (glycopeptide α-N-acetylgalactosaminidase)	3.2.1.97		hydrolyzes terminal D-galactosyl-N acetyl-D-galactosaminidic residues linked to Ser or Thr	6.0		Umomoto et al. (1977)
Phosphatases						
alkalone phosphatase	3.1.3.1		hydrolyzes orthophosphoric monoesters, broad specifitcy also transphosphorylations	7.0-9.0	fluoride, molybdate, orthophosphate	Chaconase et al. (1980)
acidic phosphatase	3.1.3.2		Hydrolyzes orthophosphoric monoesters, broad specifity, also transphosphorylations	3.0-6.0		Verjee (1969)

lower specificity like chymotrypsin, thermolysin or pepsin cleave the protein adjacent to several amino acid residues, thereby yielding more and shorter peptide fragments (Table 3). **Coagulation factor Xa, thrombin** and **enteropeptidase** (formerly termed enterokinase) are endopeptidases especially used during the isolation of recombinant proteins (Table 3). Their highly specific cleavage sites are Ile-Glu-Gly-Arg*Xaa for factor Xa, P_4-P_3-Pro-Arg/Lys*P_1'-P_2' for thrombin with P_3 and P_4 being hydrophobic amino acids, and $(Asp)_4$-Lys*Ile for enteropeptidase (the asterisk indicates the point of cleavage). Such sites are cloned into positions, where they allow to cleave fragments or proteins of interest out of a construct or fusion protein.

Exopeptidases attack proteins at their carboxy or amino termini, respectively. For the protein chemist, these enzymes are valuable tools, to cleave off N-terminal blocking groups, like an N-acetyl or formyl group (Farries et al. 1991) or a pyroglutamate residue (Armentrout 1969). Even if a protein is denatured, the N-terminus often is not accessible to exopeptidases. The proteins then first have to be cleaved into smaller fragments, followed by the isolation of the terminal fragment, before applying exopeptidases.

Beside the sequential chemical degradation of the C-terminal amino acid residues, complementary to the N-terminal Edman-degradation (Edman and Begg 1967), **carboxypeptidases** are used to get sequence information from the C-terminal end of a protein. As this is not a sequential reaction, but a continuous degradation of the C-terminal amino acid residues, the time course of the enzymatic hydrolysis has to be recorded to identify the amount of released amino acids by amino acid analysis.

Attachment of carbohydrate side chains to proteins may hinder proteases in cleaving the protein backbone and prevent the identification of the amino acid residues involved in the protein/carbohydrate linkage during Edman-degradation. Therefore, another group of enzymes, not belonging to the proteinases, but important for protein characterization are the **glycosidases**, especially those enzymes, which are able to cleave off the carbohydrate side chain from amino acid residues, like N-glycosidase A or F (Maley 1989). The reaction products are aspartic acid, ammonia and the complete oligosaccharide. O-Oligosidase cleaves off oligosaccharides from serine or threonine (Umomoto et al. 1977) (Table 3).

    Glycoproteins

The activation of many proteins is regulated by the phosphorylation of serine, threonine or tyrosine residues. These post translational modifications can be identified by the enzymatic dephosphorylation by **phosphatases**, resulting in a change of the molecular weight, which can be determined by mass spectrometry, and in a change of the mobility during isoelectric focussing, caused by a change in the electric charge.

Experimental Considerations

The experimental parameters for enzymatic hydrolysis are: buffer, pH range, temperature, incubation time and enzyme to substrate ratio. The chosen buffer depends on the required pH-range, which is necessary for the used enzyme (Table 2). Volatile buffers are preferred, since they can be easily removed by lyophylization in order to avoid buffer induced perturbation of the subsequent separation of the peptide mixture. For most cleavage reactions a pH range around 8.0 can be achieved by 0.1 M ammonium bicarbonate or N-ethylmorpholine buffers. At lower pH values ammonium acetate may be used (pH 4.0-5.5). Many enzymes are also active in the presence of detergents (SDS 0.1%, subtilisin and endopeptidase Lys-C even up to 1%; CHAPS, octylglucoside and NP40 up to 2%). The presence of organic solvents like acetonitrile and isopropanole helps to dissolve many proteins without destroying the enzymatic activity (up to 20% acetonitrile or isopropanole are tolerated by chymotrypsin, endopeptidase Glu-C and pepsin, up to 40% by trypsin, endopeptidase Lys-C, endopeptidase Asp-N, subtilisin, thermolysin, papain and elastase). Even reducing agents like 0.5% ß-mercaptoethanol (endopeptidase Lys-C and endopeptidase Glu-C) or 1% ß-mercaptoethanol (endopeptidase Asp-N) can be used. The incubation is performed at a temperature optimum of 37°C for 4 h to 16 h. For digestion in solution an enzyme to substrate ratio of 1:100 to 1:20 (w/w) is recommended. To achieve a high substrate concentration the volume of the incubation buffer should be as small as possible. For digestion of proteins in a gel matrix or bound onto a membrane a ratio of 1:10 to 1:1 should be used. To avoid autodigestion, the proteases should only be dissolved immediately prior to use.

For the digestion of proteins bound onto membranes a pretreatment of the membranes with a quenching agent like polyvinylpyrrolidone (PVP 40) is necessary to avoid nonspecific binding of the proteases. Extensive washing should remove excess PVP 40 to avoid artifact peaks during chromatography. A one-step procedure with cleavage buffer, containing 1% hydrogenated Triton X-100 (RTX-100) has been described (Fernandez et al. 1994), Triton thereby serving as a quenching agent and helping to elute peptides from the membrane.

An important step during all cleavage procedures is the simultaneous digestion of a blank sample. This can be simple buffer, a membrane or a piece of gel matrix, all having the same origin as the real sample. This helps to identify artifact peaks from salts or detergents and to find autolytic fragments of the used enzymes during chromatography.

Chemical Fragmentation

Complementary to enzymatic degradation, chemical fragmentations are frequently employed. A mayor advantage is that the corresponding chemical reagents are almost insensitive to detergents and salts.

Many chemical cleavage methods have been described (Kasper 1975, Fontana 1986) but only a few are applied in practice. The most widely used chemical fragmentation method for proteins is the cleavage with cyanogen bromide (Gross and Wittkop 1961). It cleaves Met-Xaa bonds specifically (low yields, were Xaa is serine or threonine) and nearly quantitatively, a reaction which no known peptidase is able to catalyze selectively. As methionine residues are rare in proteins, generally a few large fragments are obtained. A further advantage is, that the reaction is performed in 70 % formic acid or TFA, which are powerful solvents. However, these acids can cause further cleavages between aspartic acid and proline.

Separation and Recovery of the Cleavage Mixture

After cleavage of a protein into several fragments, the resulting peptide mixture has generally to be separated, although there are some mass spectrometric methods (Pappin 1997, Hopp and Baktiar 1997) which allow to characterize and analyze peptides in a mixture. The separation might be done by electrophoretic methods (PAGE or CE) with subsequent blotting onto a membrane or by chromatographic methods (HPLC).

Size exclusion chromatography (Chicz and Regnier 1990), ionexchange chromatography (Chicz and Regnier 1990), affinity chromatography (Chicz and Regnier 1990, Mant and Hodges 1991) or hydrophobic interaction chromatography (HIC) are separation methods frequently used for larger peptides or proteins. The method of choice for separation and fractionation of proteolytic digests on both the analytical and preparative scale is reversed phase high pressure liquid chromatography (RP-HPLC). The column support is silica based with derivatized silanol groups. Peptides are eluted with increasing concentrations of organic solvents (Meyer 1989). Practical as well as theoretical aspects have been discussed (Chicz 1990, Dolan 1991, Mant 1991, Nugent 1991a,b).

The most popular mobile phase system is 0.1% (v/v) TFA as solvent A and 0.085% (v/v) TFA in 85% (v/v) acetonitrile as solvent B. If gradient systems of higher pH are needed, 10 mM ammonium acetate, pH 6 as solvent A and 10 mM ammonium acetate, 85 % (v/v) in acetonitrile are used. The flow rates depend on the size of the used columns. The column size is chosen, depending on the amount of the cleaved protein (Table 4). Protein amounts in the low picomole range retrieved from a 1D- or 2D-PAGE, stained with Coomassie blue or silver are sufficient for microsequencing or mass spectrometry. These amounts require special HPLC setups to run low flow rates on columns with small inner diameter (I.D.).

In modern protein chemistry both separation and analytical methods are combined in one procedure to avoid loss of material. HPLC is done on capillary columns. The effluent is split after having passed the UV-detector. A small part (i.e. 10%) of the effluent is directly led into a ion spray mass spectrometer for mass determination and MS-sequencing. About 90% of the sample are continuously collected onto a PVDF-membrane, using a special blotting device (Eckerskorn et al. 1997). The peptides can be detected by lining up the PVDF strip with the UV-trace using visible markers as reference points. These peptides can be sequenced with the classical Edman chemistry.

Procedure

Reduction and alkylation

Reduction and alkylation with 4-vinylpyridine. Dissolve protein in 1-10 µg/µl of a solution containing 6 M guanidinium chloride, 0.1 M Tris, and 1 mM EDTA, adjusted to pH 8.3 with HCl. Dithiothreitol is added to a 2 mM excess over disulfide bonds in the protein. Nitrogen or argon is passed through the solution which is then incubated for 1 h at 37°C. 4-vinylpyridine is added at a 1.1 fold molar excess over total thiol groups and the solution is incubated under nitrogen for 1 h. The reaction is quenched with mercaptoethanol and desalted immediately (dialysis or RP-HPLC).

Enzymatic fragmentation

Enzymatic fragmentation in solution. Prepare a stock solution of the protease. Dissolve the protein in a volatile buffer (use minimum volume, to achieve high substrate concentration); e.g. 0.1M ammonium bicarbonate,

Table 4. Column sizes for the separation of peptide mixtures

	Capacity (mg)	Inner Diameter I.D. (mm)	Flow Rate (µl/min)
LC	1-10	4.6	400-2000
narrow bore LC	0.25-1.5	2	50-400
microbore LC	0.05-0.3	1	20-100
capillary LC	< 0.04	0.1-1	1-100
nano LC	< 0.01	0.05-0.1	0.1-1

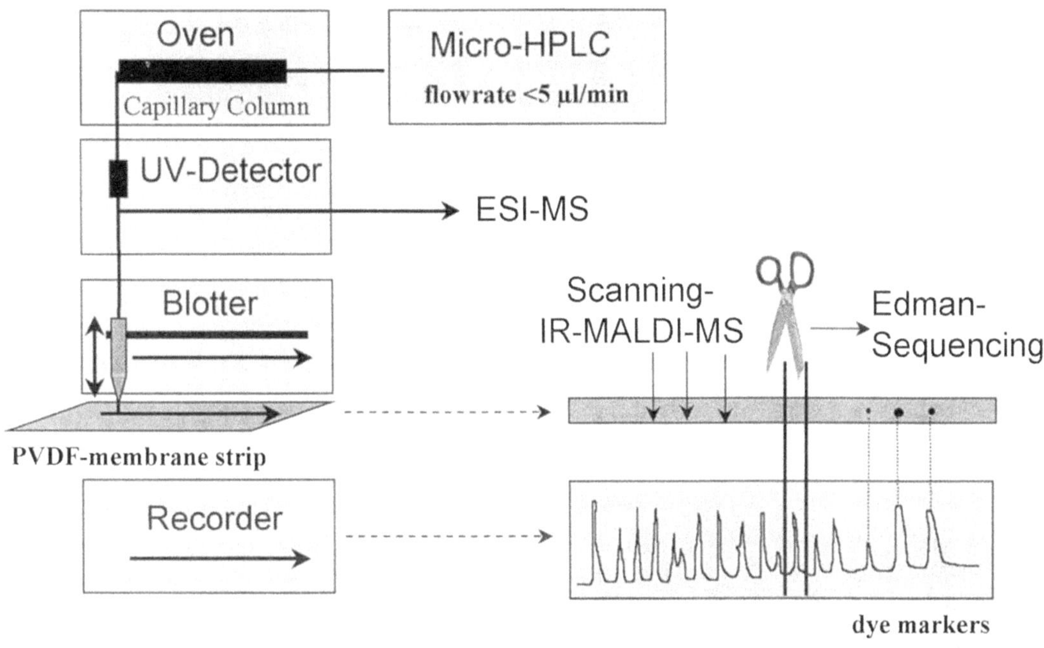

Fig. 4. Schematic diagram of the capillary HPLC/blotter/MS system. Separation of a peptide mixture by reversed phase HPLC and splitting of the effluent. Detection and characterization of the peptides by ionspray mass spectrometry (about 1/10) and simultaneous blotting of the peptides (9/10 of the total amount) on a membrane for subsequent Edman degradation or MALDI-MS.

pH 7.8-8.5 or 0.1M ammonium acetate, pH 4.0-5.5. Endopeptidase Glu-C, Asp-N and Arg-C can be used as supplied. For endopeptidase Lys-C 1 mM EDTA should be added and for trypsin, chymotrypsin, thermolysin and subtilisin the reaction mixture should include 2 mM Ca^{2+}. Incubate enzyme and substrate at a ratio of 1:20 to 1:100 for 4-8 hours at 37°C. Stop incubation by acidifying the solution with acetic acid or TFA.

Enzymatic fragmentation in a PAGE-matrix. Cut gel bands of interest into approx. 1 x 2 mm pieces and place in an 1.5 ml plastic tube. Do the same with a "blank" section of the gel. Add about 100 µl (solution should cover the gel pieces) of 50% CH_3CN/0.1M Tris/HCl, pH 8.0 to the gel pieces. Wash for 15 min at room temperature and remove supernatant. Dry washed gel pieces to about 30% of the original volume (alternatively shrink with acetonitrile). Add an enzyme (trypsin or endopeptidase Lys-C) solution in 0.1M Tris/HCl, pH 8.0, with an enzyme/protein ratio of 1:1 -1:10. Add as much buffer as the pieces need to be covered. Incubate at 37°C for 4-

8 hours. Extract peptides by elution with 0.1% TFA followed by an elution with 50% CH$_3$CN. Combine all elution steps and dry samples. Redissolve for reversed phase chromatography.

Enzymatic fragmentation on a membrane. Cut PVDF-slice of interest into approximately 1 x 2 mm pieces and place them into an 1.5 ml plastic tube. Do the same with a "blank" section of the membrane. Quench membranes with 50 µl of 0.2% polyvinylpyrrolidone (PVP 40) for 20 minutes. Remove the quenching reagent and wash extensively with H$_2$O bidest. (PVP 40 is difficult to remove completely and appears as a broad peak late in the chromatogram). Cover the pieces with trypsin or endopeptidase Lys-C solution in 0.1M Tris/HCl, pH 8.0, at an enzyme/protein ratio of 1:1 - 1:10.. Incubate at 37°C for 4-8 hours. Extract peptides by elution with 0.1% TFA (twice) followed by 50% CH$_3$CN. The addition of hydrogenated Triton RTX-100 helps to elute even more hydrophobic peptides. But if for identification of the peptide a coupled LC-MS is used, detergents interfere with mass spectrometry and should be avoided. Combine all elution steps and dry samples. Redissolve for RP-HPLC.

Chemical fragmentation with cyanogenbromide. Generally, a 100-fold molar excess of CNBr over methionine residues is used. The reaction is performed in 70% formic acid for 2-16 h in the dark. The mixture is flushed with nitrogen. Then 5-10 volumes of water are added and an excess of reagent is removed by lyophilisation. This step is repeated twice (CNBr is highly toxic - use KOH to neutralize the vapor). The same reaction can also be performed on PVDF membranes.

Enzymatic deglycosilation of N-glycosidic linked oligosacharides with N-Glycosidase F (PNGase F). 10 µg of glycoprotein are dissolved in about 50 µl 20-250 mM potassium or sodium phosphate buffer, pH 6.0-8.5. 0.5 units of N-glycosidase F are added and the mixture is incubated at 37°C over night. The presence of detergents (N-octylglucoside, Triton X-100, Nonidet P40, or CHAPS at a concentration of 0.5 2% or SDS up to 0.2%) might benefit deglycosilation of native proteins.

Enzymatic deglycosilation of O-glycosidic linked oligosacharides with O-Glycosidase. 10 µg of glycoprotein are dissolved in about 50 µl 20-250 mM Tris-maleate or potassium phosphate buffer, pH 6.0-7.6. 0.5 units of O-glycosidase are added and the mixture is incubated at 37°C over night. The concentration of SDS should not exceed 0.1% and an anionic detergent like Triton X-100 should be present at a 10-fold concentration over SDS to prevent inactivation of the O-glycosidase.

Removal of N-acetyl groups from blocked peptides with acylpeptide hydro-lase. Fragment protein of interest into smaller peptides and isolate the N-terminal peptide containing the acetylation (the enzyme acylpeptide hydrolase does not work efficiently on peptides longer than 30 amino acid residues). Dissolve sample (2µg) in 10 µl 0.1M sodium phosphate buffer, pH 8.0, 5 mM DTT, 10 mM EDTA and 5% glycerol. Add an aliquot of 0.05 µg of enzyme in 10 µl buffer and incubate under nitrogen at 4°C over night. Add again 0.05 µg of enzyme and incubate again for 8 hours at room temperature. Before sequencing, the peptide has to be desalted by RP-HPLC.

Removal of N-acetyl groups

C-terminal sequence analysis using carboxypeptidase Y. Dissolve 1 nmol of peptide in 100 ml 5 mM citrate buffer pH 6.5, 1% SDS, containing 0.01 mM norleucine as internal standard. The amount of the internal standard should be adjusted to the sensitivity of your analyzer and should allow for compensation of handling loss. Remove 10 µl as the zero time sample and add 0.1 nmol of carboxypeptidase Y (or a mixture of several carboxypeptidases). Incubate the sample at room temperature and remove 10 µl aliquots at T=1, 2, 5, 10, 20, 30, 60 minutes. Add immediately 5 µl acetic acid to each fraction, to stop the reaction and subject each fraction to amino acid analysis. A graph is plotted showing the amount of each amino acid released with time. The C-terminal sequence is deduced from the relative rate of release of each amino acid.

Sequence analysis

For peptides smaller than 4 kD, the sequence can also be deduced by the masses of the remaining degraded peptide, using matrix assisted laser desorption ionisation mass spectrometry (MALDI-MS).

References

Aebersold RH, Leavitt J, Saavedra RA, Hood LE, Kent SBH (1987) Internal amino acid sequence analysis of proteins separated by one- or two-dimensional gel electrophoresis after in situ protease digestion on nitrocellulose. Proc Natl Acad Sci USA 84: 6970-6974

Allen G (1989) Sequencing of Proteins and Peptides. Elsevier, Amsterdam

Amons R (1987) Vapor-phase modification of sulfhydryl groups in proteins. FEBS Lett 212: 68-72

Armentrout, R.W., Doolittle, R.F. (1969): Pyrrolidonecarboxylyl peptidase: stabilisation and purification. Archives Biochem. & Biophys. 132, 80-90.

Bailey JL and Cole RD (1959) Studies of the reaction of sulfite with proteins. J Biol Chem 234: 1733-1739

Barrett AJ (1981) Leukocyte elastase. Meth Enzymol 80: 581-588

Baugham DJ (1970) The serine proteinases - Thrombin assay. Meth Enzymol 19: 145-157

Brune DC (1992) Alkylation of cysteine with acrylamide for protein sequence analysis. Anal Biochem 207: 285-290

Chaconas G, van de Sande JH (1980) 32P labeling of RNA and DNA restriction fragments. Methods Enzymol, Academic Press, New York 65: 75-85

Chicz RM, Regnier FE (1990) High performance liquid chromatography; effective protein purification by various chromatographic methods. Methods Enzymol 182: 392-421

Cleland WW (1964) Dithiothreitol, a new protective reagent for SH groups. Biochemistry 3: 480-482

Cleveland DW, Fischer SG, Kirschner MK, Laemmli UK (1977) Peptide mapping by limited proteolysis in sodium dodecyl sulfate and analysis by gel electrophoresis. J Biol Chem 252: 1102-1106

Cole EG, Mecham DK (1966) Cyanate formation and electrophoretic behavior of proteins in gels containing urea. Anal Biochem 14: 215-222

Crestfield AM, Moore S, Stein WH (1963) The preparation and enzymatic hydrolysis of reduced and S-carboxymethylated proteins. J Biol Chem 238: 622-627

Darbre A (1986) Practical Protein Chemistry. Wiley, Chichester

Dolan JW (1991) Preventive maintainance and trouble shooting LC instrumentation. In: High-Performance Liquid Chromatography of Peptides and Proteins: Separation, Analysis and Conformation. (Mant CT, Hodges RS, eds.) CRC Press, Boca Raton, pp. 23-29

Drapeau GR (1980) Substrate specificity of a proteolytic enzyme isolated from a mutant of *Pseudomonas fragi*. J Biol Chem 255 (3): 839-840

Drapeau, G.R. Boily Y. Houmard, J. (1972) purification and properties of an extracellular protease of *Staphylococcus aureus*. J. Biol. Chem. 247, 6720-6726.

Ebeling W, Hennrich N, Klockow M, Metz M, Orth HD, Lang H (1974) Proteinase K from *Tritirachium album* limber. Eur J Biochem 47: 91-97

Eckerskorn C, Lottspeich F (1989) Internal amino acid sequence analysis of proteins separated by gel electrophoresis after tryptic digestion in polyacrylamide matrix. Chromatographia 28: 92-94

Eckerskorn C, Mewes W, Goretzki H, Lottspeich F (1988) A new siliconized-glass fibre as support for protein-chemical analysis of electroblotted proteins. Eur J Biochem 176: 509-519

Eckerskorn C, Strupat K, Kellermann J, Lottspeich F, Hillenkamp F. (1997) High-sensitive peptide mapping by micro-LC with on-line membrane blotting and subsequent detection by scanning-IR-MALDI mass spectrometry. J Prot Chem 16: 349-362

Edman P, Begg G (1967) A Protein Sequenator. Eur J Biochem 1: 80-91

Farries TC, Harris A, Auffret AD, Aitken A (1991) Removal of N-acetyl groups from blocked peptides with acylpeptide hydrolase. Eur J Biochem 196: 679-685

Fernandez J, Andrews L, Mische SM (1994) A one-step enzymatic digestion procedure for PVDF-bound proteins that does not require PVP-40. Techniques in protein chemistry V (Crabb JW, ed.) Academic Press, San Diego, pp 215-222

Folk, JE (1970) Carboxypeptidase B (porcine panreas). Meth Enzymol 19: 504

Fontana A (1972) Modification of tryptophan with BNPS-skatole (2-(2-nitrophenylsulfenyl)-3-methyl-3-bromoindolenine). Meth Enzymol 25: 419-423

Friedman M, Krull LH, Cavins JF (1970) The chromatographic determination of cystine and cysteine residues in proteins as S-β-(pyridylethyl)cysteine. J Biol Chem 245: 3868-3871

Fruton JS (1977) Specificity and mechanism of pepsin action on synthetic substrates. Adv Exp Medicine & Biology 95: 131-140

Glazer AN, De Lange, RJ, Sigman DS (1975) Chemical modification of proteins. North-Holland, Amsterdam.

Gross E, Wittkop B (1961) Selective cleavage of the methionyl peptide bonds in ribo-nuclease with cyanogen bromide. J Am Chem Soc 83: 1510-1511

Hayashi, R., Moore, S., Stein W.H. (1973) Carboxypeptidase from Yeast. Large scale pre-peration and the application to COOH-terminal analysis of peptides and proteins. J. Biol. Chem. 248, 3889-3892

Heinrickson, R.L. (1977) Methods Enzymol. 47, 189-193

Hirs CHW (1967a) Performic acid oxidation. Meth Enzymol 11: 197-199

Hopp CE, Bakhtiar R (1997) An introduction to electrospray ionisation and matrix-assisted laser desorption/ionisation mass spectrometry - essential tools in a modern biotechnology environment. Biospectroscopy 3(4): 259-280

Kaspar CB (1970) Protein Sequence Determination. (Needleman SB ed.) Springer-Ver-lag, Berlin, Heidelberg, New York, pp 114-161

Kaspar CB (1975) Fragmentation of proteins for sequence studies and separation of the peptide mixture. In: Protein sequence determination. (Needleman SB ed.) Springer, Heidelberg, pp 114-161

Kellner R (1994) Chemical and Enzymatic Fragmentation of Proteins. In: Microcharac-terisation of Proteins. (Kellner R, Lottspeich F, Meyer H ed) VCH, Weinheim, pp 11-27

Kennedy TE, Gawinowicz MAQ, Barzilai A, Kandel ER, Sweatt JD (1988) Sequencing of proteins from two-dimensional gels by using in situ digestion and transfer of peptides to polyvinylidene difluoride membranes: Application to proteins associated with sen-sitization in *Aplysia*. Proc Natl Acad Sci USA 85: 7008-7012

Light, A. & Janska, H. (1989) Enterokinase (enteropeptidase): Comparative aspects. Trends Biochem. Sci. 14, 110-112

Lindley H (1956) A new synthetic substrate for trypsin and ist application to the deter-mination of the amino acid sequence of proteins. Nature 178: 647-648

Lynn KR, Labow RS (1984) A comparison of four sulfhydryl cathepsins (B,C,H, and L) from porcine spleen. Can J Biochem Cell Biol 62: 1301-1308

Maley F (1989) Characterisation of Glycoproteins and their associated oligosacherides through the use of Endoglycosidases. Anal Biochem 180: 195-204

Mant CT, Hodges RT (1991a) Mobile phase preparation and column maintenance. In: High-Performance Liquid Chromatography of Peptides and Proteins: Separation, Analysis and conformation (Mant CT, Hodges RS eds) CRC Press Boca Raton, pp37-45

Mant CT, Hodges RT (1991b) Mobile phase preparation and column maintenance. In: High-Performance Liquid Chromatography of Peptides and Proteins: Separation, Analysis and conformation (Mant CT, Hodges RS eds) CRC Press Boca Raton, pp 69-94

Matsudeira PT (1993):A practical guide to protein and peptide purification for micro-sequencing. Academic Press, San Diego

Meyer HE (1994) HPLC. In: Microcharacterisation of Proteins (Kellner R, Lottspeich F, Meyer H eds) VCH, Weinheim, pp 11-27

Neugebauer JM (1990) Detergents: An overview. In: Protein Purification (Deutscher MP ed.) Academic Press, San Diego, pp239-253

Nugent KD (1991) Commercially available columns and packings for reversed phase HPLC of peptides and proteins. In: High-Performance Liquid Chromatography of Peptides and Proteins: Separation, Analysis and conformation (Mant CT, Hodges RS eds) CRC Press Boca Raton, pp279-287

Owen WG, Esmo CT, Jackson, CM(1974) The conversion of prothrombin to thrombin. J Biol Chem 249: 594-605

Pappin DJC (1997), Peptide mass fingerprinting using MALDI-TOF mass spectrometry. In: Methods in Molecular Biology, Vol.64. Protein Sequencing Protocols. pp 165-173. Humana Press Inc.: Totowa, New Jersey, USA.

Petra PH (1970) Carboxypeptidase A. Meth Enzymol 19: 460

Pluskal MG, Przekop MB, Kavonian MR, Vecoli C, Hicks DA (1986) Immobilon PVDF transfer membrane: a new membrane substrate for western blotting of proteins. Biotechniques 4: 272-283

Raftery MA, Cole RD (1966) On the aminoethylation of proteins. J Biol Chem 241: 3457-3461

Riviere LR, Fleming M, Elicone C, Tempst P (1991) Study and applications of the effects of detergents and chaotropes on enzymatic proteolysis. Techniques in protein chemistry II (Villafranca JJ ed) Academic Press, San Diego, pp171-179

Sambrook J, Fritsch EF, Maniatis T (1989) Molecular Cloning. A Laboratory Manual B16. Cold Spring Harbour Press

Schenkein I, Levy M, Franklin EC, Frangione B, (1977 Jul) Proteolytic enzymes from the mouse submaxillary gland. Specificity restricted to arginine residues. Arch Biochem Biophys 182(1): 64-70

Serano L, Avila J, Maccioni RB (1984) Controlled proteolysis of tubulin by subtilisin. Localisation of the site for MAP_2 interaction. Biochemistry 23: 4675

Tschesche, H. Kupfer S. (1972) C-terminal-sequence determination by carboxypeptidase C from orange levels. Eur. J. Biochem. 26, 33-46

Umomoto J, Bhavanandan VP, Davidson EA (1977) Purification and properties of an Endo-alpha-N-acetyl-D-galactosaminidase from *Diplococcus pneumoniae*. J Biol Chem 252: 8609-8614

Verjee ZHM (1969) Isolation of three acid phosphatases from wheat germ. Eur J Biochem 9: 439-444

Yokoyama S, Oobayashi, Tanabe O, Ichishima E (1981) Agr Biol Chem 45:311-319

Abbreviations

SDS	sodium dodecylsulfate
PAGE	polyacrylamide gel electrophoresis
RP-HPLC	reversed phase high performance liquid chromatography
LC-MS	liquid chromatography coupled to mass spectrometry
MALDI-MS	Matrix assisted laser desorption ionisation mass spectrometry
$\alpha_2 M$	α_2-macroglobulin
DFP	diisopropylfluorophosphate
PMSF	phenylmethysulfonylfluoride
4-NA	4-nitronilide
*	point of cleavage
TLCK	L-1-chloro-3-(4-tosylamido)-7-amino-2-heptanonhydrochloride
TPCK	L-1-chloro-3-(4-tosylamido)-4-phenyl -2-butanone
ZPCK	carbobenzoxy-L-phenylalanin-chloromethylketone
Xaa	any amino acid resiue
Uaa	bulky hydrophobic residue
pGlu	pyroglutamate

Chapter 15

Limited Proteolysis in the Study of Protein Conformation

ANGELO FONTANA, PATRIZIA POLVERINO DE LAURETO,
VINCENZO DE FILIPPIS, ELENA SCARAMELLA AND MARCELLO ZAMBONIN

Introduction

The methods of choice for determining the three-dimensional structure of globular proteins are X-ray crystallography and two-dimensional NMR (Lecomte, 1991; MacArthur et al., 1994). Currently we are observing a flow of reports describing the structural analysis of proteins, since significant theoretical and methodological advances have been made recently in utilizing both X-ray and NMR techniques. However, every experimental technique for protein structure determination has strengths and weaknesses and, in particular, we need protein crystals for X-ray crystallography and a non-aggregating protein solution at a millimolar concentration for NMR. Since these requirements are not always fulfilled, perhaps alternative methods can be employed, even if these will provide structural information at a lower level resolution. In this chapter we will show that a classical biochemical method such as limited proteolysis can be used to probe structure. and dynamics of proteins in solution, providing experimental results which are easy to obtain and well complement those derived from the use of other more classical physicochemical methods and approaches.

The proteolytic cleavage of a polypeptide chain occurs only if the site of cleavage can bind and adapt itself in a specific way to the stereochemistry of the active site of the protease (Shechter and Berger, 1967). This is difficult to achieve with native globular proteins, since the tight conformation of the polypeptide chain renders most of the peptide bonds inaccessible to the active site of the protease and consequently native proteins are usually quite resistant to proteolysis. On the other hand, this binding and adaptation at

Correspondence to: Angelo Fontana, Patrizia Polverino de Laureto, Vincenzo De Filippis, Elena Scaramella, Marcello Zambonin, University of Padua, CRIBI Biotechnology Centre, Viale G. Colombo 3, Padua, 35121, Italy (*phone* +39-49-8276156; *fax* +39-49-8276159; *e-mail* fontana@cir.bio.unipd.it)

the active site of the protease is easily achieved with fully unfolded proteins. It is conceivable to suggest that protein degradation requires the unfolded species as a substrate and that the actual equilibrium between the native (N) and unfolded (D) protein species is controlling the rate of protein degradation to small peptides. However, evidence has shown that the native protein can also be attacked by the protease. In this case, limited proteolysis occurs at the level of only one or very few peptide bonds, leading to the formation of "nicked" proteins, constituted by rather large fragments remaining associated in a stable (and often functional) complex (Richards and Withayathil, 1959; Mihalyi, 1978; Price and Johnson, 1990; Wilson, 1991; Vita et al., 1985; Signor et al., 1990).

A nicked protein, usually, is much more labile than the intact form and its unfolding leads to a suitable substrate for an easy and extensive proteolytic degradation to small peptides. Further proteolysis therefore is much faster than the initial peptide bond fission at the level of the native protein and, consequently, during proteolysis the intact protein and small peptides are present in the reaction mixture, without intermediate size products. On the other hand, if nicked proteins are sufficiently stable, they can resist further extensive degradation by proteolysis and can be isolated and characterized. As shown in Figure 1, a dual mechanism of protein degradation can be pro-

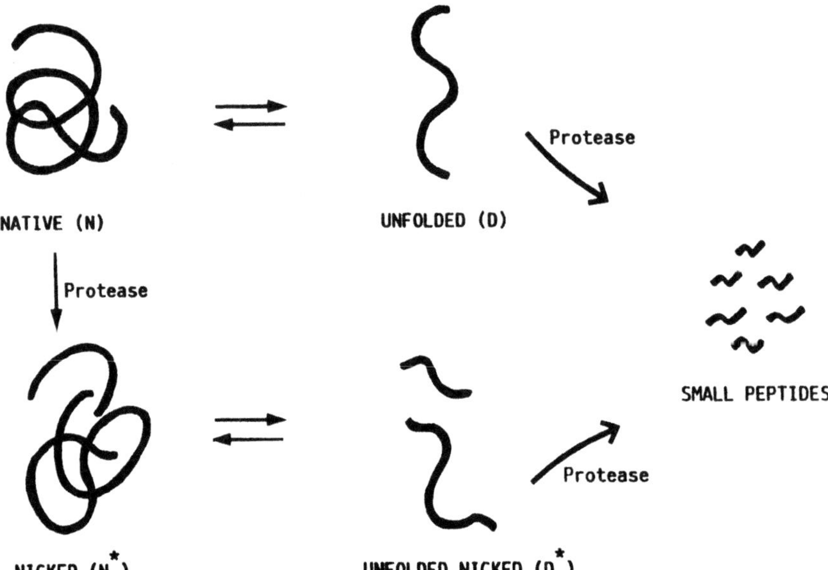

Fig. 1. Proposed scheme of the proteolytic degradation process of a globular protein. N, native protein; D, unfolded protein; N*, nicked protein; D*, unfolded nicked protein.

posed, the one in which the substrate for the protease is the unfolded protein and the other in which nicking of the native, folded protein is the initial key step in the degradation process (Fontana et al., 1993).

It is plausible to suggest that the limited proteolysis phenomenon derives from the fact that a specific chain segment of the compact, folded protein substrate is sufficiently exposed and flexible to fit the active site of the protease for an efficient and selective hydrolysis (Hubbard et al., 1992). This seems quite reasonable, but it is important to experimentally verify this general notion in more precise and quantitative terms. To this aim, we have carefully analyzed the results of the numerous limited proteolysis experiments conducted over the years on globular proteins of known three-dimensional structure in order to unravel the specific key features of the peptide bond(s) suffering limited proteolysis in respect to those of the many other peptide bonds of the protein (Fontana et al., 1993). First of all, this analysis clearly revealed that the sites of limited proteolysis are located at the surface of the protein and not embedded in the rigid protein interior. However, the notion of exposure/protrusion/accessibility (Novotny and Bruccoleri, 1987; Hubbard et al., 1991) is clearly not sufficient to explain the limited proteolysis phenomenon, since it is evident that even in a small globular protein there are many exposed sites. Therefore, surface accessibility of the proteolysis site is a required property, but not sufficient to explain limited proteolysis.

Enhanced chain flexibility (segmental mobility) appears to be the key feature of the site of proteolysis, as clearly demonstrated by the results of a systematic study on the limited proteolysis and autolysis of the thermophilic protease thermolysin (Fontana et al., 1986). The results of this study have demonstrated that there is a clear-cut correlation between sites of proteolytic attack and sites of enhanced chain flexibility, this last reflected by the crystallographic temperature factors (B-values) (see also below). The proposal was advanced therefore that limited proteolysis of a globular protein occurs at flexible loops and, in particular, that chain segments in a regular secondary structure (such as helices) are not sites of limited proteolysis (Fontana et al., 1986, 1993). In a more recent study, Hubbard et al. (1994) proposed that large local motions near the scissile peptide bond are required for proteolysis, leading to significant changes in the conformation of a chain segment constituted by up to 12 residues (see also Hubbard, 1993).

A possible explanation of the fact that helices and, in general, elements of regular secondary structure are not easily hydrolyzed by proteolytic enzymes can be given also on the basis of energetic considerations. If proteo-

Secondary structural of substrates

lysis occurs at the centre of the helical segment, it is likely that the helix will be fully destroyed by end-effects and consequently all hydrogen bonds, which cooperatively stabilize it, are broken. On the other hand, a peptide bond fission at a disordered flexible site probably does does not substantially change the energetics of that site, since the peptide hydrolysis can easily be compensated by some new hydrogen bonds with water. The proposal can be advanced therefore that proteolysis of rigid elements of secondary structure is thermodynamically very disadvantageous (Krokoszynska and Otlewski, 1996).

That secondary structure elements and, in particular, helices behave as rigid rods not amenable to proteolysis is substantiated by a number of limited proteolysis experiments (see Fontana et al., 1993). As an example, the results obtained with porcine citrate synthase (Lill et al., 1984) serve to illustrate that indeed helices are not attacked by proteases (for the calmodulin case see below). The structure of this enzyme is quite unusual for a protein of a size of 50 kDa, in that it consists almost entirely of α-helices. The subunits of this dimeric enzyme are each constituted by two domains, a large one and a small one of about 100 amino acid residues accounting for about a quarter of the subunit (Remington et al., 1982). Proteolysis occurs selectively at the level of the small helical domain (region 270-390), as schematically shown in Figure 2. It is seen that all sites of cleavage occur outside the helical segments of the polypeptide chain. Moreover, inspection of the three-dimensional model of the enzyme reveals that these sites are located at loops connecting helical segments.

The concept of flexibility/loose structure in dictating the limited proteolysis event can be used to explain the fact that, in relatively large proteins, cleavage often occurs at chain segment(s) connecting the structural domains (Wetlaufer, 1981; Neurath, 1980, 1986). These "hinge" regions between domains are expected to be much more flexible than the rest of the coiled polypeptide chain forming the globular units of the domains. Indeed, limited proteolysis is a classical procedure to produce from large proteins individual domains for further structural and functional characterization (Wetlaufer, 1981; Fontana, 1990). Finally, the chain flexibility notion explains the fact that, in certain proteins, the rather loose and flexible parts can be removed by limited proteolysis, allowing the isolation of the "core" of the protein as the most proteolysis-resistant moiety and thus as a stable protein entity (Dalzoppo et al., 1985; Vindigni et al., 1994).

The considerations outlined above emphasize that limited proteolysis can be used as a simple technique to unravel features of structure and dynamics of proteins. The technique is modest in demands for protein sample requirements, instrumentation and experimental efforts. Usually, the results of limited proteolysis experiments can elucidate specific conformational aspects of a protein, identify sites of flexibility along the polypeptide chain and locate structural domains. A specific advantage of the technique is that it provides data on the features of structure and dynamics of a protein in solution, even if these data do not reach the high resolution level provided by other physicochemical techniques (X-ray, NMR). In the following, we will discuss some experimental aspects of the limited proteolysis approach and we will illustrate examples of specific applications. The reader may find herewith some personal selections of issues raised and examples, but no attempt is made to provide a complete coverage of the limited proteolysis phenomenon nor of its applications in protein structure research, since this would require writing a book rather than a (short) chapter. The interested reader can find additional information on the limited proteolysis technique, including some experimental procedures, in the articles of Price and Johnson (1990) and Wilson (1991) or in the book of Mihalyi (1978).

Experimental Outline

Choice of the Protease

The limited proteolysis approach for probing protein conformation implies that the proteolytic event should be dictated by the stereochemistry and flexibility of the protein substrate and not by the specificity of the attacking

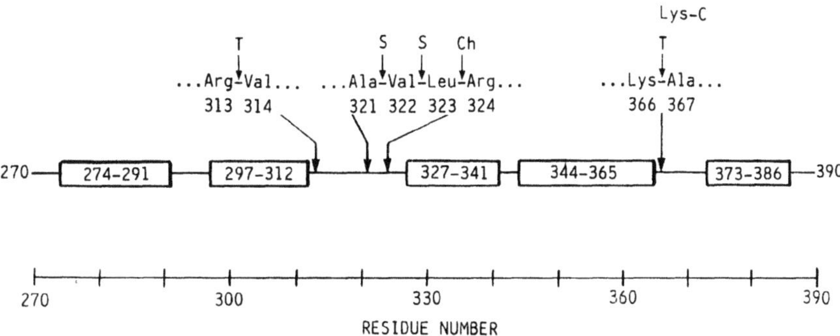

Fig. 2. Schematic representation of the secondary structure of the small domain of citrate synthase, as derived from the X-ray structure (Remington et al., 1982). Number inside the boxes refer to the starting and ending residues of helices. Arrows indicate sites of proteolytic attack by trypsin (T), subtilisin (S), chymotrypsin (Ch) and Lys-C endoproteinase (Lys-C) (experimental data taken from Lill et al., 1984).

protease. To this aim, the most suitable proteases are those displaying broad substrate specificity, such as subtilisin (Morihara et al., 1970; Ottesen and Svendsen, 1970), thermolysin (Heinrikson, 1977), proteinase K and pepsin (see Bond, 1990, for additional references). These endopeptidases display a moderate preference for hydrolysis at hydrophobic or neutral amino acid residues, but often cleavages occur at other residues as well (Bond, 1990; see also Keil, 1982, for a comparative analysis of the kinetics of hydrolysis of different peptide bonds by several proteases). Usually, exopeptidases, such as amino- or carboxy-peptidases, are not used for limited proteolysis, even if few specific applications of these proteases have been described in the literature to remove short chain segment(s) at the amino- or carboxy-terminal ends of proteins.

The recommended approach is to perform trial experiments of proteolysis of the protein of interest utilizing several proteases of broad substrate specificity, such as subtilisin and thermolysin. Nevertheless, initial experiments can be conducted also utilizing proteases of more restricted specificity, such as trypsin, Glu-C specific protease from *Staphylococcus aureus* V8 (Drapeau, 1977), Lys-C protease, Arg-C protease or chymotrypsin. Nowdays, 30 different proteases are available from commercial sources (Bond, 1990) and their purity usually is satisfactory for the specific needs of the limited proteolysis experiments. In some instances, the samples of commercial proteases can be treated with inhibitors that inactivate the contaminants. For example, chymotrypsin preparations can be treated with tosyl-lysine chloromethylketone (TLCK) to inhibit contaminating trypsin or trypsin with tosylamido-2-phenyl-ethylchloromethyl ketone (TPCK) to inactivate chymotrypsin. Note, TLCK-chymotrypsin and TPCK-trypsin are both commercially available.

Since proteases often are unstable due to autolysis, it is recommended to check the proteolytic activity of the proteases samples and usually fresh solutions of the proteases should be used. Alternatively, protease solutions can be stored frozen in order to minimize autolysis. In general, it is recommended to prepare protease solutions immediately prior to use.

In few studies, protease immobilized onto solid supports (*e.g.*, Sepharose) have been used for proteolysis experiments, since after digestion the protease can be easily removed from the proteolysis mixture. It is important, however, to check if the immobilized protease behaves as the one in solution, since proteolysis can be controlled also by diffusion effects of the protein substrate to the immobilized protease.

Monitoring and Controlling the Digestion

The usual way to analyze the time-course proteolysis is by SDS-PAGE
(Laemmli, 1970; Schägger and von Jagow, 1987). This is a method of choice,
due to its simplicity, high sensitivity, high resolving power and capability to
analyze many samples in a single gel. Usually, 1-5 µg of a protein digest can
be analyzed by SDS-PAGE. Since the protein fragments produced by limited
proteolysis can remain associated in a stable complex at neutral pH, usually
the electrophoretic analysis is conducted in the presence of SDS, which permits dissociation of the fragments and their separation on the basis of their
molecular masses. The procedure utilizing the discontinuous buffer system
of Laemmli (1970) or the Tricine buffer system of Schägger and von Jägow
(1987) are used for SDS-PAGE analysis, this last providing excellent separations of polypeptides in the range from 1 kDa to 100 kDa masses. A quantitative analysis of the proteolytic fragments produced and separated in the
gel can be achieved by densitometry of the bands stained with the Coomassie Blue dye. Before the electrophoretic analysis of the proteolytic mixture, it
is essential to stop the proteolysis reaction by inactivating the protease (see
below).

The protein fragment mixture generated by limited proteolysis can
be separated by classical chromatographic procedures, as well as by
reverse-phase(RP)-, ion exchange-, gel filtration-HPLC. The RP-HPLC
technique on columns with a hydrophobic stationary phase, such as C_8-
or C_{18}-silica, can be employed with success, even if the rather large fragments produced by limited proteolysis usually are not efficiently separated
as small peptides. Usually, the fragment mixture is applied to a C_8- or C_{18}-
column under acidic conditions, usually in the presence of 0.05-0.1% trifluoroacetic acid (TFA), and then the column is eluted with a gradient of
aqueous acetonitrile. Advantages of this solvent mixture is that it has a low
background absorbance at 200-220 nm and is volatile. The peptide/protein
fragments can be collected manually as they elute from the column and then
recovered by lyophilization for further analysis (amino acid analysis after
hydrolysis, sequencing, mass spectrometry).

The SDS-PAGE or HPLC analyses can provide the first indication on the
extent and time-course of the proteolysis reaction. Since the limited proteolysis approach is a kinetic process and involves a bimolecular reaction
between the protein substrate and the protease, it is possible to predict, on
the basis of the first trial experiments, how to change the enzyme:substrate
(E:S) ratio, reaction time, concentration and incubation temperature in order to limit or to enhance the extent of protein digestion. Usually, the de-
finition of the initial peptide bond fissions of the protein substrate is the

more interesting and useful, for example for the identification of the flexible sites of the protein chain (see above).

Protease con-
centr.
Temperature

Possible ways to control proteolysis is by using a low concentration of protease, short reaction times and low temperature. It is not easy to predict in advance the most useful experimental conditions for conducting a limited proteolysis experiment, since these actually depend on the structure, dynamics, stability/rigidity properties of the protein substrate and on the actual aim of the experiment, *i.e.*, isolation of the rigid core of the protein or location of the sites of protein flexibility. The most commonly used enzyme:substrate (E:S) ratio to be employed is 1:50 or 1:100, but E:S ratios as low as 1:1000 or 1:5000 can be necessary, if the protein is very labile and thus proteolysis is very fast. Similarly, the length of reaction time can be from 1 min up to several hours or even days (see examples below). Usually, the temperature of reaction has a relatively modest effect and that expected from the general rule that the speed of a reaction roughly doubles for a ~10°C increment. On the other hand, the temperature of proteolysis can be very critical, if the protein substrate suffers heat-mediated unfolding. It has been calculated that the rate of chymotryptic digestion of unfolded myoglobin in respect to that of the folded native species is ~10^{19} higher (McLendon and Radany, 1978).

Inhibition and/or Inactivation of Proteases

The limited proteolysis approach implies that the hydrolytic reaction is conducted under carefully controlled conditions, since if one waits long enough extensive proteolysis of a globular protein is expected to occur, thus minimizing the utility of the approach. Moreover, during the manipulation of the proteolytic mixture for the isolation and analysis of "nicked" proteins or fragment species, there is the risk that the mixture is exposed to denaturating solvent conditions which would render the products of initial proteolysis vulnerable to further proteolysis. It is desirable therefore to inhibit rapidly and irreversibly the protease at the end of the proteolytic reaction by the use of specific inhibitors or to inactivate the protease by its denaturation in acid or alkaline solution or in the presence of a detergent.

The easiest way to stop proteolysis is to add to the mixture trifluoroacetic acid (TFA) to a final concentration of 0.5% (v/v). An aliquot of this acid solution can be directly analyzed by RP-HPLC. It should be considered that some proteases, such as trypsin, are inhibited in acid but recover their activity if the pH of the solution is brought to neutrality, thus causing further proteolysis. Pepsin is fully active at pH 1-4, including in aqueous TFA, and can be inactivated by making the digestion mixture alkaline (pH ~9) by the addition of aqueous ammonia (see Lin et al., 1993, for references). Alternatively, in order to stop proteolysis, the mixture can be added to the sample buffer of SDS-PAGE and heated at 100°C. However, this procedure is not necessarily the most appropriate, since some proteolytic enzymes (trypsin, proteinase-K, V8-protease) are not fully inactivated in the presence of SDS and there is the risk therefore that upon mixing the proteolytic mixture with the SDS-containing buffer proteolysis proceeds further. Proper controls therefore should be conducted in order to check the irreversible inactivation of the protease before the SDS-PAGE analysis.

A great variety of inhibitors of proteases of natural and synthetic origin have been described over the years and thus an ample choice of them is available for stopping proteolysis (Salvesen and Nagase, 1990). For inhibition of serine proteases (*e.g.*, chymotrypsin, trypsin, V8-protease), diisopropylphosphofluoridate (DFP), phenylmethylsulfonyl fluoride (PMSF) and 3,4-dichloroisocoumarin (DCI) are the most widely used inhibitors. Peptide aldehydes (*e.g.*, leupeptin), containing an aldehyde moiety in place of the usual α-carboxylate, can be used to inhibit serine proteases. However, the most classical irreversible inhibitors of serine protease are peptide chloromethyl ketones, such as tosyl-lysine chloromethyl ketone (TLCK) and tosylamido-2-phenyl-ethylchloromethyl ketone (TPCK).

    Several protein inhibitors form tight, but reversible, complexes with a number of serine proteases and can be utilized to stop proteolysis (avian ovomucoids, soybean trypsin inhibitor, Kunitz inhibitor). Besides inhibition in situ by adding these protein inhibitors, proteases can be selectively removed from the proteolysis mixture by affinity techniques, *i.e.*, by utilizing affinity adsorbents constituted by inhibitors immobilized onto a solid support (*e.g.*, Sepharose). Some of these adsorbents are commercially available and the choice of them depends on the type of protease present in the proteolysis mixture.

Metallo-proteases can be rapidly inhibited by adding a metal-chelating agent. For example, thermolysin is a zinc-dependent and calcium-stabilized protease and can be inhibited by EDTA, 1,10-phenanthroline, tetraethylenepentamine and phosphoramidon (Heinrikson, 1977). Of note, 1,10-phenantroline is the preferred inhibitor of thermolysin, since this compound has a higher stability constant for zinc than for calcium and inactivates the protease even in the presence of 10 mM Ca^{2+}. Excess zinc ions (in the mM range) inhibit thermolysin by binding at active site residues.

Serine proteases

Metallo proteases

Isolation and Characterization of Protein Fragments

The usual applications of the limited proteolysis technique is to produce nicked proteins, to isolate structural domains or to locate the loose parts in a polypeptide chain (see below). To this aim, it is mandatory to isolate the proteolysis products for both identification purposes and for subsequent studies. For analytical purposes, the separation of fragments by SDS-PAGE, their blotting onto a suitable polymeric matrix (Matsudaira, 1987) and subsequent N-terminal sequencing could be an easy way to identify the fragments produced, if the amino acid sequence of the protein is known. The usual work-up, however, of a proteolytic mixture is the analysis of aliquots of this mixture by RP-HPLC, which, even at the analytical level, allows the separation of protein fragments in sufficient quantities for their characterization by N-terminal sequencing, amino acid analysis after acid hydrolysis, as well as molecular mass determination by mass spectrometry. Since RP-HPLC is usually conducted utilizing an aqueous acetonitrile solvent mixture and in the presence of 0.05-0.1% (v/v) trifluoroacetic acid (TFA), these solvent conditions are likely to denature and dissociate the fragment complex of a nicked protein. Thus, RP-HPLC is not suitable for isolating a nicked protein, which can instead be isolated by gel filtration or ion exchange chromatography in aqueous buffers at neutral pH.

The recent advances in mass spectrometry, in terms of both methodology and instrumentation, are dramatically changing the usual, classical procedures of isolation and characterization of proteins and peptides. Electrospray(ES)-mass spectrometry(MS) provide accurate molecular mass analysis of protein fragments even when they are present in a mixture and in minute amounts (Andersen et al., 1996; Hess et al., 1993; Mann and Wilm, 1995). In particular, the high accuracy (0.01% or better) of the ES-MS analysis can lead to the unambigous identification of the fragments and thus of sites of cleavage, if the amino acid sequence of the protein is known. A specific advantage of ES-MS is that the separated fragments being eluted from the HPLC column can be directly injected into the ion source of the instrument. Matrix-assisted laser-desorption ionization(MALDI)-MS can be effectively used for protein fragment identification (Karas and Hillenkamp, 1988). The MALDI-MS technique, even if it is less accurate (~0.05-0.1%) than ES-MS, offers the advantages of permitting analyses even if salts of moderate concentrations are present in the sample. In particular, fragments of relatively high molecular masses can be analyzed by MALDI-MS. It is likely that the recent and extremely important developments of the MS techniques will prompt a more general and wide-spread application of the limited proteolysis approach in protein research. Indeed, the combined use of limited proteolysis and mass spectrometry for the elucidation of aspects of protein structure has been described in a number of recent papers (Brockerhoff et al., 1992; Mohsen et al., 1995; Seielstad et al., 1995; Cohen et al., 1995; Zappacosta et al., 1996; Fontana et al., 1997; Polverino de Laureto et al., 1995a,b, 1997).

Selected Applications

Monitoring the Protein Unfolding Process

The unfolding of a globular protein to the random-coil polypeptide chain can be monitored by a variety of spectroscopic techniques, but also by proteolytic probes (Imoto et al., 1986). This possibility has been demonstrated by Pace and Barrett (1984) by determining the kinetics of the selective trypsin hydrolysis of the Arg^{77}-Val^{78} peptide bond in the folded versus unfolded ribonuclease T_1. This study was made possible by the fact that (i) ribonuclease T_1 contains only two peptide bonds (Lys^{41}-Thr^{42} and Arg^{77}-Val^{78}) susceptible to tryptic hydrolysis and cleavage at Lys^{41} can be blocked by selective N-acetylation and (ii) this protein unfolds completely when the two disulfide bonds are broken by reduction. The results of this elegant study indicated that the Arg^{77}-Val^{78} peptide bond is cleaved at least 1700 times more rapidly in the unfolded than in the folded species of ribonuclease T_1.

The ND equilibrium of a globular protein is shifted towards the native, folded (N) species if a protein is stable and, conversely, towards the denaturated (D) species if a protein has marginal stability. Considering that the ND equilibrium can control the rate of protein degradation by proteolysis (see above), it is clear that proteases can be used to monitor the unfolding process. If the unfolded state (D) is the only true substrate for the attacking protease, digestion of the protein occurs by an all-or-none process (Imoto et al., 1986). Indeed, there is a correlation between protein turnover in vivo, *i.e.*, proteolytic degradation, and thermodynamic stability of proteins (McLendon and Radany, 1978). Moreover, the highly stable proteins and enzymes isolated from thermophilic microorganisms not only show noteworthy stability to heat and other common protein denaturants, but are also more resistant to proteolytic degradation than their mesophilic counterparts (Fontana et al., 1991).

Proteolytic enzymes can be used also to follow the local unfolding of a protein molecule when exposed to a denaturating environment. For example, the effect of a low concentration of guanidine hydrochloride on ribonuclease A is to cause local conformational changes near the active site (Yang and Tsou, 1995). Similarly, the thermal unfolding of ribonuclease A leads to the local unfolding of a quite restricted region of the protein, as shown by the fact that a variety of proteases initially cleave at chain segment 31-46 (Arnold et al., 1996, and references cited therein).

Protein Dissection and Location of Protein Domains

The word protein domain refers to a relatively compact and folded element of protein sub-structure which can be identified by visual inspection of the protein whole structure (Goldberg, 1969; see Wetlaufer, 1981; and references cited therein). Domains are almost invariably seen in relatively large proteins constituted by more than ~100 amino acids (Fontana et al., 1985). Domains can be rather well separated from each other, so that large protein molecules often appear constituted by lobes. Protein fragments containing an entire protein domain are expected to maintain in solution the structure they have in the intact native protein (Vita and Fontana, 1982; Fontana, 1990), since essentially all stabilizing interactions and forces leading to the structure in the native protein are likely to be maintained, even after cleavage of the connecting peptide segment (hinge) between domains. Limited proteolysis proved to be the method of choice to dissect multi-domain proteins into fragments capable of an independent folding. In fact, the hinge peptides between domains are usually flexible and sites therefore of preferential proteolysis, whereas the individual domains have a tight and rigid conformation preventing extensive proteolysis (Ahmed et al., 1986; Tasayaco and Carey, 1992; Wetlaufer, 1981). Several examples of limited proteolysis occurring at flexible hinges, poorly resolved in the electron density map, have been described in literature, the most classical case being the proteolytic dissection of the immunoglobulin molecule into the Fab and Fc pieces (Neurath, 1980; Wetlaufer, 1981; Bennett and Huber, 1984). We will herewith illustrate in detail just one example of proteolysis of the two-domain protein calmodulin. This example helps us to reinforce the notion that chain flexibility is a major parameter in dictating selective proteolysis and, moreover, that helices are not targets for proteolysis.

The crystal structure of the 148-residue chain of calmodulin (CaM) has been solved and shown to contain two domains, each binding two calcium ions (Babu et al., 1985, 1988). The two domains are connected by a long, solvent-exposed α-helix of about eight turns (Chattopadhyaya et al., 1992). In the presence of calcium, trypsin cleaves selectively CaM at the level of this helix into two fragments of about equal size, residues 1-77 and 78-148

(Draibikowski et al., 1982). This tryptic cleavage of a helix appears to be unusual and to contradict the proposal that only flexible loops are the favorite sites of limited proteolysis (Fontana et al., 1986). However, studies on the structure of CaM in solution provided evidence of the dynamic features of the central helix. For example, X-ray solution scattering data gave indications of differences between the solution and crystal structure of CaM and, in particular, that the interconnecting helix is bent in solution, thereby bringing the two calcium-binding domains in closer contact (Heidorn and Trewhella, 1988). The most recent and clear-cut evidence of the dynamic nature of the long central helix of CaM in solution came from both NMR (Ikura et al., 1992; Barbato et al., 1992) and X-ray (Meador et al., 1992) studies. The three-dimensional structure of CaM complexed with a 26-residue synthetic peptide (residues 577 to 602 of myosin light chain kinase) has been solved by heteronuclear NMR spectroscopy. In the peptide complex, CaM maintains in solution its two-domain topology of the crystalline state, but "the long central helix (residues 65 to 93) which connects the two domains is disrupted into two helices connected by a long flexible loop (residues 74 to 82)" (Ikura et al., 1992). Even the crystal structure of the peptide-CaM complex solved at 2.4 <Å resolution revealed that the "central helix is unwound and expanded into a bent between residues 73 and 77" (Meador et al., 1992). Therefore, the selective peptide bond fission of CaM by trypsin (Draibikowski et al., 1982), leading to the dissection of the protein into two domains, occurs at a flexible site of the "central helix".

The CaM case convincingly documents the great potential of proteolytic enzymes as probes of structure and dynamics of globular proteins and adds strength to the proposal (Fontana et al., 1986, 1993) that flexible loops, and never helices, are the sites of proteolytic attack. There has been considerable controversy on the presence, role and structure of the central helix of CaM in current literature. The results of a simple and clear-cut experiment of selective tryptic hydrolysis of CaM (Draibikowski et al., 1982) has anticipated the key conclusions on the structure of CaM achieved by NMR, X-ray, small angle X-ray scattering, time-resolved fluorescence, chemical crosslinking experiments, as well as molecular dynamics calculations (Persechini and Kretsinger, 1988; Ehrhardt et al., 1995; James et al., 1995).

Removing Loose, Flexible Parts from a Protein

The N- and C-terminal ends of proteins usually are more flexible than the rest of the chain, as shown by the results of X-ray analysis of proteins in the solid state, as well as in solution by utilizing spectroscopic methods. This explains the fact that several proteins possess ragged N- and C-terminal ends, as a result of a relatively easy proteolysis at these chain sites mostly by the action of exopeptidase. The possibility of removing the loose parts from a protein moiety by proteolysis has been tested in a number of cases. Below we will comment a few specific examples.

Hirudin is the most potent and specific inhibitor of the blood-clotting enzyme thrombin. A number of hirudin variants have been isolated mostly from the blood sucking leach *Hirudo medicinalis* (Stringer and Lindenfed, 1992; Tapparelli et al., 1993). All of them consist of a single polypeptide chain of 63-66 amino acid residues with three disulfide bridges. Both X-ray (Priestle et al., 1993) and NMR (Folkers et al., 1989; Haruyama and Wüthrich, 1989) were used to elucidate the structure of hirudin from *H. medicinalis* and evidence was provided that the hirudin's polypeptide is constituted by a N-terminal rigid core and a C-terminal disordered tail. Limited proteolysis was used to probe the structure of hirudin from *H. medicinalis* (Chang, 1990) and from *H. manillensis* (Vindigni et al., 1994). The disordered C-terminal tail of hirudin, variant HM2 from *H. manillensis* constituted by a 64-residue chain, has been removed by proteolysis using a variety of proteases (subtilisin, thermolysin, trypsin, V8-protease), thus producing fragments containing the well-structured N-terminal core domain (fragments 1-41 and 1-47) (Vindigni et al., 1994).

Several C-terminal fragments of thermolysin are capable of folding into a native-like structure independently from the rest of the polypeptide chain, thus possessing protein domain properties. These fragments of thermolysin were prepared by cyanogen bromide cleavage of the protein at the level of methionine residues in position 120 and 205 of the 316-residue chain of thermolysin (Fontana, 1990; Vita and Fontana, 1982; Dalzoppo et al., 1985). Of course the dimensions of the fragments, for example the most studied fragment 206-316, were dictated by the location of the Met residue in the protein chain. With the aim to define the minimum size of a C-terminal fragment capable to acquire a stable, native-like conformation, fragment 206-316 was digested by several proteases (Dalzoppo et al., 1985). From the kinetics of proteolysis digestion and analysis of the isolated subfragments, it was found that the rather short fragment 255-316 (62-residues, lacking disulfide bridges) was quite resistant to further proteolysis, implying a tightly folded conformation. Indeed, both circular dichroism (Dalzoppo et al., 1985) and, in particular, NMR measurements (Rico et al., 1994) provided a clear-cut evidence of a stable , native-like structure of this small thermolysin fragment.

In a recent study, Darby et al. (1996) followed the same rationale utilized in identifying the minimum size of a folded fragment of thermolysin. The putative structural domains of protein disulfide isomerase (PDI) have been prepared using protein engineering methods (fragments 1-157 and 148-257) and then they were subjected to limited proteolysis in order to remove the surplus residues. The proteolytic trimming of the recombinant PDI fragments was conducted utilising a variety of proteases. The procedure allowed the production and isolation of the most stable protein moieties of the fragments and thus to define the boundaries of the PDI domains.

Location of Sites of Flexibility Along the Protein Chain

Limited proteolysis can be used to detect flexible sites of the polypeptide chain of a globular protein, thus representing a useful technique to study protein mobility, complementing experimental (NMR, X-ray diffraction, fluorescence, hydrogen exchange) and theoretical approaches in more common use (Gurd and Rothgeb, 1979; Frauenfelder et al., 1979; Sternberg et al., 1979). Indeed, a native globular protein is not a static entity, but rather one which undergoes conformational fluctuations about its most stable conformation, as clearly emphasized by numerous studies using a variety of physical methods. Crystallographic results have provided direct evidence of intramolecular mobility in proteins or of a static/dynamic disorder, as indicated by the fact that certain parts of the protein molecule are poorly resolved in the electron density maps. Specific evidence for motion in proteins derives from the determination of the crystallographic temperature factors (B-values), which represent the mean-square displacement of each atom. When plotted against residue number, the profile of B-values provides a graphic image of the degree of mobility along the protein chain (in the solid state). In general B-values are low in helical regions and high at loops and chain termini (Frauenfelder et al., 1979; Sternberg et al., 1979; Ringe and Petsko, 1985; Holmes and Matthews, 1982).

The most clear-cut evidence of the possibility to probe protein mobility by proteolysis came from the results of a systematic study of limited proteolysis and autolysis of thermolysin (Fontana et al., 1986; Vita et al., 1988). This enzyme is an exceptionally stable metallo-endopeptidase isolated from *B. thermoproteolyticus* which contains a functional zinc ion and four calcium ions bound to the 316-residue polypeptide chain of known three-dimensional structure at 1.6 Å resolution (Holmes and Matthews, 1982). Nicked thermolysin species constituted by two- (Vita et al., 1985) as well as three-fragments (Fassina et al., 1986) associated in a stable complex were isolated and characterized. The isolation of several nicked species was made possible by the fact that, even after nicking of the thermolysin polypeptide chain, the resulting protein species is stable enough to resist further proteolysis and degradation, as normally observed with the labile proteins of mesophilic origin. A close examination of the three-dimensional model of the thermolysin molecule (Holmes and Matthews, 1982) reveals that all sites of limited proteolysis occur at exposed loops and not within chain segments of regular secondary structure (helices, sheets) and, in particular, that there is a striking correlation between sites of proteolysis and sites of high segmental mobility, these last given by the temperature factors (B-values) determined crystallographically (Fontana et al., 1986) (see Fig. 3).

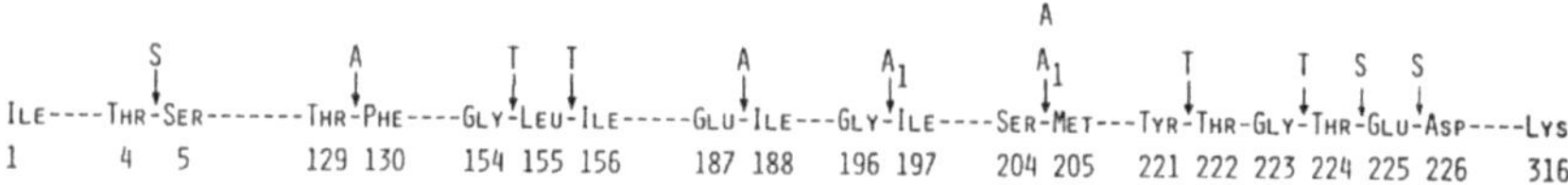

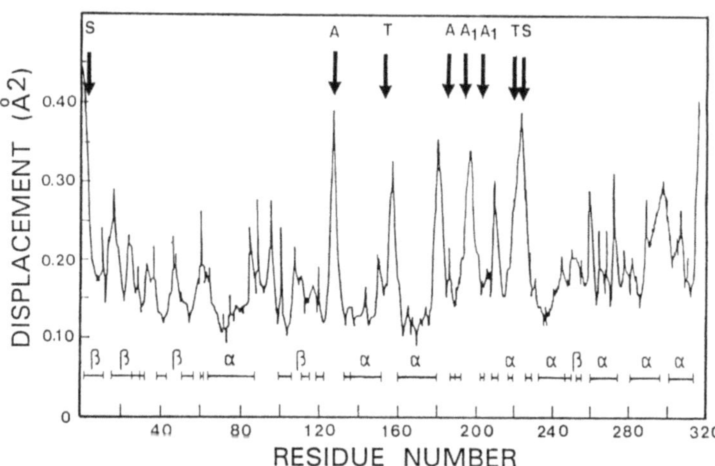

Fig. 3. (*Top*). Schematic representation of the amino acid sequence at the sites of limited proteolysis of thermolysin by subtilisin (S) and autolysis by heat (T) or in the presence of EDTA (A and A₁). S, site of cleavage by subtilisin; A, cleavage by autolysis in presence of 1.5 mM CaCl₂ and 1 mM EDTA; A₁, cleavage by autolysis in presence of 1.5 mM CaCl₂ and 10 mM EDTA; T, cleavage by thermal autolysis (see Fontana et al., 1986, for experimental details). (*Bottom*). Plot of the average main-chain temperature factors (solid line) along the polypeptide chain of thermolysin (adapted from Holmes and Matthews, 1982). Bars at the bottom of the figure indicate segments of secondary structure (helices and strands). Arrows indicate sites of limited proteolysis or autolysis of thermolysin observed under different experimental conditions.

The results of proteolysis of thermolysin are similar to those obtained with other proteins. A detailed survey of a number of cases of limited proteolysis of proteins of known three-dimensional structure revealed that, as a rule, the proteolysis sites are located at loops of the polypeptide chain and moreover that these sites are mobile, as determined by X-ray methods (Fontana et al., 1989b, 1993). The general conclusion is that flexible loops are the most favored sites for the protein-protein recognition phenomenon between the proteolytic enzyme and the globular protein substrate. Several examples of limited proteolysis occurring at flexible hinges, poorly resolved in the electron density map, have been described in literature. Indeed, the correlation between the flexibility/motility determined by X-ray methods and that determined by proteolysis experiments is quite striking, on one hand therefore demonstrating the utility of the proteolytic approach and on the other that protein mobility seen in protein crystals is relevant to motion solution. Fontana et al. (1993) have discussed several examples of protein flexibility sites detected by both X-ray methods and proteolysis. In the following, an additional example will be illustrated in detail to emphasize the utility of the proteolytic approach to pinpoint sites of protein flexibility.

Human growth hormone (hGH) was subjected to the action to proteolytic digestion by a variety of proteolytic enzymes with the aim to prepare hGH-fragments maintaining some of the biological properties of the intact hormone. These studies identified a sequence region (residues 134-147) in the hGH molecule highly susceptible to limited proteolysis. Figure 4 summarizes the sites of selective proteolytic cleavage of hGH by trypsin, plasmin, thrombin and subtilisin (Li, 1982). More recently, a kinetic analysis of the proteolytic digestion process by V8-protease has shown that among the 14 possible sites of proteolysis of the hGH chain, Glu^{33} was the site of preferential attack, followed by Glu^{56} and Glu^{66} (Polverino de Laureto et al., 1995c). The determination of the crystal structure of hGH complexed with the extracellular domain of its receptor (De Vos et al., 1992) permits a structural evaluation of the proteolysis data. Interestingly, the structure of the complex consists of one molecule of hGH per two molecules of receptor. The topography of hGH is a four-helix bundle (residues 9-34, 72-92, 106-128 and 155-184) with three short helices. Crystallographic data indicated a quite disordered segment connecting helix 3 (residues 106-128) and 4 (residues 155-184). In particular, the chain segment 147-153 was not included in the final model of hGH, since this region is not visible in the electron density map and thus disordered. It is stricking to observe therefore that all sites of limited proteolysis of hGH occur at chain segments which appear to be highly mobile in the crystal structure of the protein (see Fig. 4). Of note, the Glu-specific V8-protease (Drapeau, 1977) does cleave at a flex-

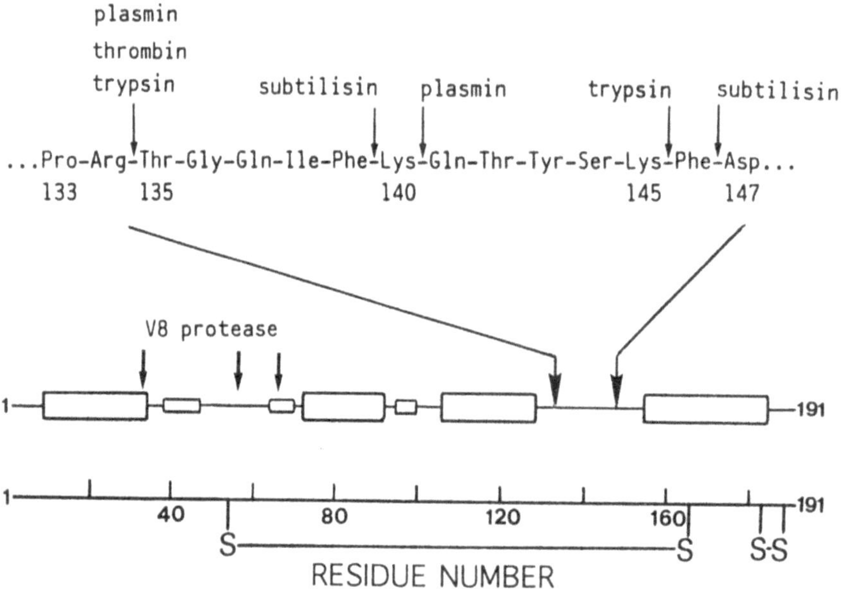

plasmin
thrombin
trypsin
subtilisin
plasmin
trypsin
subtilisin
...Pro-Arg-Thr-Gly-Gln-Ile-Phe-Lys-Gln-Thr-Tyr-Ser-Lys-Phe-Asp...
133   135   140   145   147
V8 protease
1
191
1
40
80
120
160
191
S
S
S·S
RESIDUE NUMBER

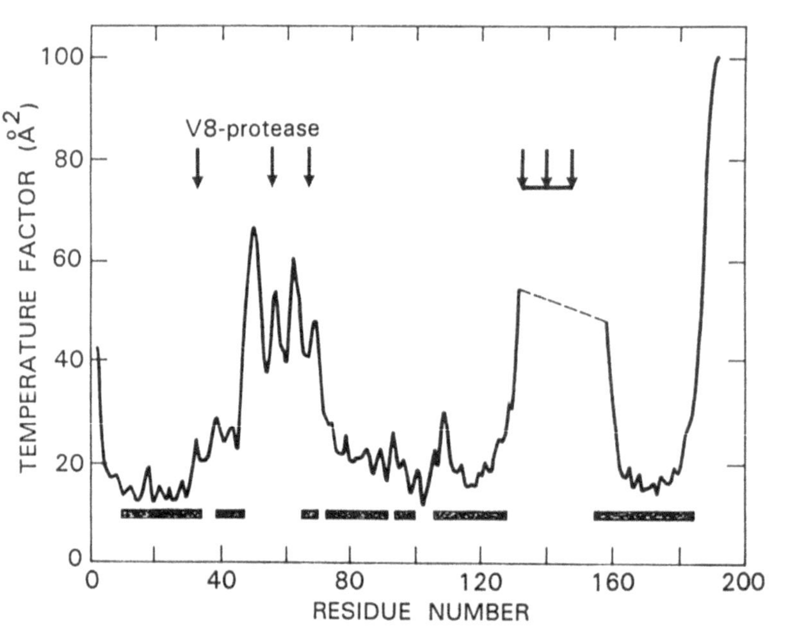

V8-protease
TEMPERATURE FACTOR (Å²)
100
80
60
40
20
0
0   40   80   120   160   200
RESIDUE NUMBER

ible region, but not at the most disordered/flexible chain segment 133-147, since in this last region there is no Glu residues to be attacked by the Glu-specific V8-protease.

Probing of Structure of Partly Folded States of Proteins

There is nowadays a general consensus that the folding of a polypeptide chain of a relatively large globular protein into its unique three-dimensional and functionally active structure occurs via folding intermediates (Creighton et al., 1996). These partly folded states of proteins are difficult to characterize, because they are usually short-lived or exist as a distribution of possible conformers. A variety of experimental techniques and approaches have been utilized in recent years for characterizing these partly folded states, such as spectroscopic techniques (circular dichroism, NMR, fluorescence), calorimetry, gel filtration chromatography, as well as genetic methods and theoretical calculations. More recently, proteolytic enzymes have been used as probes of the structural features of partly folded states of proteins, leading to experimental results complementing those obtained by other most commonly used techniques (Polverino de Laureto et al., 1995 a,b; 1997; Fontana et al., 1997).

A protein species which appears to possess the properties of a kinetic folding intermediate is the "molten globule state" of a protein, which can be generated from a globular protein under specific and mildly denaturant solvent conditions and occurs at the equilibrium, thus allowing the analysis of its structural features (Kuwajima, 1989; Ptitsyn, 1992, 1995). The

---

Fig. 4. (*Top*). Scheme of the secondary structure and sites of limited proteolysis along the polypeptide chain of human growth hormone (hGH). Boxes indicate helical segments seen in the crystal structure of hGH in the complex with its receptor (De Vos et al., 1992). Major boxes refer to the helical segments of the four-helix boundle of the hormone, whereas the small boxes are short helical segments seen in the complex of hGH with its receptor. The sites of limited proteolysis (initial nicking) are those observed by treating the hormone with various proteases (Li, 1982) or V8-protease (Polverino de Laureto et al., 1995c). At the bottom is shown the location of disulfide bonds along the polypeptide chain of the hormone. (*Bottom*). Profile of the temperature factor along the polypeptide chain of hGH, as determined by X-ray analysis of the complex of hGH with its receptor (De Vos et al., 1992). Arrows indicate sites of proteolytic cleavages by V8 protease (region 30-70) (Polverino de Laureto et al., 1995c) and by a variety of other proteases (region 134-147) (Li, 1982). The chain segment 134-155 of hGH is flexible and disordered in the crystal of the hGH-receptor complex (De Vos et al., 1992). Solid bars at the bottom of the figure indicate the helical segments along the chain of hGH determined crystallographically.

molten globule state is nowdays a controversial issue, but nevertheless some key characteristics of this state appear to be a well-defined and native-like secondary structure, lack of specific tertiary interactions and a more expanded and flexible structure in respect to that of the native protein. The most extensively characterized molten globule is the one obtained by acid-unfolding of α-lactalbumin at pH 2.0 (A-state). Limited proteolysis by pepsin has been used to probe the A-state of α-lactalbumin and it was shown that pepsin cleaves rapidly the Ala^{40}-Ile^{41} peptide bond and subsequently Leu^{52}-Phe^{53} (Polverino de Laureto et al., 1995b). This fast initial nicking at region 40-53 of α-lactalbumin implies that this region in the A-state of the protein is highly flexible, as required for a productive interaction with the active site of pepsin. Indeed, the absence of secondary structure at region 40-53 (β-sheet in the native protein at pH 7.0) in the A-state of α-lactalbumin was also previously inferred from NMR measurements (Alexandrescu et al., 1994; see Kuwajima, 1996, for additional references).

In a recent study (Fontana et al., 1997), limited proteolysis experiments have been used to probe the structural and dynamic differences between the holo and apo form of horse myoglobin (Mb). In the past, for simplicity, the structure of apoMb has often been assumed to be similar that of the native, heme-containing holoMb. However, circular dichroism data indicate a partial loss (20%) of the helical content of the protein upon removal of the heme. Moreover, at neutral pH apoMb is less compact and less stable than holoMb and seems to possess some molten globule-like properties (Lin et al., 1994), even if the results of a recent NMR study have indicated that the conformational state of apoMb does not adhere to the classical description of a molten globule (Eliezer and Wright, 1996). A variety of proteases (subtilisin, thermolysin, chymotrypsin and trypsin) initially cleave apoMb at the level of chain segment 89-96, whereas holoMb is fully resistant to proteolysis if treated under identical experimental conditions (Fontana et al., 1997). Such selective proteolysis implies that the F-helix of native holoMb (residues 82-97) is disordered in apoMb, thus enabling binding and adaptation of this chain segment at the active site of the proteolytic enzymes for an efficient peptide bond fission. That essentially only the F-helix in apoMb is largely disrupted was earlier inferred from spectroscopic measurements (Lecomte et al., 1996; Eliezer and Wright, 1996) and molecular dynamics simulations (Brooks, 1992; Tirado-Rives and Jorgensen, 1993; Yang and Honig, 1994). Therefore, the results of the proteolysis experiments conducted on apoMb provide direct experimental evidence for this, again emphasizing that limited proteolysis is a useful and reliable method for probing structure and dynamics of proteins.

Proteins when dissolved in aqueous trifluoroethanol (TFE) acquire a stable, partially folded state with a high content of α-helical conformation, but lacking the specific tertiary interactions of the native protein (Buck et al., 1993, 1995; Alexandrescu et al., 1994; Shiraki et al., 1994). This TFE-state appears to be a non-compact, expanded conformational state characterized by an ensemble of fluctuating helices (Buck et al., 1995). Recently, the conformational features of proteins dissolved in aqueous TFE was analyzed by a limited proteolysis approach utilizing thermolysin (Polverino de Laureto et al., 1995a,b, 1997). This protease appeared to be a most suitable proteolytic probe, because of its noteworthy stability under relatively harsh solvent conditions, including organic solvents (Welinder, 1988), and broad substrate specificity (Heinrikson, 1977). A variety of model proteins have been reacted with thermolysin in 50% (v/v) TFE at pH 7.0, 20-52°C, for several hours and then the protein fragments were isolated and analyzed for their identity (Fontana et al., 1995; Polverino de Laureto et al., 1995a,b, 1997). The results obtained with ribonuclease, lysozyme, cytochrome c and α-lactalbumin indicated that thermolysin cleaves these proteins in their TFE-state selectively at very few peptide bonds. In each case it was possible to show that a single peptide bond was cleaved first, followed by additional cleavages of a couple of other peptide bonds only. For example, bovine pancreatic ribonuclease A was digested with thermolysin in 50% TFE at 42°C for 30 min and from the proteolytic mixture a nicked ribonuclease selectively cleaved at the Asn^{34}-Leu^{35} peptide bond was isolated to homogeneity (Polverino de Laureto et al., 1997). When proteolysis was prolonged up to six hours, a nicked protein with the additional peptide bond Thr^{45}-Phe^{46} was isolated. Similarly, horse cytochrome c (Fontana et al., 1995) is initially cleaved by thermolysin in 50% TFE at the Gly^{56}-Ile^{57} peptide bond, hen egg-white lysozyme at Lys^{97}-Ile^{98} (Polverino de Laureto et al., 1995b) and bovine α-lactalbumin at Ala^{40}-Ile^{41} (Polverino de Laureto et al., 1995a). Proteolysis by thermolysin occurs at the amino-side of Ile, Leu and Phe, even if this protease in aqueous solution cleaves at a variety of other residues (Heinrikson, 1977; Keil, 1982).

Nevertheless, even if the proteins investigated contain many Ile, Leu and Phe residues, thermolysin cleaves specifically at only very few sites. Accepting our view that helical chain segments are not cleaved by the proteolytic probe (see above), the selective proteolytic digestion of the model proteins in their TFE-state by thermolysin is consistent with the fact that this conformational state is highly helical (for a discussion on the mechanism of selective proteolysis of proteins by thermolysin in 50% aqueous TFE see Polverino de Laureto et al., 1995a, 1997).

Concluding Remarks

The ability of a protease to interact and cleave at a specific chain segment of an intact folded protein is best understood as a dynamic process in which the chain segment must fit the precise stereochemistry of the active site of the protease in order to properly align functional/reactive groups of both enzyme and substrate polypeptide chain (induced fit mechanism; Herschlag, 1988). This process requires a significant distortion (local unfolding) of the polypeptide region suffering proteolysis (Hubbard et al., 1994) and consequently high chain flexibility (Fontana et al., 1986, 1989a,b, 1993). This is the reason why proteolytic enzymes can be used to probe both the overall and the local unfolding of the protein. The examples above discussed clearly show that proteases can pinpoint in globular proteins flexible sites (loops) with a high degree of confidence, thus complementing other physicochemical (X-ray, NMR) or computing methods in use. Specific advantages of the use of limited proteolysis for probing structure and dynamics of proteins are that the technique is simple to use, requires minute amounts of protein material, can be performed in a general biochemical laboratory and therefore does not necessarily require expensive instrumentation. The selected applications outlined in this chapter illustrate the tremendous power of the technique in analyzing conformational and even dynamic aspects of proteins.

In the past, the limited proteolysis technique was somewhat difficult to apply, since the analytical methods required to isolate and characterize protein fragments were labor-intensive and not sensitive enough. The availability of automatic, efficient and highly sensitive techniques of protein sequencing and, in particular, the recent dramatic advances of mass spectrometry (Andersen et al., 1996; Wilm et al., 1996) in analyzing peptides and proteins, are likely to prompt a more systematic use of the limited proteolysis approach as a simple first step in the elucidation of structure-dynamics-function relationships of a novel and rare protein, especially if available in minute amounts.

References

Ahmed SA, Fairwell T, Dunn S, Kirschner K, Miles EW (1986) Identification of three sites of proteolytic cleavage in the hinge region between the two domains of the $_2$ subunit of tryptophan synthase of *Escherichia coli* and *Salmonella typhimurium*. Biochemistry 25:3118-3124

Alexandrescu AT, Ng Y-L, Dobson CM (1994) Characterization of a TFE-induced partially folded state of α-lactalbumin. J Mol Biol 235:587-599

Andersen JS, Svensson B, Roepstorff P (1966) Electrospray ionization and matrix assisted laser desorption/ionization mass spectrometry: Powerful analytical tools in recombinant protein chemistry. Nature Biotech 14:449-557

Arnold U, Rücknagel KP, Schierhorn A, Ulbrich-Hofman R (1996) Thermal unfolding and proteolytic susceptibility of ribonuclease A. Eur J Biochem 237:862-869

Babu YS, Bugg CE, Cook WJ (1988) Structure of calmodulin refined at 2.2 <Å resolution. J Mol Biol 204:191-204

Babu YS, Sack JS, Greenhough TJ, Bugg CE, Means AR, Cook WJ (1985) Three-dimensional structure of calmodulin. Nature 315:37-40

Barbato G, Ikura M, Kay LE, Pastor RW, Bax A (1992) Backbone dynamics of calmodulin studied by ^{15}N-relaxation using inverse detected two-dimensional NMR spectroscopy: The central helix is flexible. Biochemistry 31:5269-5278

Bennett WS, Huber R (1984) Structural and functional aspects of domain motions in proteins. CRC Crit Rev Biochem 15:291-384

Bond JS (1990) Commercially available proteases. In: Beynon RJ, Bond JS (eds) Proteolytic enzymes: A practical approach. IRL Press, Oxford, pp 232-240

Brockerhoff SE, Edmonds CG, Davis TN (1992) Structural analysis of wild-type and mutant yeast calmodulins by limited proteolysis and electrospray ionization mass spectrometry. Protein Sci 1:504-516

Brooks CL (1992) Characterization of "native" apomyoglobin by molecular dynamics simulation. J Mol Biol 227:375-380

Buck M, Radford SE, Dobson CM (1993) A partially folded state of hen egg-white lysozyme in trifluoroethanol: Structural characterization and implications for protein folding. Biochemistry 32:669-678

Buck M, Schwalbe H, Dobson CM (1995) Characterization of conformational preferences in a partly folded protein by heteronuclear NMR spectroscopy: Assignment and secondary structure analysis of hen egg-white lysozyme in trifluoroethanol. Biochemistry 3:13219-13232

Chang JY (1990) Production, properties and thrombin inhibitory mechanism of hirudin amino-terminal core fragments. J Biol Chem 265:22159-22166

Chattopadhyaya R, Meador WE, Means AR, Quiocho F (1992) Calmodulin structure refined at 1.7 Å resolution. J Mol Biol 228:1177-1192

Cohen SL, Ferrè-D'Amarè AR, Burley KS, Chait BT (1995) Probing the solution structure of the DNA-binding protein Max by a combination of proteolysis and mass spectrometry. Protein Sci 4:1088-1099

Creighton TE, Darby NJ, Kemmink J (1996) The role of partly folded intermediates in protein folding. FASEB J 10:110-118

Darby NJ, Kemmink J, Creighton TE (1996) Identifying and characterizing a structural domain of protein disulfide isomerase. Biochemistry 35:10517-10528

Dalzoppo D, Vita C, Fontana A (1985) Folding of thermolysin fragments: Identification of the minimum size of a carboxyl-terminal fragment that can fold into a stable native-like structure. J Mol Biol 182:331-340

De Vos AM, Ultsch M, Kossiakoff AA (1992) Human growth hormone and extracellular domain of its receptor: Crystal structure of the complex. Science 255:306-312

Draibikowski W, Brzeska H, Venyaminov SY (1982) Tryptic fragments of calmodulin. J Biol Chem 257:11584-11590

Drapeau GR (1977) Cleavage at glutamic acid with staphylococcal protease. Methods Enzymol 47:189-191

Ehrhardt MR, Urbauer JL, Wand AJ (1995) The energetics and dynamics of molecular recognition by calmodulin. Biochemistry 34:2731-2738

Eliezer D, Wright PE (1996) Is apomyoglobin a molten globule? Structural characterization by NMR. J Mol Biol 263:531-538

Evans SV, Brayer GD (1990) High-resolution study of the three-dimensional structure of horse heart metmyoglobin. J Mol Biol 213:885-897

Fassina G, Vita C, Dalzoppo D, Zamai M, Zambonin M, Fontana A (1986) Autolysis of thermolysin: Isolation and characterization of a folded three-fragment complex. Eur J Biochem 156:221-228

Folkers PJM, Clore GM, Driscoll PC, Dodt J, Kohler S, Gronenborn AM (1989) Solution structure of recombinant hirudin and Lys47Glu mutant: A nuclear magnetic resonance and hybrid geometry-dynamical simulated annealing study. Biochemistry 28:2601-2617

Fontana A (1989b) Limited proteolysis of globular proteins occur at exposed and flexible loops. In: Kotyk A, Skoda J, Pacek V, Kostka W (eds) Highlights of modern biochemistry. Zeist, Amsterdam, The Netherlands, VSP Int Publ, pp 1711-1726

Fontana A (1990) Independent folding of protein domains. In: Rivier JE, Marshall G (eds) Peptides: Chemistry, structure and biology. ESCOM, Leiden, the Netherlands, pp 557-565

Fontana A (1991) How nature engineers protein (thermo)stability. In: Di Prisco G (ed) Life under extreme conditions. Springer-Verlag, Berlin-Heidelberg, pp 89-113

Fontana A, Fassina G, Vita C, Dalzoppo D, Zamai M, Zambonin M (1986) Correlation between sites of limited proteolysis and segmental mobility in thermolysin. Biochemistry 25:1847-1851

Fontana A, Polverino de Laureto P, De Filippis V (1993) Molecular aspects of proteolysis of globular proteins. In: Van den Tweel W, Harder A, Buitelaar M (eds) Protein stability and stabilization. Elsevier Science Publ, Amsterdam, pp 101-110

Fontana A, Vita C, Dalzoppo D (1985) Hierarchic organization of globular proteins: Experimental studies on thermolysin. J Biosci 8:57-66

Fontana A, Vita C, Dalzoppo D, Zambonin M (1989) Limited proteolysis as a tool to detect structure and dynamic features of globular proteins: Studies on thermolysin. In: Wittman-Liebold B (ed) Methods in protein sequence analysis. Springer-Verlag, Berlin, pp 315-324

Fontana A, Zambonin M, De Filippis V, Bosco M, Polverino de Laureto P (1995) Limited proteolysis of cytochrome c in trifluoroethanol. FEBS Lett 362:266-270

Fontana A, Zambonin M, Polverino de Laureto P, De Filippis V, Clementi A, Scaramella E (1997) Probing the conformational state of apomyoglobin by limited proteolysis. J Mol Biol 266:223-230

Frauenfelder H, Petsko GA, Tsernoglou D (1979) Temperature-dependent X-ray diffration as a probe of protein structural dynamics. Nature 280:558-563

Goldberg ME (1969) Tertiary structure of *Escherichia coli* β-galactosidase. J Mol Biol 46:441-446

Gurd FRN, Rothgeb TM (1979) Motions in proteins. Adv Protein Chem 33:73-165

Haruyama H, Wütrich K (1989) Conformation of recombinant desulfatohirudin in aqueous solution determined by nuclear magnetic resonance. Biochemistry 28:4301-4312

Heidorn DB, Trewhella J (1988) Comparison of the crystal and solution structures of calmodulin and troponin c. Biochemistry 27:909-915

Heinrikson RL (1977) Applications of thermolysin in protein structural analysis. Methods Enzymol 47:175-189

Herschlag D (1988) The role of induced fit and conformational changes of enzymes in specificity and catalysis. Bioorganic Chem 16:62-96

Hess D, Covey TC, Winz R, Brownsey RW, Aebershold R (1993) Analytical and micro-preparative peptide mapping by high performance liquid chromatography/electrospray mass spectrometry of proteins purified by gel electrophoresis. Protein Sci 2:1342-1351

Holmes MA, Matthews BW (1982) Structure of thermolysin refined at $1.6 < \text{Å}$ resolution. J Mol Biol 160:623-639

Hubbard SJ (1998) The structural aspects of limited proteolysis of native proteins. Biochem Biophys Acta 1382:191-206

Hubbard SJ, Campbell SF, Thornton JM (1991) Molecular recognition: Conformational analysis of limited proteolytic sites and serine proteinase protein inhibitors. J Mol Biol 220:507-530

Hubbard SJ, Eisenmenger F, Thornton JM (1994) Modeling studies of the change in conformation required for cleavage of limited proteolysis sites. Protein Sci 3:757-768

Hubbard SJ, Thornton JM, Campbell SF (1992) Substrate recognition by proteinases. Faraday Discuss 93:13-23

Ikura M, Clore GM, Gronenborn AM, Zhu G, Klee CB, Bax A (1992) Solution structure of a calmodulin-target peptide complex by multidimensional NMR. Science 256:632-638

Imoto T, Yamada H, Ueda T (1986) Unfolding rates of globular proteins determined by kinetics of proteolysis. J Mol Biol 190:647-649

James P, Vorherr T, Carafoli E (1995) Calmodulin-binding domains: Just two-faced or multi-faceted? Trends Biol Sci 20:38-42

Karas M, Hillenkamp F (1988) Laser desorption ionization of proteins with molecular masses exceeding 10,000 Daltons. Anal Chem 60:2299-2301

Karplus PA, Schulz GE (1985) Prediction of chain flexibility in proteins: A tool for the selection of peptide antigens. Naturwissenschaften 72:212-214

Keil B (1982) Enzymic cleavage of proteins. In: Elzinga M (ed) Methods in protein sequence analysis, vol IV. Humana Press, Clifton, NJ, pp 291-304

Krokozsynska I, Otlewski J (1996) Thermodynamic stability effects of single peptide bond hydrolysis in protein inhibitors of serine proteases. J Mol Biol 256:793-802

Kuwajima K (1989) The molten globule state as a clue for understanding the folding and cooperativity of globular-protein structure. Proteins Struct Funct Genet 6:87-103

Kuwajima K (1996) The molten globule state of α-lactalbumin. FASEB J 10:102-109

Laemmli UK (1970) Cleavage of structural proteins during the assembly of the head of bacteriophage T_4. Nature 227:680-685

Lecomte JTJ (1991) NMR/X-ray: An overview. In: Villafranca JJ (ed) Techniques in protein chemistry, San Diego, CA (USA), Academic Press, pp 337-345

Lecomte JTJ, Kao YH, Cocco MJ (1966) The native state of apomyoglobin described by proton NMR spectroscopy: The A-B-G-H interface of wild-type sperm whale apomyoglobin. Proteins: Struct Funct Genet 25:267-285

Li CH (1982) Human growth hormone:1974-1981. Mol Cell Biochem 46:31-41

Lill U, Schreil A, Henschen A, Eggerer H (1984) Hysteretic behaviour of citrate synthase: Site-directed limited proteolysis. Eur J Biochem 143:205-217

Lin L, Pinker RJ, Forde K, Rose GD, Kallenbach NR (1994) Molten globular characteristics of the native state of apomyoglobin. Nature Struct Biol 1:447-451

Lin X, Lay JA, Sussman F, Tang J (1993) Conformational instability of the N- and C-terminal lobes of porcine pepsin in neutral and alkaline solutions. Protein Sci 2:1383-1390

MacArthur MW, Driscoll PC, Thornton JM (1994) NMR and crystallography: Complementary approaches to structure determination. Trends Biotechnol 12:149-153

Mann M, Wilm M (1995) Electrospray mass spectrometry for protein characterization. Trends Biol Sci 20:219-224

Matsudaira P (1987) Sequence from picomole quantities of proteins electroblotted onto polyvinylidene difluoride membranes. J Biol Chem 262:10035-10038

McLendon G, Radany E (1978) Is protein turnover thermodynamically controlled? J Biol Chem 253:6335-6337

Meador WE, Means AR, Quiocho FA (1992) Target enzyme recognition by calmodulin: 2.4 Å structure of a calmodulin-peptide complex. Science 257:1251-1255

Mihalyi E (1978) Application of proteolytic enzymes to protein structure studies. CRC Press, Boca Raton, FL

Mohsen AA, Aull JL, Payne DM, Daron HM (1995) Ligand-induced conformational changes of thymidylate synthase detected by limited proteolysis. Biochemistry 34:1669-1677

Morihara K, Oka T, Tsuzuki H (1970) Subtilisin BPN': kinetic study with oligopeptides. Arch Biochem Biophys 138:515-525

Neurath H (1980) Limited proteolysis, protein folding and physiological regulation. In: Jaenicke R (ed) Protein folding. Elsevier/North Holland Biomedical Press, Amsterdam, New York, pp 501-504

Neurath H (1986) Limited proteolysis, domains, and the evolution of protein structure. Chemica Scripta 27B:221-229

Novotny J, Bruccoleri RE (1987) Correlation among sites of limited proteolysis, enzyme accessibility and segmental mobility. FEBS Lett 211:185-189

Ottesen M, Svendsen I (1970) The subtilisins. Methods Enzymol 19:199-215

Pace N, Barrett AJ (1984) Kinetics of trypic hydrolysis of the arginine-valine bond in folded and unfolded ribonuclease T_1. Biochem J 219:411-417

Persechini A, Kretsinger RH (1988) The central helix of calmodulin functions as a flexible tether. J Biol Chem 263:12175-12178

Polverino de Laureto P, De Filippis V, Bertolero F, Orsini G, Fontana A (1994) Probing the structure of a human interleukin-6 mutant by limited proteolysis. In: Crabb JW (ed) Techniques in protein chemistry, vol V. Academic Press, New York, pp 381-388

Polverino de Laureto P, De Filippis V, Di Bello M, Zambonin M, Fontana A (1995a) Probing the molten globule state of α-lactalbumin by limited proteolysis. Biochemistry 34:12596-12604

Polverino de Laureto P, De Filippis V, Scaramella E, Zambonin M, Fontana A (1995b) Limited proteolysis of lysozyme in trifluoroethanol. Isolation and characterization of a partially active enzyme derivative. Eur J Biochem 230:779 787

Polverino de Laureto P, Toma S, Tonon G, Fontana A (1995c). Probing the structure of human growth hormone by limited proteolysis. Int J Peptide Protein Res 45:200-208

Polverino de Laureto P, Scaramella E, De Filippis V, Bruix M, Rico M, Fontana A (1997) Limited proteolysis of ribonuclease with thermolysin in trifluoroethanol. Protein Sci 6:860-872

Price NC, Johnson CM (1990) Proteinases as probes of conformation of soluble proteins. In: Beynon RJ, Bond JS (eds) Proteolytic enzymes. A practical approach. IRL Press, Oxford, pp 163-180

Priestle JP, Rahuel J, Rink H, Tones M, Grütter M (1993) Changes in interactions in complexes of hirudin derivatives and human -thrombin due to different crystal forms. Protein Sci 2:1630-1642

Ptitsyn OB (1992) The molten globule state. In: Creighton TE (ed) Protein Folding. Freeman, New York, pp 243-300

Ptitsyn OB (1995) Molten globule and protein folding. Adv Protein Chem 47:83-229

Remington S, Wiegand G, Huber R (1982) Crystallographic refinement and atomic models of two different forms of citrate synthase. J Mol Biol 158:111-129

Richards FM, Vithayathil PJ (1959) The preparation of subtilisin-modified ribonuclease and the separation of the peptide and protein components. J Biol Chem 234:1459-1464

Rico M, Jiménez MA, González C, De Filippis V, Fontana A (1994) NMR solution structure of the C-terminal fragment 253-316 of thermolysin: A dimer formed by subunits having the native structure. Biochemistry 33:14834-14847

Ringe D, Petsko GA (1985) Mapping protein dynamics by X-ray diffraction. Prog Biophys Mol Biol 45:197-235

Salvesen G, Nagase H (1989) Inhibition of proteolytic enzymes. In: Beynon RJ, Bond JS (eds) Proteolytic enzymes: A practical approach. IRL Press, Oxford, pp 83-104

Sandmeier E, Christen P (1980) Mitochondrial aspartate aminotransferase 27/32-410: Partially active enzyme derivative produced by limited proteolytic cleavage of native enzyme. J Biol Chem 255:10284-10289

Schägger H, von Jagow G (1987) Tricine sodium dodecyl sulfate polyacrylamide gel electrophoresis for the separation of proteins in the range from 1 kDa to 100 kDa. Anal Biochem 166:368-379

Schechter I, Berger A (1967) On the size of the active site in proteases. I. Papain. Biochem Biophys Res Commun 27:157-162

Seielstad DA, Carlson KE, Kushner PJ, Greene GL, Katzenellenbogen JA (1995) Analysis of the structural core of the human estrogen receptor ligand binding domain by selective proteolysis/mass spectrometric analysis. Biochemistry 34:12605-12615

Shiraki K, Nishikawa K, Goto Y (1994) Trifluoroethanol-induced stabilization of the α-helical structure of α-lactoglobulin: Implications for non-hierarchical protein folding. J Mol Biol 245:180-194

Signor G, Vita C, Fontana A, Frigerio F, Bolognesi M, Toma S, Gianna R, De Gregoriis E, Grandi G (1990) Structural features of neutral protease from *Bacillus subtilis* deduced from model-building and limited proteolysis experiments. Eur J Biochem 189:221-227

Sternberg MJE, Grace DEP, Phillips DC (1979) Dynamic information from protein crystallography: An analysis of temperature factors from refinement of the hen egg-white lysozyme. J Mol Biol 130:231-253

Stringer KA, Lindenfeld JA (1992) Hirudins: Antithrombin anticoagulants. Ann Pharmacother 26:1535-1540

Tapparelli C, Metternich R, Ehrhardt C, Cook NS (1993) Synthetic low molecular mass thrombin inhibitors: Molecular design and pharmacological profile. Trends Pharmacol Sci 14:366-376

Tasayaco ML, Carey J (1992) Ordered self-assembly of polypeptide fragments to form native-like dimeric Trp repressor. Science 255:594-597

Tirado-Rives J, Jorgensen WL (1993) Molecular dynamics simulations of the unfolding of apomyoglobin in water. Biochemistry 32:4175-4184

Vindigni A, De Filippis V, Zanotti G, Visco C, Orsini G, Fontana A (1994) Probing the structure of hirudin from *Hirudinaria manillensis* by limited proteolysis: Isolation, characterization and thrombin-inhibitory properties of N-terminal fragments. Eur J Biochem 226:323-333

Vita C, Dalzoppo D, Fontana A (1985) Limited proteolysis of thermolysin by subtilisin: Isolation and characterization of a partially active enzyme derivative. Biochemistry 24:1798-1806

Vita C, Dalzoppo D, Fontana A (1988) Limited proteolysis of globular proteins: Molecular aspects deduced from studies on thermolysin. In: Chaiken I, Chiancone E, Fontana A, Neri P (eds) Macromolecular Biorecognition. Humana Press, Clifton NJ (USA), pp 57-67

Vita C, Fontana A (1982) Domain characteristics of the carboxyl-terminal fragment 206-316 of thermolysin: Unfolding thermodynamics. Biochemistry 21:5196-5202

Welinder KG (1988) Generation of peptides suitable for sequence analysis by proteolytic cleavage in reversed-phase high-performance liquid chromatography solvents. Anal Biochem 174:54-64

Wetlaufer DB (1981) Folding of protein fragments. Adv Protein Chem 34:61-92

Wilm M, Shevchenko A, Houthaeve T, Breit S, Schweigerer L, Fotsis T, Mann M (1996) Femtomole sequencing of proteins from polyacrylamide gels by nano-electrospray mass spectrometry. Nature 379:466-469

Wilson JE (1991) The use of monoclonal antibodies and limited proteolysis in elucidation of structure-function relationships in proteins. Methods Biochem Anal 351:207-250

Yang A-S, Honig B (1994) Structural origins of pH and ionic strength effects on protein stability: Acid denaturation of sperm whale apomyoglobin. J Mol Biol 237:602-614

Yang HJ, Tsou CL (1995) Inactivation during denaturation of ribonuclease A in guanidinium hydrochloride is accompanied by unfolding of the active site. Biochem J 305:379-384

Zappacosta F, Pessi A, Bianchi E, Venturini S, Sollazzo M, Tramontano A, Marino G, Pucci P (1996). Probing the tertiary structure of proteins by limited proteolysis and mass spectrometry. The case of Minibody. Protein Sci 5: 802-813

Chapter 16

Limited Proteolysis in the Study of Membrane Proteins

HASSAN Y. NAIM

Introduction

A substantial proportion of glycoproteins of mammalian cells is found associated with the surface membrane. Several modes of association are known. For example, transmembrane or integral proteins are linked with the membrane by specific stretches of the protein consisting predominantly of hydrophobic amino acids that function as membrane anchors and span the membrane at least once. Some other proteins are covalently attached to a phospholipid in the membrane through an oligosaccharide (denoted glycophosphatidyl inositol-(GPI)-anchored proteins). Association of glycoproteins with the cell surface and even the types of these associations could be determined by treatment of intact cells or vesicular membranes with limited dilutions of proteolytic enzymes, such as trypsin. For example, the cell surface expression of recombinant pro-lactase-phlorizin hydrolase (pro-LPH) in transfected COS-1 cells (Naim et al., 1991) or the hemagglutinin of influenza virus (HA) in infected CV-1 cells (Doyle et al., 1985; Naim and Roth, 1993) has been assessed by the susceptibilty of these proteins to limited dilutions of trypsin. The appearance of specifically cleaved molecular species of pro-LPH and HA is indicative for the presence of these molecules at the cell surface. The mode of association of angiotensin-converting enzyme (ACE) (Naim, 1992) and the Lyt-2/3 (CD-8) antigen of cytotoxic T-lymphocytes (Lüscher et al., 1985) has been characterized by incubation of inside-out membrane vesicles with limited amounts of trypsin. A reduction in the apparent molecular weight on SDS-gels of ACE or Lyt-2/3 upon limited proteolysis with trypsin is indicative of the presence of cytoplasmic tails and hence of a transmembrane orientation of these proteins.

Hassan Y. Naim, School of Veterinary Medicine Hannover, Institute of Physiological Chemistry, Bünteweg 17, Hannover, 30559, Germany (*phone* +49-511-9538780; *fax* +49-511-9538585; *e-mail* hnaim@biochemie.tiho-hannover.de)

Limited proteolysis with enzymes such as trypsin provides also an easy, cheap and powerful tool in the analysis of protein folding. Transmembrane, GPI-anchored and secretory proteins undergo contranslational and post-translational conformational and structural modifications in the ER which ultimately lead to the acquisition of correct tertiary and quaternary structure (for reviews see Rose and Doms, 1988; Hurtley and Helenius, 1989). Some of the most significant alterations are cotranslational glycosylation, disulfide bond formation and oligomerisation. Protease sensitivity assays could be used to discriminate between correctly folded and malfolded proteins. The rationale is that the exposure of specific sites accessible to proteolytic digestion is dependent on the specific conformation of the protein. A different proteolytic band pattern will be expected when two different conformations of a protein are subjected to proteolytic digestion. Malfolded proteins are even more susceptible to proteolytic digestion and, in contrast to correctly folded proteins, could be readily degraded by proteolytic enzymes.

In this chapter I shall present protocols in which trypsin has been used at limited dilutions to examine

- the cell surface expression of lactase-phlorizin hydrolase (LPH),

- the membrane association of angiotensin-converting enzyme (ACE), and

- the folding of LPH.
  The protocols are of general use and could be followed to address similar questions with other membrane proteins.

Outline

Outlines of the procedures used to study cell surface expression of a membrane surface (Subprotocol 1), the mode of association of a protein with the membrane (Subprotocol 2) and the folding of a membrane protein (Subprotocol 3) are shown in Figure 1, 2 and 3.

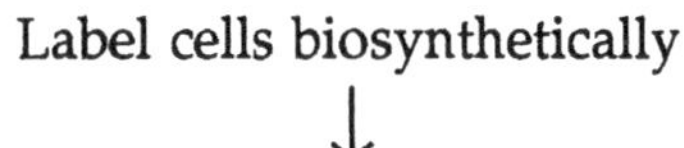

Fig. 1. An outline of the procedure used to examine the cell surface expression of a membrane protein

Label cells biosynthetically

Wash medium and incubate

in FCS-free medium

↓

Add trypsin and incubate

for 30 min

↓

Collect medium and prepare

cellular extracts

↓

Immunoprecipitate protein

of interest from medium and

cell extracts

↓

SDS-PAGE and

fluoro- /autoradiography

Fig. 2. An outline of the procedure used to study the mode of association of a protein with the membrane.

Label cells or tissue biosynthetically

↓

Prepare membrane vesicles

of the cell surface

(for example brush border membranes)

↓

Separate inside-out from right-side out

vesicles by lectin chromatography

using wheat germ agglutinin

↓

Treat half of the inside-out

vesicles with trypsin

↓

Make cellular lysates of trypsin-treated

and untreated vesicles

↓

Immunoprecipitate protein of interest

↓

SDS-PAGE and

fluoro-/autoradiography

Immunoprecipitate a control native protein

or other conformations (e.g. mutants)

of the same protein using protein A-Sepharose

↓

Resuspend immunoprecipitates in PBS

↓

Digest immunoprecipitates with trypsin

↓

Denature samples by boiling

in SDS-PAGE sample buffer

↓

Submit to SDS-PAGE and

fluoro-/autoradiography

Fig. 3. An outline of the procedure used to investigate the folding of a membrane protein.

Subprotocol 1
Cell Surface Expression of Lactase-Phlorizin Hydrolase (LPH)

Materials

- COS-1 cells (ATCC)   **Reagents**
- cDNA encoding LPH in pCMV-2 expression vector (Naim et al., 1991)
  (denoted pHI-6)
- Protein A-Sepharose (Pharmacia)
- DEAE-Dextran (Pharmacia)

- Dulbecco's modified Eagle's Medium (DMEM)   **Culture media**
- penicillin/streptomycin (PS)
- DMEM + PS + 10 mM Hepes (all from Gibco)
- Fetal calf serum (FCS)

Buffers and solutions

- 10 mM chloroquine in sterile water
- Lysis buffer
  25 mM Tris-HCl, pH 8.1
  50 mM NaCl
  0.5% sodium deoxy cholate
  0.5% Triton X-100
- Protease inhibitors
  Solution A: Dissolve 5 mg leupeptin, 1 mg pepstatin, 1 mg aprotinin in 5ml methanol. 5 µl of this mixture are added to 1 ml lysis buffer to lyse cells corresponding to one 100 mm plate. The final concentrations of these inhibitors are: 5 µg leupeptin, 1 µg pepstatin. Make appropriate aliquots and keep these at – 20°C. Usually 25 µl aliquots are suitable that could be used in the lysis procedure of 5 samples.
  Solution B: Dissolve 10 mg of soybean trypsin inhibitor in 1 ml of either PBS or DMEM + PS + 10% FCS. Make 20 µl aliquots and keep these at -20°C. Usually soybean trypsin inhibitor is used at 4-fold the concentration of trypsin.
  Solution C: Prepare a 100 mM solution of phenylmethylsulfonyl fluoride (PMSF) in methanol and keep at 4°C. Use PMSF at 1 mM final concentration.
- Washing buffer I
  Phosphate-buffered saline (PBS)
  0.5% Triton X-100
  0.05% sodium deoxy cholate
  0.01% sodium dodecyl sulfate
- Washing buffer II
  125 mM Tris-HCl, pH 8.1
  500 mM NaCl
  0.5% Triton X-100
  10 mM EDTA
- Trypsin stock solutions
  Solution A: Trypsin/Tris: 10 mg trypsin in 1 ml 25 mM Tris-HCl, pH 7.5, 50 mM NaCl
  Solution B: Trypsin/DMEM: 10 mg trypsin in 1 ml DMEM
- 2-fold SDS-PAGE sample buffer
  40 mM Tris-HCl, pH 6.8
  4% SDS
  20% glycerol
  bromophenol blue (to color, 0.002%)
- 1M dithiothreitol

Procedure

Cell culture and incubation of cells should be made in a humidified CO_2 incubator.

1. On day 0 plate 1 x 10^6 COS cells on 100 mm plate in 10 ml DMEM + 10% FCS + 1% PS.

2. On day 1 wash cells with 5 ml DMEM + PS + 10 mM Hepes.

3. Mix 1.5 ml DMEM + PS + Hepes with 5 µg DNA (pHI-6) and 9 µl DEAE-Dextran (50 mg/ml), add the solution to the cells and incubate for 1 h at 37°C.

4. Gently shake every 20 minutes.

5. Wash with 5 ml DMEM + 10 mM Hepes

6. Add 10 ml of DMEM + 10% FCS containing 100 µM chloroquine (from 10 mM stock freshly made in sterile water).

7. Incubate for 4 h at 37°C.

8. Wash cells twice with DMEM + PS + 10% FCS and incubate in the same medium for 40 - 46 hours at 37°C. Change medium once.

9. Wash twice with DMEM without FCS and incubate with 5 ml of DMEM containing 50 µg trypsin (5 µl from trypsin/DMEM solution) for 30 min at 37°C. Control cells are processed similarly but without trypsin.

10. Collect incubation medium, add immediately 200 µg soybean trypsin inhibitor (from 10 mg/ml solution made with DMEM + PS + FCS) and leave on ice for 30 min.

11. Add immediately to the cells 5 ml of ice-cold FCS and leave cells on ice for 5 min. Repeat the same step once.

12. Aspirate FCS, place the cells on ice, add 1 ml of ice-cold lysis buffer and 5 µl of mixture of protease inhibitors (solution A), 20 µl of soybean trypsin inhibitor (solution B) and 10 µl of PMSF (solution C) and leave on ice for 10 min. Scrape cells while still on ice with a rubber man and transfer to an Eppendorf tube and leave on ice for 30 min with occasional vortexing.

13. Centrifuge cell extracts at 100,000 x g for 1 h at 4°C.

14. Immunoprecipitate LPH molecules from the supernatant of the centrifugation and the incubation medium (step 10) as follows. Add to each sample 1 µl of monoclonal anti-LPH and rotate at 4°C for 2h. Add to each sample 40 µl of 50% (v/v) Protein A-Sepharose in PBS and rotate for 1h at 4°C.

15. Wash the immunoprecipitates (Protein A-Sepharose/antigen (LPH)/antibody) first with washing buffer I (4 times, 1 ml each time) and then twice with washing buffer II. For beads washing, the appropriate buffer (I or II) is added to the beads and the suspension is gently shaken and centrifuged in an Eppendorf centrifuge for 2 seconds. The supernatant is aspirated and the same procedure is repeated.

16. After final wash aspirate buffer completely and denature the samples by boiling in 25 µl 2-fold sample buffer, 20 µl H_2O (and for reduced samples 5 µl 1M DTT).

17. Submit the samples to 5% or 6% slab gels.

18. Stain gel with Coomassie blue, treat it with a fluorographic reagent and expose on Kodak X-Omat film.

Results

Studies on the biosynthesis and processing of LPH in organ culture of small intestinal biopsy samples have revealed that the first detectable form of LPH is a single chain mannose-rich polypeptide precursor (pro-LPH, 215-kDa) (Naim et al., 1987). This form is processed and cleaved in the Golgi apparatus to the 160-kDa mature brush border LPH. Initial cleavage of pro-LPH takes place at a trypsin-sensitive site (Jacob et al., 1996). The intriguing question raised in this respect is that of the role of the cleavage process in the acquistion of enzymatic activity and transport competence of LPH to the cell surface. One way to examine this is to express uncleaved pro-LPH in cells and determine the fate of such forms in the cell. For this purpose a full length cDNA encoding pro-LPH was expressed in COS-1 cells. In these cells pro-LPH is not cleaved (Naim et al., 1991) rendering this system suitable for analyses of the transport and function of pro-LPH. By virtue of the sensitivity of pro-LPH to trypsin, the cell-surface expression of pro-LPH could be probed by trypsin treatment of intact transfected COS-1 cells. In this case cleavage of pro-LPH to a brush border-like molecule is anticipated.

Figure 4, shows that LPH immunoprecipitated from transfected COS-1 cells is composed exclusively of the uncleaved pro-LPH polypeptides, corresponding to the mannose-rich (215-kDa) and the complex glycosylated (230-kDa) forms (lane 1) (for more details see Naim et al., 1991). Practically no cleaved forms representing the 160-kDa mature LPH could be detected. Addition of limited dilutions of trypsin to intact cells revealed a 160-kDa polypeptide (Fig. 4, lane 3). This band appeared concomitant with the disappearance of the 230-kDa polypeptide. The data indicate therefore that the 160-kDa is derived from the complex glycosylated 230-kDa pro-LPH and that this species is expressed at the cell surface. By contrast, the mannose-rich 215-kDa pro-LPH was unaffected and represents an intracellular

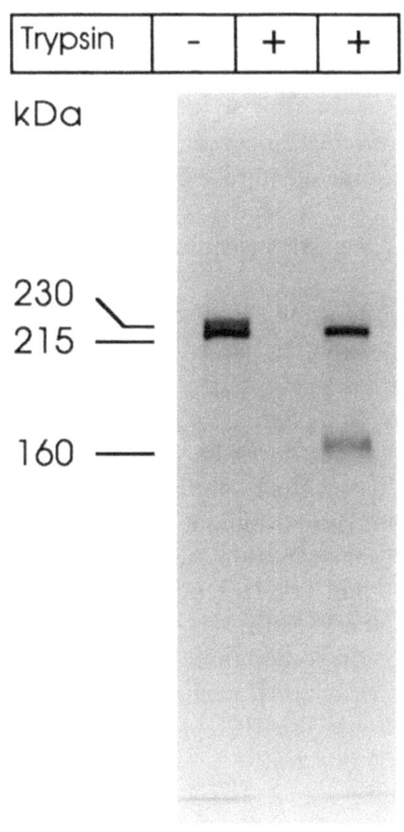

Fig. 4. Cell-surface expression of pro-LPH. Transfected COS-1 cells were labeled with ^{35}S-methionine for 6 h and subsequently incubated for 30 min at 37°C with DMEM containing trypsin at a 10μg/ml final concentration (lanes 2 and 3) or without the protease (lane 1). The cell extracts (lanes 1 and 3) and the culture-medium (lane 2) were immunoprecipitated with anti-LPH and analyzed by SDS-PAGE and fluorography.

form of pro-LPH as has been previously shown (Naim et al., 1987). Cleaved products could not be found in the incubation medium, since monoclonal anti-LPH recognizes epitopes found on the 160-kDa species (Fig. 4, lane 2). Taken together the data indicate that pro-LPH is a transport-competent molecule since it has egressed the ER, processed to complex glycosylated 230-kDa species in the Golgi apparatus and reached the cell surface. In this respect trypsin has provided an easy and powerful tool to delineate this issue.

Subprotocol 2
Membrane-Association of Angiotensin-Converting Enzyme (ACE)

Materials

Reagents
- Intestinal biopsy samples (to study ACE) (but any other cell line containing the membrane protein of interest could be used)
- Protein A-Sepharose (Pharmacia)
- Wheat germ agglutinin (WGA)-Sepharose (Pharmacia)

Buffers and solutions
- Buffers and solutions
- Lysis buffer
- Protease inhibitors
- Washing buffer I
- Washing buffer II
- Trypsin stock solutions
  (for composition of all these solutions refer to Subprotocol 1)
- Homogenisation buffer
  10 mM Tris-HCl, pH 7.4
  150 mM NaCl
- Lectin-elution butter
  2.5% N-acetyl D-glucoseamine
  10 mM Tris-HCl, pH 7.4
  150 mM NaCl

Procedure

All steps are performed at 4°C unless otherwise stated.

1. Label biopsy samples or cells biosynthetically with ^{35}S-methionine (or other radiolabelled amino acids) at 37°C for an appropriate period of time, during which mature forms of the protein of interest are generated and transported to the cell surface. In the case of ACE 4 - 6 h of labeling are sufficient.

2. Homogenise intestinal biopsy samples in a Teflon homogeniser by adding 1 ml of homogenisation buffer to one biopsy sample (about 5 mg wet weight) and 5µl protease inhibitors mix, 20 µl soybean trypsin inhibitor and 10 µl PMSF. When cells are used about 2 - 4 x 10^6 cells are disrupted and homogenised in 1 ml of homogenisation buffer.

3. The cellular homogenate of the intestinal tissue is fractionated to brush border membranes (BBM) and intracellular/basolateral membranes (IM/BLM) by $CaCl_2$ (Naim et al., 1988 and references therein). This step is not necessary when non-polarized cells are used.

4. Homogenise BBM in 0.5 ml of homogenisation buffer. These homogenates contain right-side out and inside out membrane vesicles.

5. During steps 1 - 4 equilibrate a slurry of 0.5 ml of WGA-Sepharose by washing 10 times with homogenisation buffer, 1 ml each time.

6. Mix the BBM vesicles with the WGA-Sepharose beads and rotate gently for 2 h.

7. Spin the BBM/beads suspension for 2 seconds and retain the supernatant (Sup 1).

8. Wash the beads 3 times, 1 ml each with homogenisation buffer.

9. Elute the bound material, which contains the right side-out vesicles with 1 ml of lectin elution buffer for 1 h at 4°C.

10. Wash the beads once again 3 times with homogenisation buffer.

11. Mix Sup 1 (from step 7) with the beads. This step is necessary to ensure complete depletion of right-side out vesicles.

12. Spin for 2 seconds and retain the supernatant (Sup 2), which should contain predominantly inside-out membrane vesicles. Repeat steps 8 - 10.

13. At this stage Sup 2 is virtually free of right-side out vesicles. However, depending on the experiment, the cell type, the concentration of the membrane vesicles, it is possible that some contamination with right-side vesicles may still be present in Sup 2. It is therefore recommended to repeat step 11 to obtain a third supernatant (Sup 3).

14. Divide the inside-out vesicles in Sup 2 (or Sup 3) to two equal parts and treat one part with 50 μg of trypsin at 37°C for 1 h. Incubate the other half similarly but without trypsin.

15. Stop the reaction by adding 200 μg of soybean trypsin inhibitor and centrifuge trypsin-treated and non-treated probes for 1 h at 4°C.

16. Retain the pellets, solubilize in 1 ml of lysis buffer and subject to immunoprecipitation with anti-ACE antibodies and electrophoresis as described in protocol 1 for LPH.

Results

The main goal of this experiment is to examine the mode of association of a protein with the membrane. Here, the presence of a transmembrane domain, composed of a membrane anchor and a cytoplasmic tail, could be demonstrated. For this purpose right-side out and inside-out vesicles were prepared. In right-side out vesicles of the cell surface, membrane proteins have the right orientation with the ectodomain being oriented to the external milieu. The ectodomains of glycoproteins are usually glycosylated and the right-side out vesicles are thus expected to be retained by the lectin (WGA-Sepharose). Inside-out vesicles expose existing cytoplasmic tails of proteins. These tails are not glycosylated and the inside out vesicles therefore are not retained by the lectin and are recovered in the supernatant. The exposure of a putative cytoplasmic tail in inside-out vesicles to the external milieu could be demonstrated by the accessibility to proteolytic enzymes and subsequent reduction in the apparent molecular weight of the protein of interest. In this case the result is indicative of the existence of a cytoplasmic tail and thus a transmembrane domain.

Figure 5 shows the results obtained with ACE. ACE immunoprecipitated from trypsin-treated inside-out vesicles revealed a molecular species of slightly smaller apparent molecular weight (approximately 3-kDa) (lane 2) than the non-treated probe (lane 1). The observed reduction in the apparent molecular weight provides evidence for the existence of a cytoplasmic tail in ACE that is cleaved off during trypsin treatment. The results are in line with the suggested length of the cytoplasmic portion of ACE deduced from cDNA cloning of the endothelial enzyme (Soubrier et al., 1988).

Fig. 5. Mode of association of ACE with the membrane. Biopsy samples were labeled for 6 h with ^{35}S-methionine, homogenized and brush border vesicles were prepared. Right-side out and inside-out vesicles were separated by lectin affinity chromatography using wheat germ agglutinin-Sepharose. The inside-out vesicles were treated (lane 2) or not treated (lane 1) with trypsin. The vesicles were solubilized, immunoprecipitated with anti-ACE antibodies and the immunoprecipitates were analyzed by SDS-PAGE and fluorography.

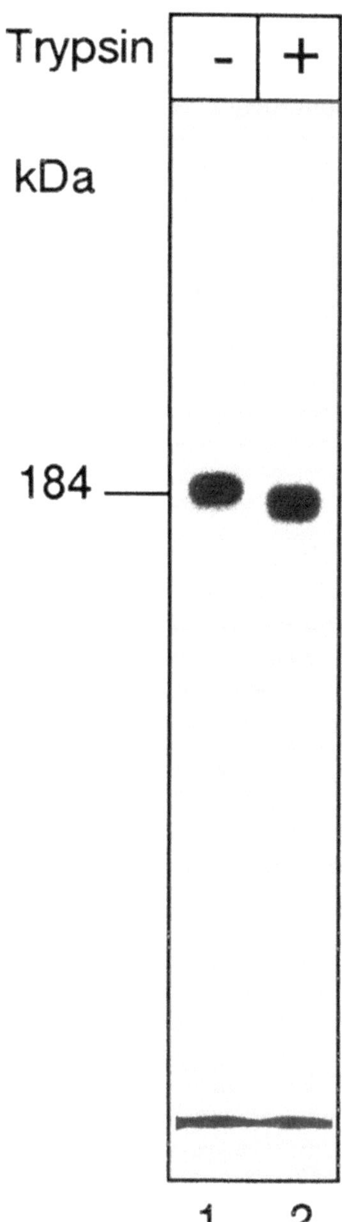

Subprotocol 3
Assessment of the Folding State of a Membrane Protein by Probing its Protease Sensitivity

Materials

Buffers and solutions
- Lysis buffer
- Protease inhibitors
- Washing buffer I
- Washing buffer II
- Trypsin stock solutions
  (for composition of these solutions refer to Subprotocol 1)

Procedure

Two forms of pro-LPH were prepared for this experiment, both of which were synthesized in a cell-free transcription/translation system (for details see Jacob et al., 1995). One form was produced in the presence of GSSG, which oxidises existing thiol groups and therefore facilitates the formation of disulphide bonds in proteins. The other pro-LPH species was synthesized in the absence of GSSG and does not contain intramolecular disulphide bonds. The folding of these two forms was assessed in the experiment described below. The procedure could be used, however, to study the folding of purified proteins from cellular extracts or other sources.

1. Isolate the protein of interest (in this case pro-LPH) by immunoprecipitation using a suitable antibody conjugated to Protein A-Sepharose (as described in Subprotocol 1).

2. Resuspend the immunoprecipitates in 35 µl PBS.

3. Add 5 µl of trypsin (50 µg) from trypsin stock solution (solution A) and incubate at 37°C for 30 min.

4. Arrest the reaction by addition of 25 µl of 3-fold concentrated SDS-PAGE sample buffer (+ 5 µl 1 M DTT for reduced samples) and boiling for 5 min.

5. Submit to SDS-PAGE and fluorography or autoradiography as described in subprotocol 1.

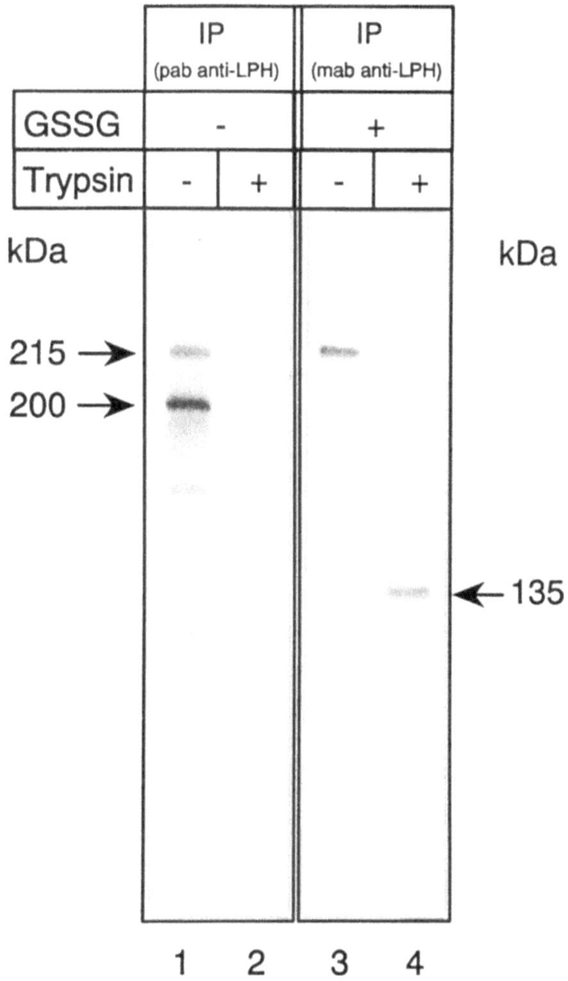

Fig. 6. Trypsin sensitivity assay. Pro-LPH was synthesized in the presence of microsomal membranes and in the absence (lanes 1 and 2) or presence (lanes 3 and 4) of 2 mM of the oxidizing agent GSSG. The translation products were immunoprecipitated with polyclonal anti-LPH (pab anti-LPH) (lanes 1 and 2) or monoclonal anti-LPH (mab anti-LPH) (lanes 3 and 4). The immunoprecipitates (IP) were treated with trypsin (lanes 2 and 4) or not treated (lanes 1 and 3). The samples were analysed by SDS-PAGE on 7% slab gels.

Results

The role of disulphide bond formation in the folding of pro-LPH was assessed in a cell-free transcription/translation system using the oxidising agent GSSG. In the presence of GSSG pro-LPH could be immunoprecipitated by a conformation-specific monoclonal anti-LPH antibody (Fig. 6, lane 3). Pro-LPH generated in the absence of GSSG could only be isolated using a polyclonal antibody directed against a denatured form of the enzyme (Fig. 6, lane 1) (see also Jacob et al., 1995, Fig. 3). The conformation of both forms was probed by trypsin. As shown in Figure 6, trypsin treatment of pro-LPH translated in the absence of GSSG resulted in a complete degradation of this species (lane 2). By contrast, pro-LPH immunoprecipitated from GSSG-treated samples with a monoclonal antibody was cleaved specifically into a major 135-kDa polypeptide in the presence of trypsin and a faint 80-90-kDa species (Fig. 6, lane 4). The reactivity with a conformation-specific monoclonal antibody and the specific cleavage by trypsin indicate that pro-LPH synthesized in the presence of GSSG is a correctly folded molecule. By contrast, the complete degradation of pro-LPH generated in the absence of GSSG indicates that this polypeptide has a different folding pattern and an increased susceptibility to trypsin. Together, the results strongly support the notion that disulphide bond formation is crucial in the acquisition of pro-LPH to its native conformation.

References

Doyle C, Roth MG, Sambrook J, Gething MJ (1985) Mutations in the cytoplasmic domain of the influenza virus hemagglutinin affect different stages of intracellular transport. J Cell Biol 100:704-714

Hurtley SM, Helenius A (1989) Protein oligomerisation in the endoplasmic reticulum. Annu Rev Cell Biol 5:277-307

Jacob R, Bulleid NJ, Naim HY (1995) Folding of human intestinal lactase-phlorizin hydrolase. J Biol Chem 270:18678-18684

Jacob R, Radebach I, Wüthrich M, Grünberg J, Sterchi EE, Naim HY (1996) Maturation of human intestinal lactase-phlorizin hydrolase. Generation of the brush border form of the enzyme involves at least two proteolytic cleavage steps. Eur J Biochem 236:789-795

Lüscher B, Rousseaux-Schmid M, Naim HY, Macdonald HR, Bron C (1985) Biosynthesis and maturation of the lyt-2/3 molecular complex in mouse thymocytes. J Immunol 135:1937-1944

Naim HY, Sterchi EE, Lentze MJ (1987) Biosynthesis and maturation of lactase-phlorizin hydrolase in the human small intestinal epithelial cells. Biochem J 241:427-434

Naim HY, Sterchi EE, Lentze MJ (1988) Biosynthesis of the human sucrase-isomaltase complex. Differential O-glycosylation of the sucrase subunit correlates with its position within the enzyme complex. J Biol Chem 263:7242-7253

Naim HY, Lacey SW, Sambrook JF, Gething MJH (1991) Expression of a full-length cDNA coding for human intestinal lactase-phlorizin hydrolase reveals an uncleaved, enzymatically-active, and transport-competent protein. J Biol Chem 266:12313-12320

Naim HY (1992) Angiotensin-converting enzyme of the human small intestine. Subunit and quaternary structure, biosynthesis and membrane association. Biochem J 286:451-457

Naim HY, Roth MG (1993) Basis for selective incorporation of glycoproteins into influenza virus envelope. J Virol 67:4831-4841

Rose JK, Doms RW (1988) Regulation of protein export from the endoplasmic reticulum. Annu Rev Cell Biol 4:257-288

Soubrier F, Alhenc-Gelas F, Hubert C, Allegrini J, John M, Tregear G, Corvol P (1988) Two putative active centers in human angiotensin I-converting enzyme revealed by molecular cloning. Proc Natl Acad Sci USA 85:9386-9390

Chapter 17

Segregating Cells - Proteases in Tissue Culture

ULRICH N. WIESMANN

Introduction

Rous and Johns first introduced the use of trypsin to detach growing cells from explanted tissue pieces (Rous and Johns 1916). For more than 80 years trypsin has remained a favorite enzyme for the primary dissociation of tissues and for detaching cells in monolayers for subsequent replating. The trypsin that was used by Rous and Johns and in many instances still is used is a mixture of various pancreatic enzymes. Low concentrations of crude trypsin usually did not harm cells, but higher concentrations would kill the cells. Tryptic isolation of individual cells from tissue for primary cultures were used to dissociate chick embryonic tissues for cell cultures (Weymouth 1974). Moskona (1952) was using a 3 % crude trypsin solution in a calcium and magnesium free solution. The use of calcium and magnesium free solution was based on the evidence that these cations play a role in stabilising the intercellular matrix and thus in controlling mutual cellular adhesiveness. A first exhaustive review on the isolation of cells has been published by Rinaldini (Rinaldini 1958). He also reported that purified trypsin was less effective than crude trypsin preparations pointing out the existence of other enzymes in the mixture of crude trypsin. The intercellular material which has to be digested in order to segregate the cells are mucopolysaccharides and proteins of which collagen is one of the major intercellular proteins. Elastins are another kind of intercellular mucoproteins which are digested by the specific enzyme elastase.

Tissue dissociation requires a combination of enzymatic, chemical and mechanical procedures which are all damaging to various degrees. Furthermore, isolation by cell sorting through centrifugation results in structural

Ulrich N. Wiesmann, Unversity of Berne, Childrens Hospital, Freiburgstr. 15, Bern, 3010, Switzerland (*phone* (031) 632 95 43; *fax* (031) 632 47 72; *e-mail* UlrichWiesmann@dk-f2.unibe.ch)

alteration whose magnitude depends on the type and the speed of the rotor as well as on the chemical composition of the medium. The procedure suitable for one given tissue of a given species may not be appropriate for another tissue or even for the same tissue in another species. Thus no universal method exists and each must be worked out for each particular problem (Romrell et al 1975).

Procedure

Mechanical Dissociation

For embryonic tissues the **mechanical dissociation** methods may offer a significant advantage in preparing cell suspensions greatly enriched for one type of cell such as myoblasts making additional preplating steps unnecessary (Teppermann et al 1975). After dissociation of muscle with a **mechanical method** utilizing shearing forces obtained in a Vortex, the extent of myotubformation were greatly improved (Bullaro and Brookmann 1976) when compared to treatment with crude trypsin.

An interesting method has been described for the isolation of large numbers of viable cells from several human tissues by combining **mechanical desintegration** with 0.1 mg/ml **trypsin** or 0.5 mg/ml **collagenase** and 0.1 mmol **EGTA**.

Pieces of human fetal intestine, kidney, liver, lung and skin were placed in a sterile bag, containing 60 ml of calcium/ magnesium-free phosphate buffered saline with the appropriate enzymes, EGTA and 0.1 % methylcellulose.

The bag was then placed into a blender (Stomacher model 80, lab blender) and mixed at low speed for 3 to 20 minutes at room temperature.

After single cell suspension was observed by phase contrast microscopy 10 ml of bovine calf serum were added to the cell suspension.

Thereafter 30 ml of cold Hank's balanced salt solution containing 5 % bovine calf serum was added and the entire cell suspension passed through a cloth sieve (100 mesh, 140 micron) and the cells collected by centrifugation.

The method describes shorter incubation times with relatively low concentrations of proteolytic enzymes and the yield was 2 to 3 fold greater than with many other enzymatic and mechanical methods. The valiability of the cells was between 86 to 93 % (Gibson-D'Ambrosio et al, 1986).

Proteases and Antiproteases in Cultured Cells

Cultured cells such as fibroblasts and keratinocytes **produce proteases** (metallo-proteases and several types of gelatinase as well as plasminogen activators (Varani et al 1995).

Aortic endothelial cells also **secrete proteases** such as elastase (West and Kumar 1986). Articular cartilage explants synthesise and **secrete proteases** into the medium which are able to degrade matrix proteins and proteoglycans. Stimulation of protease secretion occurs by cytokins like IL-1 alpha, beta and TNF-alpha (Arsenis and Mc Donnell 1989). The action of proteases in cultured cells including fibroblasts may be responsible for the detachment of confluent cultures from the substratum as an entire sheeth resulting in loss of the cultures.

On the other hand chondrocytes in primary culture **produce the collagenase inhibitor** TIMP (tissue inhibitor of metallo proteases) in the absence of detectible procollagenase. (Lefebvre et al 1990). Sometimes addition of antiproteases may help to preserve cultured cells, especially in the absence of serum. Surface structures on guinea pig ventricular myocytes could be **protected** from endogenous proteolytic enzyme activity with **aprotinin** during mechanical disaggregation. Myocytes aggregated by this procedure were shown to retain many of the morphological and metabolic characteristics of intact cardiac muscle cells (Bailey et al 1987).

Trypsin inhibitors themselves may however damage dissociated cells. Dissociation of epithelia and connective tissue of rat oral mucosa by **elastase sojabean** trypsin inhibitor causes greater ultrastructural damage among dissociated cells than elastase alone, reflected in damage to mitochondria and resulting in decline of oxydative metabolism (Fine et al 1981).

Proteases and Mycoplasm Infections

Mycoplasms infections still cause severe problems in cell cultures particulary in permanent lines and despite rapid detection . Serum, crude trypsin and contamination through laboratory workers may be the cause of **mycoplasma** infections. On the other hand the tissue used for cell culture and dissociation may already be contaminated by mycoplasms. Infections with mycoplasma usually first go unrecognized in cell cultures. There may be disturbances of growth rythms and increased utilisation of aminoacids from the medium (Draghici et al 1969). Latent **mycoplasma** infections may mimic metabolic changes in cultured cells by direct interference of a microbial enzyme with the assay system. A rapid and simple bioassay to detect and distinguish particle associated and soluble phosphorlyase activity by (^{3}H) TdR degradation may be a useful screening assay for mycoplasma contamination in tissue culture (Merkenschlager et al 1988). The only methods proposed for the elimination of the mycoplasma are either laborious or unsatisfactory. Treatment with antibiotics often leads to the development of resistance.

Serum and trypsin can be **sterilized by beta-propiolacton (BPL)** which, even in small concentrations can inactivate several types of bacteria, mycoplasms and viruses.

500 ml of serum are mixed with 5 ml of pure BPL together with 70 ml of 1 M tris solution. After incubating at room temperature for 18 to 24 hours the serum is warmed to 56° C for 30 minutes to remove nonhydrolisable traces of the reagent (Döhner and Kiessig 1975).

An interesting approach has been communicated that the passage of cells through nude mice results in the eliminiation of mycoplasm. In vitro a brief cocultivation of contaminated cells with mouse macrophages in the presence of antibiotics may also do the same job (Schimmelpfeng et al 1980).

Crude trypsins have been found generally to be more effective than purified crystaline trypsin to disaggregate cells indicating that other proteases contained in the preparations are operating. The optimal pH of trypsin is high (pH 8 to 9) and may not be optimal for the other enzymes in the mixture nor physiological for the cells, thus the pH used for trypsinization is between 7.2 to 7.8.

Proteolytic enzymes have been shown to **adsorbe** to cell surfaces and persist in an active form as long as 24 hours thereafter. They were found to prevent the formation of the glycoprotein cell coat material at the surface and to interfere with the attachment, spreading and growth of the cells on glass (Poste 1971). Snow and Allan 1970 have shown that crude trypsin (0.1 %) removed more sialic acid from cell surfaces than crystaline trypsin (0.04 or 0.001 %). Crystaline trypsin released only minimal amounts of nucleic acids suggesting that trypsin itself may be acting mainly at the cell surface, although an intracellular action of trypsin could be shown by degradation of polyribosomes (Hosick et al 1971). Antigenic determinants may also be partially lost after trypic proteolysis using EDTA / trypsin (Korver et al 1995).

During dissociation of (^{14}C) **linoleic acid labeled endothelial** cell monolayers with 0.25 % trypsin or with 0.125 trypsin + 0.01 % EDTA more radioactivity was released than with 0.01 % **collagenase + 0.01 % EDTA**. Morphological studies failed to reveal any distinctive differences of surface morphology following the various enzyme treatments. Collagenase treatment of

Trypsin (EC 3.4.4.4)

endothelial cell monolayers seems to be the least dramatic harvesting method for subcultures as far as the integrity of the lipids in the cell membrane is concerned (Kirkpatrick et al 1985).

The intracellular damage may be repaired by the cells so that normal function of the cells is resumed within hours and up to 1 day. Other kind of damage requires a longer time suggesting synthesis of new cellular elements in the process of cell growth. Because of the progressive nature of the deleterious effects of trypsin upon cell viability it is important that trypsinization is followed as quickly as possible by addition of a trypininhibitor to stop trypsin action. As serum is customarily included in the culture media the addition of serum is the most common method for termination trypsin digestion. In cases where defined media are to be used the continuing action of trypsin is a problem. Washing the cells proves to be insufficient because some of the trypsin remains firmly bound to cell structures. The use of trypsin inhibitors of plant sources may stop trypsin action but may also have undesirable effects by themselves.

The time for trypsinization in some instances seems to be responsible for the type of cells that are obtained. In rat brain tissue the protoplasmic astrozytes which are the major type of neuroglial cells in the brain seem to be highly susceptible to degradation during tissue trypsinization. After trypsinization for less than 10 minutes mostly protoplasmic astrozytes were obtained while after prolonged trypsinization fibrous astrozytes were the predominant cells. The trypsin concentration was 0.1 % (Sigma, twice crystalized) (Chatterjee and Sakar 1984).

Longterm trypsinisation (Difco 1:250) procedure at low temperature (4° C), pH 7,5 for 12 to 20 hours may in representative tissues (tumors) yield large numbers of viable cells. The trypsinisation may be completed by warming to 37° C and by mechanical disintegration. The cell suspension was passed through a double layer of gauze into minimal essential medium containing 30 % bovine serum (Engelholm et al 1985).

Effects of combining EDTA with trypsinization

The time for tryptic cell dissociation may be considerably reduced using trypsin (0.025 %) with EDTA (5 mmol/l). It seems to be important whether trypsin and EDTA are given simultaneously or as consecutive steps following each other. EDTA treatment may modify the surface components involved with intercellular adhesion and facilitate cell dissociation by trypsin (Tokiwa et al 1979).

Retinal pigmented epithelium cells most efficently and gently could be dissociated into single viable cells when **trypsinized** in **EDTA** solution at low temperature and low enzyme concentration followed by incubation in culture medium up to 4 hours. The completely dissociated cells had a much higher plating efficiency and a more uniform pattern of colony growth and differentiation than a dose obtained under any other tested conditions. Complete dissociation, separating the basal lateral cell borders, induced however dedifferentiation of retinal pigmented epithelium (Vielkind and Crawford 1988). This observation may also apply for other epithelial cells such as the intestine and the kidney.

Combined or consecutive enzymatic dissociation using trypsin and other proteases may reduce time and enhance viability and cell growth. Primary cultures of pulp cells from bovine permanent incisores isolated by **trypsin, collagenase** with and without **trypsin pretreatment** and by a **trypsin collagenase mixture** showed that trypsin pretreatment followed by collagenase digestion was preferable in regard to the high number of isolated cells, satisfactory adhesion, and good growth. Using this technique the cells were responsive to parathyreoid hormone at the poliferating stage and had high alkaline phosphatase activity (Nakashima 1991).

Chorionic villus cells have been obtained from first and second trimester pregnancies by treating cells either with trypsin-**EDTA** for 2 hours followed by collagenase over night or by **trypsin-EDTA** for 1 hour and **collagenase** for 2 further hours. Using short term enzymatic digestion the cultivation time was reduced from 14 days to 6 days. The number of viable cells and their attachement was more efficient than after prolonged treatment with collagenase. With this method sufficient amounts of metaphases of good quality were present in 93 % of primary cultures (Smidt-Jensen et al 1989).

Mouse hair follicles could be completely dissociated into a suspension of single cells by **sequential digestion** with **trypsin** and **chondroitinase ABC** under mild conditions (Frater 1976).

Epithelial cells in culture that develop intercellular desmosomes are difficult to dissociate by trypsin without damaging the cells. Incubation with **dithiotreitol** (TDT) (1.5 - 24 mg/ml) **followed** by subsequent **trypsinisation** (1 mg crystalin tryspin, trypure Novo Copenhagen/ml) destroyed the desmosomes, whereas cell membranes, organels and nucleids remained undamaged (Hentzer and Kobayashi 1976).

An alternative method for dispersing strongly adhesive cells in tissue culture has been suggested by Sanford by combining **trypsin** (0.5 mg of 1:250 trypsin and 0.2 mg of **EDTA**/ml at 20 C°) with a short **ultrasonic** treatment (75 second) in a Branson model 220 ultrasonic cleaner (100 watt output). This resulted in an abrupt removal of all cells from the plastic surface without any damage to the viability of the cells (Sanford 1974).

Trypsin covalently bound to sephadex beads (6-25 superfine) does not bind to isolated cells. Neuronal cells isolated from rat brain showed that the histiolytic efficiency of the bound enzyme measured in terms of cell yields was approximately 100-fold greater than that of the free enzyme (Ginns and Guarnieri 1976).

Collagenase (clostridio peptidase A,E,C 3.4.4.19)

The ability of some anaerobic microorganisms to digest tendons and muscles was observed clinically in the last century. A collagenase from clostridium velchii was isolated by Okley et al 1946. The current commercially available preparations of **Collagenase** from clostridium histolyticum (clostridio peptidase A, EC 3.4.24.3) cleaves collagen and other proline containing peptides at glycin residues. Crude type I or V collagenase from Sigma (St Louis Missouri) or Worthington Biochemical Corporation (Freehold, New Jersey) give excellent results. Collagenase is dependent on the presence of calcium which is activating, and is sensitive to other divalent cations which are either activators (Zn_2) or inhibitors (magnesium). Calcium chelators such as EDTA should not be used simultaneously with collagenase. Since collagen is predominantly an extracellular protein collagenase may be regarded as relatively innocuous to cell structures. Crude bacterial **collagenases** contain nonspecific proteases. Purified preparations of collagenase are quite ineffective in dissociating tissues. They can be substituted by a combination of **purified collagenase** and 0.01 % **w/v purified trypsin**.

When **trypsin** was present during **collagenase** treatment of the tissue the performance of the collagenases was improved and stabilised.

Cells from human and bovine endrometrium have been isolated by incubation with a mixture of 0.5 % **collagenase**, 0.1 % **hyaluronidase** and 0.1 % **pronase** in balanced salt solution containing 1 % chicken serum at 37°C. After a digestion period of 10 minutes large numbers of structurally and metabolically intact dispersed cells were obtained (Marcus et al 1984). Hepatocytes were isolated from human fetal livers by **mechanical fragmentation** followed by dissociation of the liver tissue with **collagenase/dispase** mixture. This resulted in high yield and viability of hepatocytes (Hoshi et al 1987).

Not always is cell dissociation by collagenase optimal. Comparison of crude collagenase with other proteases showed that the choice of enzymes for harvesting depends upon the cell type and purpose of the cultures.

Cells from primary human breast tumors are difficult to isolate because of the high content of stromal tissue. Comparing **collagenase** type IV (2 mg/ml) in the presence of 5 % serum for 24 hours, with **pronase** 0,075 % for 1 hour, showed that **low dose collagenase** treatment for 24 hours gave the best results with the highest number of valiable cells per gram of tissue and with very high plating efficiency of 4 isolated tumor cells (Besch et al 1983).

Morphologically and functionally intact mast cells were isolated from the lung and mesentery of normal or actively sensitized dogs using the **pronase** or **collagenase** tissue dissociation methods. Dissociation by **collagenase** (type) yielded about 6 times as many metachromatically staining cells. The ratio of mast cells and basophils in all samples were independant of the enzyme used for the tissue dissociation but the average histamine content of the cells obtained with the pronase method was significantly higher than after collagenase treatment (Heymanns et al 1982).

The choice of proteases for tissue dissociation may influence the properties of the subsequent cultures. The electrophysiological and pharmacological properties of aggregates prepared from cells of 7 day old chick embryo heart ventricles depends on the enzyme used for the cell dissociation. Half of the aggregates formed from the **collagenase** dissociated cells were tetrodotoxin insensitive. Aggregates prepared with **trypsin** dissociated cells displayed properties which more closely resembled those of intact 7 day embryonic ventricle tissue (Colizza et al 1983).

Pronase is not a well defined enzyme but rather a mixture for several proteinases isolated from streptomyces griseus with a wide spectrum of proteolytic activity. In the presence of the calcium chelator the calcium independent esterase activity is largely predominant. The broad substrate specificity of the pronase, recommends its use for separating cells. Good quality lyophilized pronase can be obtained from Boeringer (Mannheim, Federal Republic of Germany) or Merck (Darmstadt, Federal Republic of Germany). It is recommended that the enzyme activity be tested first on a sample and if found suitable, amounts large enough to allow for work over an extended period of time should be ordered. Serum contains no pronase inhibitors so thorough washing to remove the enzyme is recommended. However, pronase may not be removed completely from the surface of cells by washing and may continue to damage the cells (Poste 1971).

Pronase (EC 3.4.-)

Various tissues from chick embryos (kidney, sensory ganglia, heart cells) have been dissociated with **pronase** followed by filtration through a porous plastic filter. Chick pectoral muscle is a tissue that has proven resistant to crystaline trypsin or chelating agents but Wilson and Lau (1963) found that stirring with pronase at 0.05 % in Hank's solution at 37°C released single cells in 30 minutes. Sullivan and Schafer (1966) have successfully and routinely used a lower concentration (0,025 %) for transfer of cell lines derived from human skin, and Weinstein (1966) has used 0.05 % pronase for the regular transfer of human diploid fibroblast cultures. Viability (trypan blue exclusion) was 90 % after the pronase treatment and no adverse effects upon chromosome morphology were noted. Gastric mucosal cells from the rat can be separated and collected using proteolytic enzymes. **Pronase** (1 %) achieved better results than did **trypsin** (2 %) in collecting single isolated cells with higher yields and viability. The cells, however, dissociated with trypsin retained glandular structure as in situ. The most effective method to obtain dissociated cells from the generative zone of the mucosa was to collect the cells dissociated with 1 % pronase continuously for a period of 50 to 45 min after the start of dissociation (Kurokawa et al 1975).

Cell suspensions prepared from pancreatic islets by different enzymatic dissociations proved to be superior to mechanical dissociation. Enzymatic dissociation was performed by fractionated treatment of the tissue with low concentration of enzymes in Hank' solution for 2 to 3 minutes at room temperature. Concentrations of trypsin were 0.01 %, of pronase 1.0 mg/ml and of dispase 1.2 mg/ml. The percentage of single cells was consistently higher with dispase and pronase treatment with a very high (90 %) cell viability by dye exclusion. The cells were better preserved after **pronase** and **dispase** treatment (Kohnert and Hehmke 1986).

When comparing **collagenase, trypsin** and **pronase** to obtain cell suspension from human prostates the dissociation using pronase gave both the largest number of nucleated cells and the largest proportion of viable cells. One third of these cells contained histochemically detectable acid phosphatases (Helms et al 1975).

When comparing different methods of cell dissociation for the isolation of goblet cells from cat trachea the most successful method was to use **EDTA** to loosen the basement membrane followed by **1 % pronase** to dissociate the epithelial cells (Sherman et al 1988).

Elastase (pancreato peptidase E, EC 3.4.4.7)

Elastase is the only pancreatic enzyme able to digest the fibrous connective tissue protein elastin. Elastase contains no prostetic group and requires no metal ions and has pH optimum of pH 8.8. Crude elastase has been found superior to trypsin for disaggregating chick embryonic cardiac ventricles (Rinaldi 1959, Levinson and Green 1965). Elastase also digests the elastic fibers of the arteries, thus myocytes and endothelial cells may be obtained for cell culture. Single cells from lung or kidney tissue also have been obtained with **elastase** dissociation (Phillips 1972). No intracellular changes attributable to elastase have been reported, however elastase has not commonly been used for tissue dissociation.

Papain (EC 3.4.22.2)

In contrast to trypsin and to bacterial proteases, papain is a protease of plant origin, obtained from the tropical plant papaia. It is a cystein protease and is not inhibited by antitrypsin antiproteases. Reports of tissue dissociation with papain are rare, but it's use may have considerable advantages.

Tissue dissociation with **papain** rather than with **trypsin** produced less cellular debris and a higher neuronal cell yield from the brain of fetal rat at

18 to 21 day gestation. Such neurons when cultured on antimitotic agent-treated astrocytes could be maintained in culture for several weeks (Fitzgerald et al 1992).

An elegant technique to isolate cells of high purity has been recently developed. Coupling of a murine monoclonal antibody, HEA 125, specific for human epithelial cells, to magnetic beads permitted positive selection of a population containing essentially only the desired cells. Binding conditions were at 0°C on ice for 1 hour. After dilution of the suspension with PBS, beads were collected with a magnet. Cells were removed from the beads by incubation with 10 U /ml **papain** (Boeringer Mannheim, Germany) for 30 to 60 minutes at 37°C). The PBS contained 1 mmol/l EDTA and 2.5 mmol/l mercaptoethanol (Kemmner et al 1992).

Ficin (EC 3.4.22.3)

Ficin is a serin protease and can be inhibited by trypsin inhibitors. A major advantage of ficin, is the use in media containing calcium and magnesium. Comparing the effects of **trypsin, papain, promelain** and **ficin** on bovine dental pulp tissue showed that ficin was the most suitable enzyme for isolation and on the growth of isolated pulp cells from various layers of the bovine pulp due to it's even rate of cell removal and the good initial viability and subsequent growth of the separated cells in monolayer culture. The results were superior to those obtained with trypsin (Miller et al 1976).

Dispase (EC 3.4.99)

Removal of the basal epidermal cells is important for longterm survival of the cultures since they contain stem cells. Harvesting of epidermal basal cells is optimal when using trypsin and the flotation method. For the isolation of the cells from the epidermal sheet **dispase** was better than **trypsin**. The cell yield was about 4 times as high as with trypsin. Cells dissociated with dispase showed a higher rate of viability and attachment to culture dishes than those obtained with trypsin (Takahashi et al 1985).

Hepatocytes from human liver were obtained as viable cells after gently dissociating with **collagenase followed by dispase**. This procedure could be used to study human hepatocytes function in vitro. The biopsy specimens were cut into smaller fragments in cold calcium free Hank's solution and stirred for 30 minutes in 4°C in small flasks. The cells and fragments were centrifuged at 50 x g for 3 minutes and the pellet suspended in 10 ml of

0.05 % collagenase IV (Worthington) in Hank's solution. The suspension was agitated in a waterbath at 90 oscillations per minute for 20 minutes at 35°C. The centrifugation procedure was repeated and the cell pellet suspended in 1000 U/ml dispase I in Hank's solution and placed in a waterbath and oscillated for 60 minutes at 37°C. After filtration through a sterilizied nylon net the suspension of cells was centrifuged 3 more times at 50 x g for 3 minutes. Each time the supernated fluid was discarded and the pellet resuspended in fresh Hank's solution. Buffer and enzyme solutions were oxygenated with 95 % O_2 and 5 % CO_2 prior to use and the pH was adjusted to 7.2 by sodium bicarbonate. There was a yield of 2 x 10^6 cells per g wet weight of liver tissues with a more than 90 % viability as judged by trypan blue exclusion tests (Miyazaki et al 1981).

Thermolysin (EC3.4.24.-)

Overgrowth of cultured keratinocyte preparations by fibroblasts could be significantly reduced by utilizing **thermolysin** since this enzyme selectively digests the dermal-epidermal junctions. Following separation of the epidermis from the dermis; the epidermis was digested with trypsin to obtain a single cell suspension. Compaired with the conventional procedure this isolation method was shorter and resulted in cells displaying a higher colony forming efficiency and reaching confluency 1 to 3 days earlier. Apart from not being contaminated by fibroblasts the cell population contained all basal-layer keratinocytes (Germain et al 1983).

Rat lungs were dissociated into individual viable cells after **thermolysin** perfusion through the vasculature and the trachea. Then the lungs were minced and further dissociated by washing with sequential addition of thermolysin. This appears to be a suitable procedure for dispersing lung tissue into it's cellular components much superior than the trypsinizing procedure. The number of cells after thermolysin treatment was 10 fold higher than with trypsin treatment (Frazier et al 1975).

References

Arsenis C, Mc Donnell J (1989) Effects of antirheumatic drugs on the interleukin-1 alpha induced synthesis and activation of proteinases in articular cartilage explants in culture. Agents-Actions, 27; 261-64

Bailey LE, Carlos H, Amian A, Moon KE (1987) Oxydative metabolism in guinea pig ventricular myocytes protected from proteolytic enzyme activity. Cardiovasc Res 21; 481-88

Besch GJ, Woleberg WH, Gilchrist KW, Völkel JG, Gould MN (1983) A comparison of methods for the production of monodispersed cell suspensions from human primary breast carcinoms. Breast Cancer Res Treat 3; 15-22

Bullaro JC, Brookman DH (1976) Comparison of sceletal muscle monolayer culture initiated with cells dissociated by Vortex and by trypsin methods. In Vitro 12; 564-70

Chatterjee D, Sakar PK (1984) Isolation of protoplasmic astrozytes: A procedure based on controled trypsin digestion. J. Neurochem 42; 1229-34

Colizza D, Guevara MR and Shrier A (1983) A comparative study of collagenase and trypsin dissociated embryonic heart cells reaggregation electrophysiology and pharmacology. Can J. Physio Pharmacol 61; 408-19

Döhner L, Kiessig R (1975) Zur Sterilisation von Serum und Trypsin für die Zellkultivierung mit Betapropiolakton. Labortech 16; 277-280

Draghici D, Schimmel D und Hubrig Th (1969) Mycoplasmen von Schwein und Zellkulturen: Die Isolierung von Mycoplasmen des Schweines in Primärzellkulturen. Arch Exp Veterinärmed 23; 101-134

Engelholm SA, Spang-Thomsen M, Brunner N, Nohr I and Vindelov LL (1985) Disaggregation of human solid tumors by combined mechanical and enzymatic methods. Br. J. Cancer 51; 93-98

Fine AS, Egnor RW, Forrester E, Stahl SS (1981) Elastase soybean trypsin inhibitor dissociation of rat oral mucosa: Ultrastructural and oxidative metabolic destructive changes in isolated epithelial and dermal mitochondria after dissociation. J. Invest Dermatol 76; 239-54

Fitzgerald SC, Willis MA, Yu C, Rigatto H (1992) In search of the central respiratory neurons: I. dissociated cell cultures of respiratory areas from the upper medulla. J. Neurosci Res 33; 579-89

Frazier ME, Hadley JG, Andrews TK and Drucker H (1975) Use of thermolysin for the dissociation of lung tissue into cellular components. Lab Invest 33; 231-38

Frater R (1976) A new technique for dissociation of hair follicles into single cells. Experientia 32; 675-76

Germain L, Rouabhia M, Guignard R, Carrier L, Bouvard V, Auger FA (1983) Improvement of human keratinocyte isolation and culture using thermolysin. Burns 19; 99-104

Gibson-D' Ambrosio RE, Samuel M, D' Ambrosio SM (1986) A method for isolation large numbers of viable disaggregated cells from various human tissues for cell culture establishment. In Vitro Cell dev Biol 22; 529-34

Ginns E, Guarnieri M (1976) Trypsin bound to sephadex beads. A tool for neuronal cell dissociation. Expl Cell Res 97; 42-46

Helms SR, Brazeal FI, Bueschen AJ, Pretlow TG (1975) Separation of cells with histochemically demonstrable acid phosphatase activity from suspensions of human prostatic cells in an isokinetic gradient of Ficoe in tissue culture medium. Am. J. Pathol 80; 79-90

Hentzer B, Kobayashi T (1976) Dissociation of human adult epidermal cells by disulfide reducing agents and subsequent trypsinization. Acta Derm Venerol (Stockholm) 56; 19-25

Heymanns J, Behrendt H and Schmutzler W (1982) Comparative studies of mast cells from normal (non immunized) and activly sensitized dogs. Agents-Actions 12; 192-8.

Hoshi H, Kan M, Mc Keehan WL (1987) Direct analysis of growth factor requirements for isolated human fetal hepatocytes. In-Vitro-Cell-Dev-Biol 23; 723-32

Hosick HL and Strohman R (1971) Changes in ribosome-polyribosome balance in chick muscle cells during tissue dissociation, development in culture, and exposure to simplified culture medium. J Cell Physiol 77; 145-156

Kemmner W, Moldenhauer G, Schlag P, Brossmer R (1992) Separation of tumor cells from a suspension of dissociated human colorectal carcinoma tissue by means of monoclonal antibody-coated magnetic beads. J. Immunol Methods 147; 197-200.

Kirkpatrick CJ., Melzner I, Goller T (1985) Comparative effects of trypsin, collagenase and mechanical harvesting on cell membrane lipids studied in monolayer-cultured endothelial cells and a green monkey kidney cell line. Biochim Biophys Acta 846; 120-126

Kohnert KD, Hehmke B (1986) Preparation of suspensions of pancreatic islet cells: A comparison of methods. J. Biochem Biophys Methods 12; 81-88

Korver WE, Cornelisse CJ, Herrmans J, Fleuren GJ (1995) Limited loss of 9 tumor associate surface antigenic determinants after trypic cell dissociation. Cytometry, volume 19, p. 267-72

Kurokawa IJ, Saito S, Kanamaru R, Sato T, Sato H (1975) Separation of gastric mucosal cells of rat with proteolytic enzymes, pronase and trypsin with special reference to the collection, morphology and viability of the generative cells. Tohoku J. Exp Med 116; 241-252

Lefebvre V, Peeters-Joris C and Vaes G (1990) Production of collagens collagenase and collagenase inhibitor during the dedifferentiation of articular chondrocytes in serial subcultures. Biochim Biophys Acta 1051; 266-275

Levinson C, and Green JW (1965) Cellular injury resulting from tissue disaggregation. Exp Cell Res 39; 309-317

Marcus GJ, Connor L, Domingo MT, Tsang BK, Downey WR, Ainsworth L (1984) Enzymatic dissociation of ovarian and uterine tissues. Endocr Res 10; 151-62

Merkenschlager M, Kardamakis D, Rawle FC, Spurr N, Beverley PC (1988) Rate of incorporation of radiolabelled nucleosides does not necessarily reflect the metabolic state of cells in culture: Effects of latent mycoplasma contamination. Immunology 63; 125-31

Miller WA, Everett MM, Freedman JT, Feagans WC and Cramer JF (1976) Enzyme separation techniques for the study of growth of cells from layers of bovine dental pulp. In Vitro 12; 580-8

Miyazaki K, Takaki R, Nakayama F, Yamauchi S, Koga A, Todo S (1981) Isolation and primary culture of adult human hepatocytes. Ultrastructural and functional studies. Cell-Tissue-Res 218; 13-21

Moscona A (1952) Cell suspension from organ rudiments of chick embryos. Exp Cell Res 3; 535-539

Nakashima M (1991) Establishment of primary cultures of pulp cells from bovine permanent incisors. Archs oral Biol 36; 655-63

Oakley CL, Warrack GH, Van Heyningen WE (1946) The collagenase (K toxin) of Cl. welchii type A. J Pathol Bacteriol 58; 229-235

Phillips HJ (1972) Dissociation of single cells from lung or kidney tissue with elastase. In Vitro 8; 101-105

Poste G (1971) Tissue dissociation with proteolytic enzymes. Adsorption and activity of enzyme at the cell surface. Exp Cell Res 65; 359-367

Rinaldini LM (1958) The isolation of living cells from animal tissues. Int Rev Cytol 7; 587-647

Rinaldini LM (1958) An improved method for isolation and quantitative cultivation of embryonic cells. Exp Cell Res 16; 477-505

Romrell LJ, Coppe MR, Munro DR, Ito S (1975) Isolation and separation of highly enriched fractions of viable mouse gastric parietal cells by velocity sedimentation. J. Cell Biol 56; 428-38

Rous P, Jones FS (1916) A method for obtaining suspensions of living cells from the fixed tissues, and for the plating out of individual cells. J Exp Med 23; 549-555

Sanford WC (1974) A new method for dispersing strongly adhesive cells in tissue culture. In Vitro 10; 281-300

Schimmelpfeng L, Langenberg U, Peters JH (1980) Macrophages overcome mycoplasma infections of cells in vitro. Nature 285; 661-62

Sherman JM Jr, Haase B, Carr T und Tandler B (1988) Goblet cell isolation from cat trachea: a comparison of methods. Exp Lung Res 14; 375-385

Smidt-Jensen S, Christensen B and Lind AM (1989) Chorionic villus culture for prenatal diagnosis of chromosome defects: reduction of the longterm cultivation time. Prenat Diagn 9; 309-19

Snow C, Allen A (1970) The release of radioactive nucleic acids and mucoproteins by trypsin and ethylenediaminetetraacetate treatment of baby-hamster cells in tissue culture. Biochem J (England), Oct 1970, 119(4) p 707-14

Sullivan JC, Schafer IA (1966) Survival of pronase-treated cells in tissue culture. Exp Cell Res 43; 676

Takahashi H, Sano K, Yoshizato K, Shioya N, Sasaki K (1985) Comparative studies on methods of isolating rat epidermal cells. Ann plast surg 14; 266-285

Tepperman K, Morris G, Essien F, Heywood SM (1975) A mechanical dissociation method for preparation of muscle cell cultures. J. Cell Physiol 86; 561-65

Tokiwa T, Hoshika T, Shiraishi M, Sato J (1979) Mechanism of cell dissociation with trypsin and EDTA. Acta Med Okayama 33; 1-4

Varani J, Perone P, Inman DR, Burmeister W, Schollenberger SB, Fligiel SE, Sitrin RG, Johnson KJ (1995) Human skin in organ culture. Elaboration of proteolytic enzymes in the presence and absence of exogenouse growth factors. Am J Pathol 146; 210-217

Vielkind U and Crawford BJ (1988) Evaluation of different procedures for the dissociation of retinal pigmented epithelium into single viable cells. Pigmented cell Res 1; 419-33

Viko H, Osnes JB, Sjetnan AE and Skomedal T (1995) Improved isolation of cardiomyocytes by trypsinisation in addition to collagenase treatment. Pharmacol Toxicol 76; 68-71

Waymouth C (1974) To disaggregate or not disaggregate injury and cell diaggregation, transient or permanent? In Vitro 10; 97-111

Weinstein D (1966) Comparison of pronase and trypsin for detachment of human cells during serial cultivation. Exp Cell Res 43; 234-236

West DC, Kumar S (1986) Elastase activity in capillary and aortic endothelial cells. Anticancer Res 6; 1069-72

Willson BW, Lau TL (1963) Dissociation and cultivation of chick embryo cells with an Actinomycete protease. Proc Soc Exp Biol Med 114; 649-651

Chapter 18

Protease-Catalyzed Peptide Synthesis

DIRK ULLMANN AND HANS-DIETER JAKUBKE

Introduction

Peptides play a fundamental role in the function of living systems. Since the demand for peptides is enormous, classical chemical synthesis in solution, solid-phase synthesis and recombinant techniques belong to the most important methods of peptide synthesis. Unfortunately, neither of these methods is completely free of drawbacks. Especially chemical synthesis of peptides often suffers from problems such as time-consuming side-chain protection/deprotection requirements, racemization and other side-reactions. Furthermore, in recombinant DNA methods, the incorporation of unnatural amino acids is limited in preparative scale. For these reasons, the use of suitable peptide ligases should provide an attractive alternative by combining the flexibility of chemical strategies with the advantages of the regio- and stereospecific enzymatic reactions (Jakubke 1987; Bongers and Heimer 1994; Jakubke 1994; Bordusa et al. 1997).

It could be demonstrated that important approaches towards suppressing competitive reactions in the reversal of proteolysis are: leaving group manipulations of the acyl donor ester in kinetically-controlled synthesis; peptide synthesis in frozen-aqueous systems and zymogen-catalyzed peptide synthesis (Jakubke 1995). In addition, in both equilibrium-controlled synthesis and in the kinetic approach (Schellenberger and Jakubke 1991) synthetic reactions in high-density media provide high peptide yields using equal concentrations of the educts and avoiding high concentrations of organic co-solvents. The use of proteases to perform selective transformations

Correspondence to: Dirk Ullmann, EVOTEC BioSystems AG, Schnackenburgallee 114, Hamburg, 22525, Germany (*phone* +49-040-56081247; *fax* +49-040-56081222; *e-mail* ullmann@evotec.de)
Hans-Dieter Jakubke, Leipzig University, Faculty of Biosciences, Pharmacy and Psychology, Institute of Biochemistry, Talstrasse 33, Leipzig, 04103, Germany (*phone* (0341) 9736901; *fax* (0341) 9736998 *e-mail* jakubke@server1.rz.uni-leipzig.de)

in peptide synthesis is advantageous since chemical ligation methods are prone to racemization and suffer from time-consuming side-chain protection/deprotection necessities.

The purpose of this contribution focuses on peptide coupling reactions on a preparative scale in which native and immobilized proteases can be used. In the following sections a series of protocols for standard techniques in protease catalyzed peptide synthesis widely used in preparative scale are compiled. Most of the chosen protocols follow the kinetic approach according to the protease catalyzed acyl transfer (Fig. 1). Furthermore, the selection of the examples was chosen with respect to the use of nonconventional reaction systems and to demonstrate the application up to semisynthesis of biologically active peptides and its analoga.

Subprotocol 1
Synthesis in organic-aqueous solvent systems

The enzyme and the reactants are dissolved in a monophasic solution consisting of water and a water-miscible organic co-solvent, such as N,N-dimethyl formamide (DMF), dimethylsulfoxide (DMSO), tetrahydrofuran, dioxane or alcohols. Mixtures of this type are mainly used for reactants which are slightly soluble in aqueous systems, the same applies to longer peptide segments. Mostly, water-miscible organic solvents can be applied in concentrations up to 10% of the total reaction volume, in some cases its amount can be increased to up to 50%. If the proportion of the solvent system exceeds the critical value, the essential bound water is removed from the protein surface leading to deactivation.

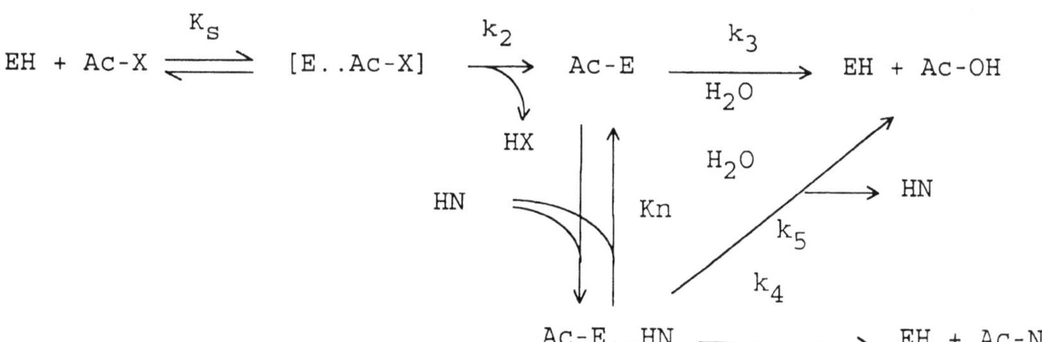

Fig. 1. Reaction pathway of protease-catalyzed acyl transfer. EH, free enzyme; Ac-X, acyl donor; HX, leaving group; Ac-E, acyl enzyme; Ac-OH, hydrolysis product; HN, nucleophile; Ac-N, aminolysis product

The advantages of synthesis in organic-aqueous solvent mixtures are promoting equilibrium-controlled synthesis and enhancement of educt solubility. This approach is still the most important enzymatic method in production scale synthesis.

Example

Chymotrypsin-catalyzed (3+7) segment synthesis of D-Phe[6]-LH RH: The fragment coupling 3+7 is a preferred strategy for the synthesis of LH-RH decapeptides. The fragments Pyr-His-Trp-OEt and H-Ser-Tyr-D-Phe-Leu-Arg-Pro-Gly-NH$_2$ can be synthesized by conventional methods without side reactions and racemization. However, the fragment condensations by chemical means is accompanied by up to 3% racemization at the activated Trp leading to a reduced peptide purity (Kemp 1979). α-chymotrypsin was examined as biocatalyst for 3+7 segment condensation in the synthesis of D-Phe[6]-LH-RH (Schellenberger et al. 1990).

Materials

α-chymotrypsin (E.C. 3.4.22.2.) is from Merck (specific activity 47 U/mg). N,N-dimethyl formamide (Chromasolv) is purchased from Fluka (Switzerland). Generally, DMF should be of high purity for all enzymatic syntheses. In addition, it can be purified by stirring it with acidic aluminium oxide (Fluka), because traces of dimethylamine are to be neutralized. The peptide fragments Pyr-His-Trp-OEt (1) and H-Ser-Tyr-D-Phe-Leu-Arg-Pro-Gly-NH$_2$ (2) were provided by Berlin-Chemie AG (Germany) and were kept at – 20°C. Reversed-phase HPLC can be performed on LiChrosorb RP8; a) analytical: 5 µm particle size, flow 1.0 ml/min, 250x4 mm I.D. (Merck) b) preparative: 10 µm particle size, flow 15 ml/min, 250x22 mm I.D. (Merck). Mobile phase: 0.05 M NaH$_2$PO$_4$, pH 2.2 / 26% acetonitrile (isocratic conditions). The peptides were detected by UV-absorption at 280 nm.

Procedure

Peptide synthesis

1. Suspend 21 g (43.7 mmol) of Pyr-His-Trp-OEt and 20 g (23.9 mmol) of H-Ser-Tyr-D-Phe-Leu-Arg-Pro-Gly-NH$_2$ in a mixture of 66 ml of DMF and 70.5 ml of water. Adjust the pH value of that mixture to 8.0 by addition of 2 M KOH under vigorous stirring.

2. Initiate the reaction by adding 50 mg chymotrypsin and stir the mixture for 6.5 hours in a 250 ml round-bottom flask with a magnetic stirrer.

3. Stop the enzymatic reaction by acidification with 10 ml of glacial acetic acid.

4. Evaporate approximately 90% of the solvents in a rotary evaporator. Take care that the temperature of the water bath is below 60°C to avoid decomposition of the peptide.

5. Precipitate the crude peptide by the addition of dry acetone. The yield should be higher than 80% (24.4g).

6. Purify the crude peptide by reversed-phase chromatography to give 16.2 g (53%) LH RH peptide and characterize it by analytical HPLC (> 99%), amino acid analysis and ESI-mass spectrometry.

The reaction time of 6.5 h was found to be optimal under the reaction con-   **Comments**
ditions given. However, in competition to the aminolysis reaction the Trp^3-Ser^4 peptide bond in the product is cleaved after 15-20 h. The chymotryptic hydrolysis at the carboxyl group of Tyr was inhibited due to the presence of D-Phe at position 6.

Subprotocol 2
Peptide Synthesis Using Immobilized Proteases

The application of proteases for peptide synthesis has been stimulated to a considerable degree by the availability of immobilized enzymes. Proteases can be immobilized without loss of function, and the potential of immobilized proteolytic enzymes for peptide synthesis has been demonstrated (Kuhl et al. 1980; Jakubke and Könnecke 1987). The simplified work-up procedure that becomes possible when immobilized proteases are used, the long-term stability of the immobilized enzyme preparations, and the successful reutilization experiments are among the advantages to such an approach. Covalently bound proteases catalyze peptide bond formation as effectively as the native enzymes, since substrate specificity and stereospecificity appear not to be altered by immobilization. Other main advantages are: The peptides synthesized are free of contamination by proteolytic activities and denatured protein. Due to the increased stability in the presence of organic solvents, higher concentrations of such solvents can be used to influence the position of the thermodynamic equilibrium.

Beside all these important advantages, several requirements must be fulfilled: The immobilization procedure should provide high and reproducible activity yields. In order to minimize leakage phenomena covalent coupling procedures should be preferred. The need of rapid attainment of the equilibrium requires high enzyme concentrations (thermodynamical approach). The support material should be easily derivatized and stable under the reaction conditions including no significant swelling capacity. Kinetically-controlled syntheses promise favorable results for use with immobilized proteases because only low concentrations of enzymes and organic solvents are needed to dissolve the reactants.

Materials

α-chymotrypsin (E.C. 3.4.22.2.) attached to macroporous acrylic beads and thermolysin (E.C. 3.4.24.4.) (protease type X-A from *Bacillus thermoproteolyticus rokko*) attached to 4% cross-linked beaded agarose are from Sigma (Germany). The assay of immobilized chymotrypsin preparations is spectrophotometrically performed with Glt-Leu-Phe-pNA as substrate (Jakubke et al. 1980). Amino acid derivatives can be obtained from Bachem (Switzerland) or can be synthesized by standard peptide synthesis protocols. Prior to use, their identity and purity are confirmed by HPLC and TLC. High-performance liquid chromatography is carried out using isocratic elution and variable-wavelength UV detection. Separations are performed with a Vydac 218TP54 peptide-column (The Separations Group).

Note: It should be stressed that immobilization is easy to perform by do-it-yourself. α-chymotrypsin could be bound to macroporous silica (20 nm pore diameter) (250 mg wet gel containing about 2.5 mg of immobilized protein). Thermolysin could be immobilized on Enzacryl AH according to standard procedures (Jakubke and Könnecke 1987).

Procedure

Kinetic approach: Chymotrypsin-catalyzed synthesis of Ac-Phe-Ala-NH$_2$

Example

1. Immobilized chymotrypsin is placed in a reaction vessel commonly used for solid-phase peptide synthesis with a G4 frit a t the bottom (10 mm in diameter), fitted with an overhead stirrer.

2. A solution of H-Ala-NH$_2$, obtained by neutralization of 49.4 mg (0.4 mmol) H-Ala-NH$_2$ x HCl with 0.1 ml cold 4 N NaOH and dilution with 1.6 ml 0.2 M carbonate buffer (pH 10.0), is added.

3. Stir the reaction mixture under avoiding any contact of the stirrer with the glass wall of the vessel.

4. Initiate the coupling reaction by addition of 0.3 ml of a stock solution of Ac-Phe-OMe (44.2 mg / 0.2 mmol) in dimethylformamide.

5. After 12 min the solution is removed by suction filtration, and the immobilized enzyme is washed successively with three 3 ml portions of water.

6. The next coupling cycle can be started by adding the substrate solutions in the order given above. The combined filtrates are worked up by addition of 1 ml 1 N HCl and rotary evaporation to dryness.

7. Dry the residue in vacuo and remove inorganic salts by extraction of the solid with three 30 ml portions of boiling ethyl acetate. The ethyl acetate extract should be subsequently rotary evaporated to dryness.

8. Dissolve the residue in a minimum amount of methanol and precipitate the peptide by addition of two volumes of diethyl ether.

9. In our synthesis the yield was 34.4 mg (62%). The melting point (m.p.) was determined 241-242°C and was unchanged after recrystallization from methanol.

Thermodynamic Approach

The thermodynamic barrier to peptide synthesis is predominantly due to the energy required for the proton transfer from the α-amino group to the α-carboxyl group. The equilibrium constant K_{con} of the conversion of the non-ionized reactants into the product is determined by the concentration of the nonionized forms in K_{ion} (Fig. 2).

Because of the unfavourable equilibrium position, relatively high concentrations of the reactants are required. In order to achieve an appreciable degree of synthesis, one needs to resort to some expedients and unusual techniques, such as using water-miscible organic solvents, biphasic systems or solvent-free micro-aqueous systems (see the review by Jakubke 1994).

$$R^1COO^- + H_3N^+R^2 \xrightleftharpoons{K_{ion}} R^1COOH + H_2NR^2 \xrightleftharpoons{K_{con}} R^1CO-NHR^2 + H_2O$$

Fig. 2. Ionization equilibrium in the thermodynamic approach

Thermolysin-Catalyzed Synthesis of Z-Phe-Leu-NH$_2$

Example

1. Add 29.9 mg (0.1 mmol) Z-Phe-OH, 21 mg (0.16 mmol) H-Leu-NH$_2$, and 2 ml 0.2 M Tris-maleate buffer, pH 7.0 to a small tube fitted for magnetic stirring.

2. Initiate the condensation by adding 50 mg thermolysin (approximately 10 U) attached to 4% cross-linked beaded agarose and allow the coupling reaction to proceed for 20 h at room temperature with constant stirring.

3. Separate the precipitated product from the immobilized thermolysin by treatment with three 3 ml portions of cold N,N-dimethylformamide followed by addition of 5 ml water.

4. Dilute the extract with 35 ml water and store it at 4°C overnight. The precipitated product is filtered and dried in vacuo, yielding 34.5 mg (84%) pure (TLC and HPLC) Z-Leu-Phe-NH$_2$, m.p. 191-192°C.

Note: After four uses of the immobilized thermolysin, the yield still amounted to 70%.

Subprotocol 3
Enzymatic Fragment Condensation / Leaving Group Manipulation

The combination of chemical solid-phase synthesis of oligopeptides or analoga and enzymatic coupling of these fragments demonstrates the state of the art in peptide synthesis and proves the potential of enzymes for the formation of peptide bonds (Jackson et al. 1994; Jakubke 1995). There is a growing interest in enzymatic semisynthesis for the chemical modification of peptides and proteins from recombinant DNA synthesis. The fundamental advantage of the kinetically-controlled method lies in the utilization of an activated ester of the acyl component. If the specificity of the enzyme for the acyl component can be raised several orders of magnitude higher than its specificity for any of the internal peptide cleavage sites by use of an appropriate ester leaving group, coupling can proceed cleanly to high yield, with high catalytic efficiency, even if the substrate contains multiple competing proteolytic cleavage sites (Schellenberger et al. 1991). This observation has important implications for protease-catalyzed condensation of long polypeptide fragments. These features can expand the practice of peptide synthesis beyond the current limits of conventional solution and solid-phase methods by facilitating the condensation of minimally protected peptide segments to produce long polypeptides and small proteins.

Enzymatic Semisynthesis of the Superpotent Analog of Human Growth Hormone Releasing Factor, [desNH$_2$Tyr^1,D-Ala^2,Ala^{15}]-GRF(1-29)-NH$_2$

The target analog [desNH$_2$Tyr^1,D-Ala^2,Ala^{15}]-GRF(1-29)-NH$_2$ (3) was synthesized by acylation of the precursor [Ala^{15}]-GRF(4-29)-NH$_2$ (1) with desNH$_2$Tyr-D-Ala-Asp-OR (2) [R=CH$_3$CH$_2$- (2a), CH$_3$- (2b), ClCH$_2$CH$_2$- (2c), C$_6$H$_5$CH$_2$- (2d), 4-NO$_2$C$_6$H$_4$CH$_2$- (2e)] catalyzed by V8 protease or the Glu/Asp-specific endopeptidase (GSE) from *Bacillus licheniformis* (Bongers et al. 1994) The latter is more stable than the V8 protease under the conditions employed in usual segment condensations and could be obtained from the commercial detergent Alcalase (Novo Nordisk) by low cost purification (Svendsen and Breddam 1992). The extent of conversion of (1) into (3) is limited by proteolysis at Asp^3-Ala^4 and Asp^{25}-Ile^{26}. However, this proteolyses is virtually eliminated by use of the appropriate ester leaving group, R. A systematic study of the kinetics of the GSE-catalyzed segment condensations of (1) and the series of tripeptide esters (2a-e) revealed that the rate of aminolysis versus proteolysis, and hence the conversion of (1) into (3), increases with increasing specificity (k_{cat}/K_m) of GSE for the tripeptide ester. The specificity varies in the order 2e>2d>2c≫2b>2a and appears to depend on an increase in the maximum turnover rate (V_{max}) with increasing basicity of R.

Example

Materials

V8 protease (EC 3.4.21.19) can be obtained from Sigma (protease type XVII-B from the V8 strain of *Staphylococcus aureus*). Usually it has a specific activity of 770 U/mg (against Boc-Glu-OPh).
The peptides [Ala^{15}]-GRF(4-29)-NH$_2$ (1) and desNH$_2$Tyr-D-Ala-Asp-OCH$_2$C$_6$H$_4$(4-NO$_2$) (2e) can be performed by solid-phase peptide synthesis (SPPS) and as reported (Bongers et al. 1992).

Note: Growth hormone releasing factor, GRF(1-44)-NH$_2$ is:
H-YADAIFTNSY^{10}RKVLGQLSAR^{20}KLLQDIMSRQ^{30}QGESNQERGA^{40}R-ARL-NH$_2$.
If the more stable GSE is intended to be used it could be purified form Alcalase (Merck 111362) as described (Svendsen and Breddam 1992).

▦ Procedure

1. Dissolve 15.8 mg (32.4 µmol) of the tripeptide ester (2e) in 100 µl DMF, followed by the addition of 32 µl 1 M triethanolamine and 120 µl water.

2. Dissolve 30.0 mg (8.1 µmol) of the nucleophile (1) x 6 trifluoroacetic acid (TFA) and 300 µg (5.2 µmol) CaO in 20 µl DMF, 5 µl water and 20 µl 1 M triethanolamine.

3. Combine the solutions and titrate the resulting mixture to pH 8.2 with 5 µl increments of 0.5 M NaOH (20 µl total).

4. Let the mixture equilibrate at 37°C and initiate the reaction by adding 15µl of the aqueous enzyme (10-30 mg/ml).

5. The conversion of (1) into (3) reached 90 % after 6 h. Stop the reaction by adding 600 µl of glacial acetic acid.

6. Purify the product by HPLC on a Vydac 218TP1022 column (2.2 x 25 cm) with the eluants: (A) 0.1 % TFA and (B) 0.1 % TFA/CH_3CN. Run a gradient such as 20-40 % (B) in 60 min at a flow rate of 25 ml/min.

7. Pool the product containing fractions and lyophilize. The yield should be approximately 20 mg (60%) of (3) x 5TFA.

Subprotocol 4
Syntheses in Frozen Aqueous-Systems

Cryoenzymatic Synthesis

Competitive reactions in enzymatic peptide synthesis can also be supressed by manipulation of the reaction conditions. Based on literature data that freezing decreases the rate of proteolysis and enhances hydroxylaminolysis (Grant and Alburn 1966) we have started enzymatic peptide synthesis in frozen aqueous systems (Schuster et al. 1990) in order to improve product yield of inefficient educts.

For this purpose, the enzyme is added to the reactants precooled to 0°C in a polypropylene tube and immediately inserted into liquid nitrogen (Fig. 3). After 20 s the tube is transferred into a cryostat. In both serine and cysteine protease-catalyzed synthetic reactions, amino components with low nucleophilic efficiency in solution at room temperature give high peptide yields in frozen aqueous systems. Secondary hydrolysis of the synthesized peptide bond is not observed in ice. In frozen aqueous systems, the endo-proteinase α-chymotrypsin is able to act as a reverse carboxypeptidase catalyzing coupling of free amino acids as amino components (Ullmann and Jakubke 1993). Using amino acid alkyl esters as amino components under these conditions the resulting peptide esters can be applied as acyl donors in further kinetically-controlled enzymatic syntheses. Frozen state enzymology opens completely new possibilities in enzymatic peptide synthesis. The simplest strategy of peptide bond formation which cannot be performed by chemical methods is using N-terminal free amino acid or peptide esters as acyl donors (Gerisch and Jakubke 1996).

The high peptide yields obtained in frozen systems can be explained by the assumption that freezing the reaction mixture results in an effective increase in concentration of the reactants in the remaining unfrozen liquid phase, which is in equilibrium with the ice crystals according to the "freeze-concentration-model" (Pincock and Kiovsky 1966).

The strategy of performing enzyme-catalyzed syntheses in the presence of organic cosolvents at low temperatures is a further way to increase the efficiency of the reactions. The avoidance of enzyme denaturation under such conditions, mainly due the high dielectric constant of the medium at subzero temperatures, is well known in cryoenzymology.

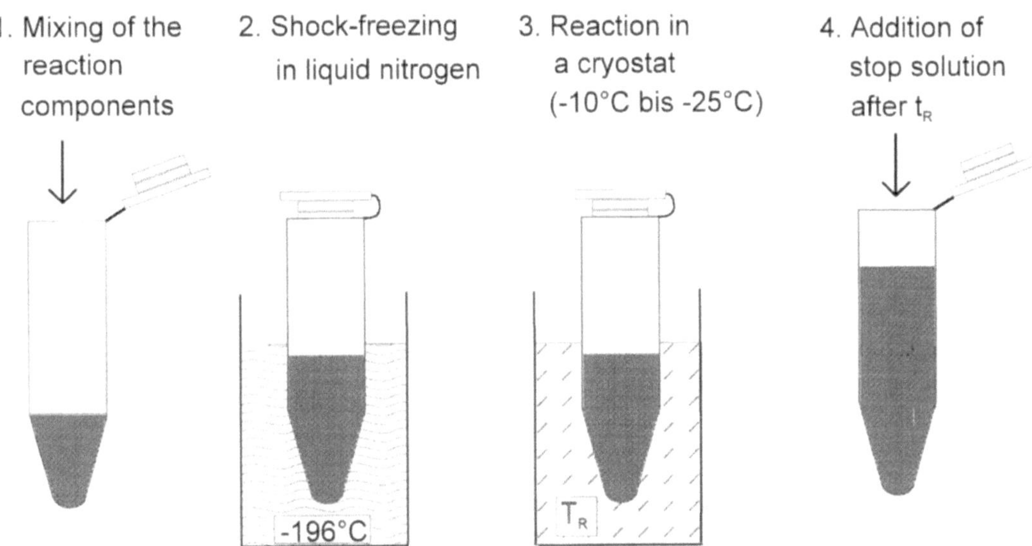

Fig. 3. Principal working scheme in frozen aqueous system

Synthesis in Frozen Aqueous System

Example α-chymotrypsin-catalyzed synthesis of α-Bet-Phe-Arg-OPr:
Under conventional reaction conditions serine and cysteine proteases require amino acid amides or peptides as nucleophiles in acyl transfer reactions because of acting as an endopeptidase. However, recently it was shown that α-chymotrypsin is able to acylate free amino acids in frozen aqueous systems (Ullmann and Jakubke 1993). The use of esters as nucleophilic components is, as mentioned above, possible but with a drastically decreased efficiency (Morihara and Oka 1977). However, acyl transfer to arginine- or lysine alkylesters in ice using α-chymotrypsin with regard to its strong preference for basic residues in the P'_1 position (subsite specificity according to Schechter and Berger 1967) was found to be the method of choice in synthesizing new potential protease substrates (for proteases with a preference for basic residues in P_1 position). Neither enzymatic synthesis at room temperature nor synthesis in organic solvents had shown to proceed in a successful manner.

Materials

α-chymotrypsin (E.C. 3.4.22.2.) can be chosen from Merck (Darmstadt) and can be used without further purification (3 times crystallized, salt-free). α-Bet-Phe-OEt was a gift from Prof. Wandrey (Institute of Biotechnology 2, Julich/Germany). However, it can be synthesized from chloracetyl-Phe-OEt by reaction with trimethylamine in an autoclave (Fischer 1994). H-Arg-OPr can be obtained from Bachem (Switzerland). Reactions were carried out in a cryostat F3 (Haake/Germany) using 5 ml polypropylene tubes (Nalgene).

Note: * In order to achieve rapid freezing in liquid nitrogen all tubes and solutions should be precooled in a refrigerator (– 20°C). This is absolutely necessary to avoid ester hydrolysis after enzyme addition and mixing the reaction components before the solution was put into nitrogen. Keep this interval as short as possible and work very quickly.

Note: * In our experiments we divided the total reaction volume into nine portions of 3 ml.
* The selection of the buffer system is critical to achive high yields. We strongly recommend the use of 0.2 M sodium carbonate-bicarbonate buffer pH 10.0 containing 0.2 M NaCl to guarantee a high ionic strength. The high pH-value results from the high pK-value of the α-amino group of H-Arg-

OPr. This fact implies also a quick working procedure to avoid non-enzymatic ester hydrolysis.

* Do not use hydrochloric acid to stop the reaction. It will not evaporate during vacuum concentration resulting in product hydrolysis.

Procedure

1. Dissolve 355 mg (1.08 mMol) Bet-Phe-OEt in 13.5 ml of water and 781 mg (2.7 mMol) H-Arg-OPr x 2 HCl in 12.2 ml 0.4 M sodium carbonate-bicarbonate buffer pH 10.0 containing 0.2 M NaCl. Store the solutions in ice/NaCl, but avoid freezing (shake from time to time).

2. Adjust the pH value of the H-Arg-OPr solution in the buffer to pH 10 by deprotonating the hydrochloride with approximately 1.3 ml 2 N NaOH (small portions, pH control).

3. Combine the solutions, shake well and divide into nine portions of each 3 ml using the 5 ml polypropylene tubes.

4. Add 30 µl α-chymotrypsin in 1 mM HCl containing 10 mM $CaCl_2$ (stock solution: 0.3 mM; 2.0 mg in 270 µl) consecutively to each reaction tube, shake, screw the tube and put it immediately into liquid nitrogen for 2 min to achieve shock-freezing.

5. Transfer the tubes from liquid nitrogen into the cryostat at – 25°C.

6. After a reaction time of 48 h stop the reaction by adding 1.5 ml of 8% trifluoroacetic acid in water to each tube.

7. After thawing the samples can be combined and concentrated on a vacuum evaporator (τ should be lower than 60°C).

8. Purify the product by preparative HPLC on a Vydac 201HS1022 column (2.2 x 25 cm) with the eluants: (A) 5 % CH_3CN in 0.1 % TFA and (B) 0.1 % TFA 95 % CH_3CN. Run a gradient of 0-65 % (B) in 45 min and 65-100 % (B) in 5 min at a flow rate of 15 ml/min.

9. Pool the product containing fractions, evaporate the acetonitrile and lyophilize. The yield should be approximately 440 mg (82%).

10. Store the product in a vacuum desiccator over P_4O_{10}.

Subprotocol 5
Cryoenzymatic Synthesis

Example α-chymotrypsin-catalyzed (3+7) segment synthesis of the luteinizing hormone releasing hormone (LH RH):

The luteinizing hormone releasing hormone Pyr-His-Trp-Ser-Tyr-Gly-Leu-Arg-Pro-Gly-NH_2 (1) is synthesized from the segments Pyr-His-Trp-OEt (2) and H-Ser-Tyr-Gly-Leu-Arg-Pro-Gly-NH_2 (3) by α-chymotrypsin catalysis (Schuster et al. 1992). The application of the cryoenzymatic approach is necessary to avoid undesired side reactions based on enzyme labile peptide bonds, e. g. ...Tyr^5-Gly^6... and ...Leu^7-Arg^8...In the D-Phe^6-analog the substrate character of these bonds is reduced (Schellenberger et al. 1990). Synthesis in frozen aqueous systems is not possible due to bad solubility of the acyl donor (2). An increase of the product yield can be achieved by saturation of the reaction mixture with KCl.

Materials

α-chymotrypsin (E.C. 3.4.22.2.) can be chosen from Merck (Darmstadt) and can be used without further purification (3 times crystallized, salt-free). The peptide fragments Pyr-His-Trp-OEt (2) and H-Ser-Tyr-Gly-Leu-Arg-Pro-Gly-NH_2 (3) were provided by Berlin-Chemie AG (Germany) and were kept at $-20°C$. However, these derivatives can be easily self-made by solution and solid-phase peptide synthesis protocols.

Procedure

1. Dissolve 62.3 mg (0.13 mmol) Pyr-His-Trp-OEt (2) and 74.8 mg (0.1 mmol) H-Ser-Tyr-Gly-Leu-Arg-Pro-Gly-NH_2 (3) in 611 µl water.

2. Add 890 µl of a saturated solution of KCl in water and 64.8 µl of 2 M NaOH to adjust the pH value.

3. Initiate the reaction by adding 10 µl of a stock solution of α-chymotrypsin (100 mg in 1 ml 10^{-3} M HCl) after equilibration at the reaction temperature.

4. Follow the reaction by RP-HPLC under isocratic elution using a mixture of 0.1% TFA and methanol (v/v = 60/40) at a flow rate of 1 ml/min with a Vydac 218TP54 column (The Separations Group).

5. After a reaction time of 5 h stop the reaction by adding 100 µl of 10% trifluoroacetic acid. The yield should be approximately 70% (83 mg).

6. Purify the product by preparative HPLC on a Vydac 201HS1022 column (2.2 x 25 cm) Use the conditions as described above for the analytical run at a flow rate of 15 ml/min.

7. Pool the product containing fractions, evaporate the methanol and lyophilize.

Subprotocol 6
Enzymatic Peptide Synthesis in High Density Media

The analysis of equilibrium position in the thermodynamic approach revealed a favourable shift towards peptide product, when educts are largely undissolved in the reaction medium and the product precipitates (Halling et al. 1995). This consideration resulted in water-based high-density media and lead to high peptide yields with equimolar supply of substrates. A further important advantage is the easy scale-up possibility from analytical to multi-mol pilot scale.

A minimal liquid phase is inevitable since enzymatic catalysis was shown to occur mostly in the liquid phase, if not exclusively (Lopez-Fandino et al. 1994). Largely undissolved substrates serve as solid phase pool from which consumed substrate can be replaced. Provided the peptide product precipitates at all, the conversion should continue until at least one of the substrate pools disappears completely. Using such a pool may shift the overall equilibrium far towards peptide product even if the reaction is performed in high-density aqueous media with equimolar supply of substrates (Eichhorn et al. 1996).

This new method attracts further attention by simple work-up procedures and its compatibility with well established chemical protection/deprotectionmethods.

Synthesis of the Sweetener Precursor Z-Aspartame

The artificial low-calorie sweetener precursor Z-Aspartame is synthesized in semi-preparative scale by thermolysin-catalyzed coupling of Z-Asp-OH and H-Phe-OMe. Since Z-Aspartame forms an adduct with H-Phe-OMe precipitating from aqueous media two equivalents of H-Phe-OMe are sup-

Example

plied. To attain a favourable pH in the reaction mixture Z-Asp-OH is neutralized with NaOH in advance. The sodium salt of Z-Asp-OH then serves to neutralize H-Phe-OMe x HCl very mildly and hydrolysis of nucleophile ester will not exceed 5% during reaction.

Materials

H-Phe-OMe x HCl and Z-Asp-OH can be obtained from Bachem (Switzerland). Thermolysin (E.C. 3.4.24.4.) (protease type X from *Bacillus thermoproteolyticus rokko*) is from Sigma (Germany).

Procedure

1. Mix intensively 267 mg (1 mmol) Z-Asp-OH with 200 µl 0.5 M Na$^+$-Hepes buffer pH 7 at 40°C in a polypropylene tube.

2. Add 240 µl 10 M NaOH and stir intensively.

3. Add 431 mg H-Phe-OMe x HCl, mix and thermostate to 40°C.

4. Start the reaction with the addition of 10 mg thermolysin in 100 µl buffer.

5. After 5-10 h suspend the precipitate in 2 ml water, adjust the pH with concentrated HCl to pH 2 and stir for 2 h.

6. The slurry can be filtered and washed with water until the filtrate shows a neutral reaction.

7. Dry the product to give an overall yield of 360 mg (84%) Z-Aspartame.

References

Bongers J, Lambros T, Liu W, Ahmad M, Campbell RM, Felix AM, Heimer EP (1992) Enzymatic semisynthesis of a superpotent analog of human growth hormone-releasing factor. J Med Chem 35:3934-3941
Bongers J, Heimer EP (1994) Recent applications of enzymatic peptide synthesis. Peptides 15:183-193
Bongers J, Liu W, Lambros T, Breddam K, Campbell RM, Felix AM, Heimer EP (1994) Peptide synthesis catalyzed by the Glu/Asp-specific endopeptidase. Int J Peptide Protein Res 44:123-129

Bordusa F, Ullmann D, Jakubke HD (1997) Peptide synthesis form N- to C-terminus: an advantageous strategy using protease catalysis. Angew Chem 1st Ed Engl 109:1099-1101

Eichhorn U, Bommarius AS, Drauz K, Jakubke HD (1997) Synthesis of dipeptides by suspension-to-suspension conversion via thermolysin catalysis - from analytical to preparative scale. J Pept Sci,3:245-257

Fischer A, Bommarius AS, Drauz K, Wandrey C (1994) A novel approach to enzymatic peptide synthesis using highly solubilizing N-alpha-protecting groups of amino acids. Biocatalysis 4:289-307

Gerisch S, Jakubke HD (1997) Enzymatic peptide synthesis in frozen aqueous solution: use of N^α-unprotected peptide esters as acyl donors. J Peptide Sci, 3:93-98

Grant NH, Alburn HE (1966) Acceleration of enzyme reactions in ice. Nature 212:194

Halling PJ, Eichhorn U, Kuhl P, Jakubke HD (1995) Thermodynamics of solid-to-solid-conversion and application to enzymic peptide synthesis. Enzyme Microb Technol 17:601-606

Jackson DY, Burnier J, Quan C, Stanley M, Tom J, Wells JA (1994) A designed peptide ligase for total synthesis of ribonuclease A with unnatural catalytic residues. Science 266:243-247

Jakubke HD, Däumer H, Könnecke A, Kuhl P, Fischer J (1980) Specificity in the hydrolysis of N-acyl-L-phenylalanine-4-nitroanilide by α-chymotrypsin. Experientia 36:1039-1040

Jakubke HD (1987) Enzymatic peptide synthesis. In: Udendriend S, Meienhofer J (eds) The Peptides: Analysis, Synthesis, Biology, vol 9. Academic Press, New York, pp 103-165

Jakubke HD, Könnecke A (1987) Peptide synthesis using immobilized proteases. In: Mosbach K (ed) Immobilized enzymes and cells (Part C). Methods Enzymol, vol 136. Academic Press, London, pp 178-188

Jakubke HD (1994) Protease-catalyzed peptide synthesis: basic principles, new synthesis strategies and medium engineering. J Chin Chem Soc 41:355-370

Jakubke HD (1995) Peptide ligases: tools for peptide synthesis. Angew Chem Int Ed Engl 34:175-177

Kemp DS (1979) Racemization in peptide synthesis. In: Gross E, Meienhofer J (eds) The Peptides: Analysis, Synthesis, Biology, vol 1. Academic Press, New York, pp 315-383

Kuhl P, Könnecke A, Döring G, Däumer H, Jakubke HD (1980) Enzyme-catalyzed peptide synthesis in biphasic aqueous-organic systems. Tetrahedron Lett 21:893-896

Lopez-Fandino R, Gill I, Vulfson EN (1994) Enzymatic catalysis in heterogeneous mixtures of substrates: The role of the liquid phase and the effects of "adjuvants". Biotechn Bioeng 43:1016-1023

Morihara K, Oka K (1977) Alpha-chymotrypsin as the catalyst for peptide synthesis. In: Nakajima H (ed) Peptide Chemistry 1976. Protein Research Foundation, Osaka, pp 9-16

Pincock RE, Kiovsky TE (1966) Kinetics of reactions in frozen solutions. J Chem Educ 43:358-360

Schechter I, Berger A (1967) On the size of the active site in proteases. I. Papain. Biochem Biophys Res Commun 27:157-162

Schellenberger V, Schellenberger U, Jakubke HD, Hänsicke A, Bienert M, Krause E (1990) Chymotrypsin-catalyzed fragment coupling synthesis of D-Phe^6-GnRH. Tetrahedron Lett 31:7305-7306

Schellenberger V, Jakubke HD (1991) Protease-catalyzed kinetically controlled peptide synthesis. Angew Chem Int Ed Engl 30:1437-1449

Schellenberger V, Görner A, Könnecke A, Jakubke HD (1991) Protease-catalyzed peptide synthesis: Prevention of side reactions in kinetically controlled reactions. Peptide Res 4:265-269

Schuster M, Aaviksaar A, Jakubke HD (1990) Enzyme-catalyzed peptide synthesis in ice. Tetrahedron 46:8093-8102

Schuster M, Aaviksaar A, Jakubke HD (1992) α-chymotrypsin-catalyzed (3+7) segment synthesis of the luteinizing hormone releasing hormone. Tetrahedron Lett 33:2799-2802

Svendsen I, Breddam K (1992) Isolation and amino acid sequence of a glutamic acid specific endopeptidase from *Bacillus licheniformis*. Eur J Biochem 204:165-171

Ullmann G, Jakubke HD (1993) Frozen aqueous systems: New efficient reaction media for enzyme-catalyzed peptide bond formation. In: Schneider CH, Eberle AN (eds) Peptides 1992. ESCOM Sci Publ, pp 36-37

Appendix

Supplement to Chapter 9:
Crystallization Conditions for Proteases

M.T. STUBBS AND M.M.T. BAUER

Crystallization Conditions for Poteinases

Table 1. Crystallization conditions are given for all proteinases present in the Protein Data Bank (PDB) at the time of writing. The proteinases are ordered according to their EC numbers, subgrouped according to species, and further subdivided according to crystal form. References given are to both the completed structure, and where available, the original crystallization conditions. When using this table to find crystallization conditions for your proteinase, please seek out the relevant original literature. Note that published crystallization conditions may deviate slightly from the 'real' ones.

pdb	nat	inhibitor	pc.	buffer	pH	precipi-tant	additive	T °C	met	spcgr	cell edges (Å)	angles (°)	res (Å)	lit
AMINOPEPTIDASES (E.C.3.4.11.x)														
Leucine aminopeptidase (E.C.3 4.11.1), *Bos taurus*, lens														
1BPM	nat	none		50mM Tris	7.8	MPD	15µM ZnSO$_4$	4	seed	P6$_3$22	130.3 130.3 121.9	90 90 120	2.4	(1)
Also 1BPN, 1BLL (amastatin)														
Aminopeptidase (E.C.3.4.11.10), *Aeromonas proteolytica*														
1AMP	nat								hd	P6$_1$22	111.0 111.0 92.1	90 90 120	1.8	(2)
CARBOXYPEPTIDASES (E.C.3.4.12-17.x)														
Carboxypeptidase A (E.C.3.4.12.2), *Sus scrofa*, pancreas														
1PCA	pro-	BA	10	1M cit	5.0	1.5M AS		4	vd	P2$_1$2$_1$2	127.1 76.9 47.5	90 90 90	2.0	(3)
Carboxypeptidase B (E.C.3.4.12.3), *Bos taurus*, pancreas														
1CPB	nat									P3$_2$	54.9 54.9 104.6	90 90 120	2.8	(4)
Serine carboxypeptidase II (E.C.3.4.16.1), *Triticum vulgaris*, wheat germ														
3SC2	nat				4-6	PEG/NaCl				P4$_1$2$_1$2	98.4 98.4 209.5	90 90 90	2.0	(5)
1WHS	nat									P4$_1$2$_1$2	95.6 95.6 208.6	90 90 90	2.3	(6)
Also 1WHT (ssi)														
D-alanyl-D-alanine carboxypeptidase / transpeptidase (E.C.3.4.16.4), *Streptomyces*														
1PTE	nat									P2$_1$2$_1$2$_1$	51.1 67.4 102.9	90 90 90	2.8	(7)
Also 2PTE, 3PTE														

Table 1. Continuous

pdb	nat	inhibitor	pc.	buffer	pH	precipi-tant	additive	T °C	met	spcgr	cell edges (Å)	angles (°)	res (Å)	lit
Serine carboxypeptidase CPY (E.C.3.4.16.5), *Saccharomyces cerevisae*														
1CPY	mut.		10	0.1M Imid.	6-8	20% PEG 6	0.3M NaAc 0.05M NaCl.		hd	$P2_13$	112 112 112	90 90 90	2.6	(8)
							Also 1YSC							
Carboxypeptidase A (E.C.3.4.17.1), *Bos taurus*, pancreas														
6CPA	nat	phosph.		20mM Tris	7.4	6% PEG 8	0.25M LiCl	4	vd	$P2_12_12_1$	61.9 67.2 76.2	90 90 90	2.0	(9)
							Also 7CPA, 8CPA							
4CPA	nat	potato-inhibitor		0.01M Tris	7.5	4% PEG6	0.12M LiCl	4	md	$P3_2$	53.45 53.45 218.5	90 90 120	2.5	(10)
2CTB	nat	none		0.02M Tris	7.5		0.2M LiCl	4	d	$P2_1$	51.6 60.27 47.25	90 97.3 90	1.5	(11)
							Also 5CPA, 1CBX, 2CTC, 1CPS, 3CPA							
SERINE PROTEINASES (E.C.3.4.21.x)														
Chymotrypsin (E.C.3.4.21.1), *Bos taurus*, pancreas														
5CHA	nat α	none	6		4	50% AS		15	soak	$P2_1$	49.3 67.2 65.9	90 101.7 901.8		(12)
							Also 4CHA, 2CHA, 6CHA							
1ACB	nat α	eglin C	22	0.05M PP	6.5	10% PEG 4/6			sd	$P2_1$	55.3 59.4 42.5	90 99.1 90	2.0	(13)
1CHO	nat α	OMTKY3		0.05M PP	5-6	50% AS			vd	$P2_1$	44.9 54.5 57.2	90 103.9 901.8		(14)
1CGI	pro-	PSTI3(rec)	10	0.05M Tris	8.2	1.4-1.6 M AS				$P4_12_12$	84.4 84.4 86.7	90 90 90	2.3	(15)

Table 1. Continuous

pdb	nat	inhibitor	pc.	buffer	pH	precipi-tant	additive	T °C	met	spcgr	cell edges (Å)	angles (°)	res (Å)	lit
1CGJ	pro-	PSTI4(rec)												
2CGA	pro-		80	0.2M PP	4.5	30% Ethanol				$P2_12_12_1$	59.3 77.1 110.1	90 90 90	1.8	(16)
1CHG	pro-		80	0.2M PP	4.5	30% Ethanol				$P2_12_12_1$	52.0 63.9 77.1	90 90 90	2.5	(17)
8GCH	nat γ	sp								$P4_22_12$	69.0 69.0 95.3	90 90 90	1.6	(18)

Also 1GCD (ssi)

pdb	nat	inhibitor	pc.	buffer	pH	precipi-tant	additive	T °C	met	spcgr	cell edges (Å)	angles (°)	res (Å)	lit
2GMT	nat γ	ssi								$P4_22_12$	69.5 69.5 97.6	90 90 90	1.8	(19)

Also 2GCH, 1GCT, 2GCT, 3GCT, 1GMC, 1GMD, 1GHA (all none); 3GCH, 4GCH, 6GCH, 7GCH, 1GMH, 1GHB (all ssi)

Trypsin (E.C.3.4.21.4), *Bos taurus*, pancreas

pdb	nat	inhibitor	pc.	buffer	pH	precipi-tant	additive	T °C	met	spcgr	cell edges (Å)	angles (°)	res (Å)	lit
2PTC	nat	PTI								I222	75.5 84.4 122.9	90 90 90	1.9	(20)

Also 1TPA (anh.)

pdb	nat	inhibitor	pc.	buffer	pH	precipi-tant	additive	T °C	met	spcgr	cell edges (Å)	angles (°)	res (Å)	lit
2TPI	pro-	PTI				2.4M MgSO$_4$				I222	75.7 84.2 122.9	90 90 90	2.1	(21)

Also 3PTI, 4PTI, 2TGP

pdb	nat	inhibitor	pc.	buffer	pH	precipi-tant	additive	T °C	met	spcgr	cell edges (Å)	angles (°)	res (Å)	lit
1TYN	nat	ssi	15	0.1M Tris		30% PEG 4	0.2M Na-cit	4	vd	$P4_32_12$	71.7 71.7 88.8	90 90 90	2.0	(22)
4PTP	nat	ssi				2.4M AS			vd	$P2_12_12_1$	54.8 58.6 67.5	90 90 90	1.34	(20)

Also 3PTB, 1TPP, 1TPO, 1NTP, 1TNJ, 1TNK, 1TNH, 1TNI, 1TNG, 1TNL, 2PTN

pdb	nat	inhibitor	pc.	buffer	pH	precipi-tant	additive	T °C	met	spcgr	cell edges (Å)	angles (°)	res (Å)	lit
1GBT	mod	BA	30	0.05M Tris	8.2	1.2M MgSO$_4$	6 mM CaCl$_2$	22	vd	$P2_12_12_1$	63.7 63.5 68.9	90 90 90	2.0	(23)

Also 1TLD, 1SMF, 1PPH, 1PPC, 1TPS

pdb	nat	inhibitor	pc.	buffer	pH	precipi-tant	additive	T °C	met	spcgr	cell edges (Å)	angles (°)	res (Å)	lit
2TLD	nat	str.inh		0.05M Tris	7.0	80% MgSO$_4$				I222	110.9 116.8 64.3	90 90 90	2.6	(24)

Table 1. Continuous

pdb	nat	inhibitor	pc.	buffer	pH	precipitant	additive	T °C	met	spcgr	cell edges (Å)	angles (°)	res (Å)	lit
1PPE	nat	CMTI-I		0.1M PP	5.0	20% PEG 6			vd	$P2_12_12_1$	59.3 55.5 74.6	90 90 90	2.0	(25)
1TAB	nat	AB-I		0.05M PP	7.2	18% AS			vd	$P4_12_12$	55.4 55.4 181.7	90 90 90	2.3	(26)
3PTN	nat	BA	60	0.05M MES	5-7	1.6M AS	1mM $CaCl_2$	20	hd	$P3_121$	55.1 55.1 109.4	90 90 120	1.7	(27)
1TGC	pro-					2.4M $MgS O_4$				$P3_121$	55.1 55.1 109.4	90 90 120	1.8	(21)
						Also 1TGT, 2TGT, 2TGA, 2TGD, 1TGB								
1TGS	nat	PPSTI	10	0.1M Ac.	6.7	1.9M $MgSO_4$			md	$P2_12_12_1$	67.1 75.5 66.9	90 90 90	1.8	(28)
Trypsin (E.C.3.4.21.4), *Sus scrofa*, pancreas														
1MCT	nat	MCTI		0.2 M $NaPO_4$	6-7	85% sat NaCl		24	hd	$P3_221$	62.7 62.7 124.3	90 90 120	1.6	(29)
1EPT	nat ε		20	10mM Ac.	6.5	0.2M Na-PP		24	vd	$P2_12_12_1$	76.9 53.4 46.6	90 90 90	1.8	(30)
1TFX	nat	TFPI-II		1M Na/ K PO4	8.0			20	hd	$P2_12_12_1$	41.9 96.2 137.7	90 90 90	2.6	(31)
Trypsin (E.C.3.4.21.4), *Homo sapiens*, pancreas														
1TRN	nat	DFP								P4	107.1 107.1 39.9	90 90 90	2.2	(32)
Trypsin (E.C.3.4.21.4), *Rattus rattus*, pancreas														
1BRB	mut	BPTI		0.1M Na-Cit	6.7	37% PEG335	0.2M AS	4	hd	$P3_221$	92.7 92.7 62.3	90 90 120	2.1	(33)
1BRC	mut	APPI							vd					
1TRM	mut	BA			8	$MgSO_4$			vd	$P2_12_12_1$	40.4 92.0 127.3	90 90 90	2.3	(34)

Table 1. Continuous

pdb	nat	inhibitor	pc.	buffer	pH	precipi-tant	additive	T °C	met	spcgr	cell edges (Å)	angles (°)	res (Å)	lit
1BRA	mut	BA								I23	124.4 124.4 124.4	90 90 90	2.2	(35)
							Also 2TRM							
Trypsin (E.C.3.4.21.4), *Salmo sclar*														
1BIT	nat	BA								$P2_12_12$	77.1 82.3 31.2	90 90 90	1.8	(36)
2TBS	nat	BA								$P2_12_12$	62.0 84.3 39.1	90 90 90	1.8	(37)
Trypsin (E.C.3.4.21.4), *Streptomyces griseus*														
1SGT	nat		13			2M AS	10 mM Ca-Ac			$C222_1$	72.3 51.0 120.1	90 90 90	1.7	(38)
Thrombin (E.C.3.4.21.5), *Homo sapiens*, plasma														
1HGT	nat α	none, hir-der	3.7	0.1M Na-PP	7.3	30% PEG 8	0.2M NaCl		vd	C2	70.8 71.7 72.6	90 100.9 902.2		(39)
		Also 1HAI, 1HAH, 1THR, 1ABI, 2HGT, 1IHS, 1IHT, 1THS, 3HAT, 1HBT, 1NRR, 1NRS, 1FPC, 1HAG												
1HLT	nat α	PPACK+TM								$P4_3$	50.9 50.9 325.8	90 90 90	3.0	(40)
1DWB	nat α	Inh.+ hir-der	10	0.1M HEPES	7.5	25% PEG 4	0.2M CaCl$_2$ 0.05M NaCl	20	vd	$P4_32_12$	90.8 90.8 132.5	90 90 90	3.2	(41)
		Also 1DWC, 1DWD, 1DWE												
1PPB	nat α	PPACK	10	0.2M Na-PP	7.0	16% PEG 6	0.2M NaCl	20	md	$P2_12_12_1$	87.7 67.8 61.1	90 90 90	1.9	(42)
		Also 1ABJ												
2HPP	nat α	PPACK+PF2								$P4_12_12$	122.7 122.7 103.7	90 90 90	3.3	(43)
		Also 2HPQ												

Table 1. Continuous

pdb	nat	inhibitor	pc.	buffer	pH	precipi-tant	additive	T °C	met	spcgr	cell edges (Å)	angles (°)	res (Å)	lit
4HTC	nat α	hirudin	5	0.1M Na-ac	4.5	30% PEG 4	0.2M MgCl$_2$	4	vd	P4$_3$2$_1$2	90.5 90.5 132	90 90 90	2.3	(44)
						Also 3HTC								
1FPH	nat α	FPA + hir-der	4	0.2M MES	6	1.5M Na-PP	10% PEG 6		hd	P4$_3$2$_1$2	90.5 90.5 132	90 90 90	2.5	(45)
1TMT	nat α	CPG	6	0.1M Na-ac	4	12% PEG6	0.1M NaCl		hd	P2$_1$2$_1$2	80.5 107.1 45.8	90 90 90	2.2	(46)
						Also 1TMU								
1HUT	nat α	PPACK+DNA-fr.								P2$_1$2$_1$2$_1$	56.5 77.4 99.5	90 90 90	2.9	(47)
1NRN	nat α	peptide								C2	129.8 51.9 63.3	90 101 90	3.1	(48)
						Also 1NRO, 1NRP, 1NRQ								
2HNT	nat γ			0.1M Tris	8.5	28% PEG 8	0.2M Na-ac		vd	C2	126.5 48.2 52.4	90 96.2 90	2.5	(49)

Thrombin (E.C.3.4.21.5), *Bos taurus*, plasma

pdb	nat	inhibitor	pc.	buffer	pH	precipi-tant	additive	T °C	met	spcgr	cell edges (Å)	angles (°)	res (Å)	lit
1HRT	nat α	hirudin								C222$_1$	59.1 102.6 143.3	90 90 90	3.3	(50)
1BBR	nat ε	FP-α			8.0	40% AS				P2$_1$	83.0 89.4 99.3	90 106.6 902.3		(51)
1ETR	nat ε	ssi	10	0.1M PP	8.0	1.9M AS		20	vd	P4$_2$2$_1$2	87.7 87.7 103.0	90 90 90	2.2	(52)
						Also 1ETT, 1ETS								
1TBA	nat α	rhodniin	10	0.3M PP	6.0	10% PEG4		0	vd	C2	114.8 112.3 92.1	90 95.0 90	2.6	(53)

Table 1. Continuous

pdb	nat	inhibitor	pc.	buffer	pH	precipi-tant	additive	T °C	met	spcgr	cell edges (Å)	angles (°)	res (Å)	lit
Coagulation factor Xa (E.C.3.4.21.6), *Homo sapiens*, plasma														
1HCG	nat		10	0.1M HEPES	7.5	2M Na/ K-PP	0.2M AS			$P4_12_12$	93.7 93.7 119.3	90 90 90	2.2	(54)
α-lytic protease (E.C.3.4.21.12), *Lysobacter enzymogenes*														
1P11	rec	ssi	20		7.2	1.3M LiSO$_4$			seed	P3$_2$21	66.6 66.6 80.1	90 90 120	1.9	(55)
Also 1P12, 2PO7, 1PO1, 1PO2, 1PO3, 1PO4, 1PO5, 1PO6, 1PO8, 1PO9, 1P10, 2PO8, 1LPR, 2LPR, 3LPR, 4LPR, 5LPR, 6LPR, 7LPR, 8LPR, 9LPR, 2ALP														
Mesentericopeptidase (E.C.3.4.21.14), *Bacillus mesentericus*														
1MEE	nat	eglin-C	5	0.025M PIPES	7.0	18% PEG 4			vd	$P2_1$	43.0 72.0 48.3	90 110 90	2.0	(56)
Coagulation factor IXa (E.C.3.4.21.22), *Sus scrofa*, plasma														
1PFX	nat	PPACK	10	0.5M ac	6.6	22% PEG6		20	hd	$P4_12_12$	128.8 128.8 77.0	90 90 90	3	(57)
Kallikrein A (E.C.3.4.21.35), *Sus scrofa*, pancreas														
2KAI	nat	BPTI								$P4_12_12$	106.2 106.2 108.6	90 90 90	2.5	(58)
2PKA	nat									$P4_12_12$	90.2 90.2 159.4	90 90 90	2.1	(59)
Elastase (E.C.3.4.21.36), *Sus scrofa*, pancreas														
7EST	nat.	ssi								$P2_12_12_1$	52.2 57.5 75.3	90 90 90	1.8	(60)
Also 1EST, 2EST, 3EST, 4EST, 5EST, 6EST, 8EST														
Elastase (E.C.3.4.21.37), *Homo sapiens*, leucocyte														
1EAU		SSI								$P2_12_12_1$	50.9 58.3 75.7	90 90 90	2.1	(61)
Also 1EAS, 1EAT, 1ESB, 1ELA, 1ELB, 1ELC, 1ELD, 1ELE, 1ESA, 1INC, 9EST, 1JIM														

Table 1. Continuous

pdb	nat	inhibitor	pc.	buffer	pH	precipitant	additive	T °C	met	spcgr	cell edges (Å)	angles (°)	res (Å)	lit
Elastase (E.C.3.4.21.37), *Homo sapiens*, neutrophil														
1HNE	nat.	ssi								$P6_3$	74.5 74.5 70.9	90 90 120	1.8	(62)
Achromobacter protease I (E.C.3.4.21.50), *Achromobacter lyticus*														
1ARC	nat	ssi								P1	37.3 42.8 48.0	120 113 69	2.0	(63)
1ARB	nat									P1	39.5 40.4 43.9	115 114 74	2.0	(63)
M-protease (E.C.3.4.21.62), *Bacillus spec.*														
1MPT	nat									$P2_12_12_1$	75.8 57.8 54.2	90 90 90	2.4	(64)
Subtilisin BL (E.C.3.4.21.62), *Bacillus lentus*														
1ST3	nat		26	50mM cit	5.8	15% PEG3.3	2mM $CaCl_2$		hd	$P2_12_12_1$	53.1 61.5 75.6	90 90 90	1.4	(65)
Subtilisin Carlsberg (E.C.3.4.21.62), *Bacillus licheniformis*														
1SCA	nat		5		5.6	17% Na_2SO_4			b	$P2_12_12_1$	76.5 55.3 53.4	90 90 90	2.0	(66)
Also 1SCB, 1SCD														
Subtilisin Carlsberg (E.C.3.4.21.62), *Bacillus subtilis*														
2SEC	nat	eglin C		0.7M PP	5.6				hd	P1	38.3 41.4 56.5	70 84 75	1.8	(67)
1CSE	nat	eglin C								P1	38.3 41.5 57.0	112 86 105	1.2	(68)
1SEL	mut.		15	0.2M IM	8	20% PEG 8		20	vd	$P2_1$	75.8 65.4 53.3	90 107 90	2.0	(69)
Subtilisin Novo BPN' (E.C.3.4.21.62), *Bacillus subtilis*														
1SIB	nat	eglin C mutant	60	0.1M K-PP	6	10% PEG 4			hd	$P3_12_1$	84.9 84.9 89.1	90 90 120	2.4	(70)
Also 1SBN														

Table 1. Continuous

pdb	nat	inhibitor	pc.	buffer	pH	precipi-tant	additive	T °C	met	spcgr	cell edges (Å)	angles (°)	res (Å)	lit
Subtilisin Novo (E.C.3.4.21.62), *Bacillus amyloliquefaciens*														
2SNI	nat	CI-2								C2	103.2 56.8 68.7	90 127.5 90	2.1	(67)
Proteinase K (E.C.3.4.21.64), *Tritirachium album*														
2PRK	nat		100	0.05M Tris	6.5	1M NaNO$_3$	10 mM CaCl$_2$		md	P4$_3$2$_1$2	68.2 68.2 108.3	90 90 90	1.5	(71)
						Also 2PKC, 1PEK, 3PRK, 1PTK								
Thermitase (E.C.3.4.21.66), *Thermoactinomyces vulgaris*														
1THM	nat		5	0.1M Na-Ac	6	22% AS	3% MPD			P2$_1$2$_1$2$_1$	73.0 64.1 47.6	90 90 90	1.4	(72)
1TEC	rec	eglin C								P2$_1$2$_1$2$_1$	63.3 72.1 89.3	90 90 90	2.2	(73)
Tissue type plasminogen activator (E.C.3.4.21.68), *Homo sapiens*, plasma														
1RTF	rec	BA	3	0.2M HEPES	7.5	20% EtOH	0.01M K-PP	4	sd	P3$_1$21	59.2 59.2 136.3	90 120 90	2.3	(74)
Glutamic acid-specific protease (Glu-SGP) (E.C.3.4.21.82), *Streptomyces griseus*														
1HPG	nat	sp	33	0.1M HEPES	7.5	50% cit	0.2M MgCl2	20	vd	C2	77.3 36.3 51.2	90 102 90	1.5	(75)
Rat mast cell protease II, *Rattus rattus*, mast cells														
3RP2	nat		10	0.02M PP	8.3	46% AS		4	b	P3$_1$	78.2 78.2 96.8	90 90 120	1.9	(76)
Sindbis virus capsid protein, *Sindbis virus*														
1SNW	nat		8	0.13M Tris	8.3	8% PEG 8			hd	P2$_1$	38.8 79.7 60.8	90 102 90	3.0	(77)
						Also 2SNV								

Table 1. Continuous

pdb	nat	inhibitor	pc.	buffer	pH	precipi-tant	additive	T °C	met	spcgr	cell edges (Å)	angles (°)	res (Å)	lit
Proteinase A, *Streptomyces griseus*														
3SGA		sp	10		4.1	1.4M Na-PP			d	$P4_2$	55.2 55.2 54.7	90 90 90	1.8	(78)
						Also 4SGA, 5SGA, 1SGC, 2SGA								
Proteinase B, *Streptomyces griseus*														
3SGB	nat	OMTKY3		0.75M PP	6.3				vd	$P2_1$	45.4 54.5 45.7	90 119 90	1.8	(79)
4SGB	nat	PCI1	12	0.2M K-PP	7	30% AS	3mM MPD	21	hd	$P2_1$	50.9 46.2 52.5	90 117 90	2.1	(80)
Tonin, *Rattus rattus*, submaxillary gland														
1TON	nat			0.01M PP	6.2	2.5M AS	1% MPD		hd	$P4_32_12$	48.6 48.6 200.2	90 90 90	1.8	(80a)
CYSTEINE PROTEINASES (E.C.3.4.22.x)														
Cathepsin B (E.C.3.4.22.1), *Homo sapiens*, lysosomal														
1HUC	nat			1.5M PP	4	1.25M AS		20	vd	$P2_1$	86.2 34.2 85.6	90 103 90	2.1	(81)
Cathepsin B (E.C.3.4.22.1), *Rattus norvegicus*, lysosomal														
1CTE	mut		8	0.1M cit	4.4	0.8M LiSO4			vd	$P2_1$	47.1 90.2 62.2	90 97.4 90	2.1	(82)
Papain (E.C.3.4.22.2), *Carica papaya*														
1PE6 1PPD		E-64C	15			64% MeOH/ EtOH (2/1)	8 mM NaCl			$P2_12_12_1$	42.8 95.7 49.7	90 90 90	2.0	(83)

Table 1. Continuous

pdb	nat	inhibitor	pc.	buffer	pH	precipi-tant	additive	T °C	met	spcgr	cell edges (Å)	angles (°)	res (Å)	lit
1STF	mod	stefinB		0.75M PP	5.0	10% PEG 1.5		20	hd	P3$_1$21	67.0 67.0 169.3	90 90 120	2.4	(84)
1PAD	nat	ssi								P2$_1$2$_1$2$_1$	45.0 104.3 50.8	90 90 90	2.8	(85)
						Also 2PAD, 4PAD, 5PAD, 6PAD, 9PAP, 1POP								
1PPP		E-64C	15	0.1M AE	9.2	64% MeOH/ EtOH (2/1)	80mM NaCl	20	vd	P2$_1$2$_1$2$_1$	43.4 102.3 50.0	90 90 90	1.9	(86)
1PIP														
1PPN			13	0.1M Na-ac	5	67% MeOH/	25mM NaCl		vd	P2$_1$	65.7 50.7 31.5	90 98.4 90	1.6	(87)
Actinidin (E.C.3.4.22.14), *Actinidia chinensis*														
2ACT			5	0.1M PP	6.0	24% AS	1mM Na$_2$S$_4$O$_6$	4	b	P2$_1$2$_1$2$_1$	78.2 81.8 33.0	90 90 90	1.7	(88)
						Also 1AEC								
Protease omega (E.C.3.4.22.30), *Carica papaya*														
1PPO	mod		18	0.05M Tris	8.0	82% EtOH	0.18M NaCl		vd	P3$_1$12	74.1 74.1 77.8	90 90 120	1.8	(89)
ASPARTIC PROTEINASES (E.C.3.4.23.x)														
Pepsin (E.C.3.4.23.1), *Sus scrofa*, stomach														
3PEP	nat									P2$_1$	55.3 73.8 36.4	90 90 103.42.3		(90)
						Also 4PEP								

Table 1. Continuous

pdb	nat	inhibitor	pc.	buffer	pH	precipi-tant	additive	T °C	met	spcgr	cell edges (Å)	angles (°)	res (Å)	lit
5PEP	nat		280		3.6	0.5M H_2SO_4		35-20tg		$P6_522$	67.4 67.4 290.1	90 90 120	2.3	(91)
1PSA	nat	RI	20		2	>20% EtOH				$P2_1$	74.4 76.5 54.1	90 100.8 90	2.9	(92)
2PSG	pro-		10	0.05M MOPS	6.1	1.8M Li2SO4			hd	C2	105.8 43.4 88.6	90 91.4 90	1.8	(93)

Progastricsin (pepsinogen C) (E.C.3.4.23.3), *Homo sapiens*, stomach

pdb	nat	inhibitor	pc.	buffer	pH	precipi-tant	additive	T °C	met	spcgr	cell edges (Å)	angles (°)	res (Å)	lit
1HTR	nat									$P4_22_12$	105.3 105.3 70.4	90 90 90	1.6	(94)

Chymosin B (E.C.3.4.23.4), *Bos taurus*

pdb	nat	inhibitor	pc.	buffer	pH	precipi-tant	additive	T °C	met	spcgr	cell edges (Å)	angles (°)	res (Å)	lit
3CMS	rec		10	0.05M PP	5.6	2M NaCl			md	I222	80.2 114.6 72.4	90 90 90	2.0	(95)
1CMS	mut		8.6	0.05M MES	6.0	30% NaCl								

Also 4CMS

Cathepsin D (E.C.3.4.23.5), *Homo sapiens*, liver

pdb	nat	inhibitor	pc.	buffer	pH	precipi-tant	additive	T °C	met	spcgr	cell edges (Å)	angles (°)	res (Å)	lit
1LYA	nat		20	0.05M NaAc	5.1	64% AS			hd	$P6_5$	125.9 125.9 104.1	90 90 120	2.5	(96)

Also 1LYB (pepstatin)

Rhizopuspepsin (E.C.3.4.23.6), *Rhizopus chinensis*

pdb	nat	inhibitor	pc.	buffer	pH	precipi-tant	additive	T °C	met	spcgr	cell edges (Å)	angles (°)	res (Å)	lit
4APR	nat	spi								$P2_12_12_1$	60.4 60.6 106.9	90 90 90	2.5	(97)

Also 5APR, 6APR, 3APR, 2APR

Table 1. Continuous

pdb	nat	inhibitor	pc.	buffer	pH	precipi-tant	additive	T °C	met	spcgr	cell edges (Å)	angles (°)	res (Å)	lit
Endothia aspartic proteinase (E.C.3.4.23.6), *Endothia parasitica*														
5ER1	nat	ssi								$P2_1$	53.5 73.9 45.7	90 109.6 902.0		(98)
Also 2ER6, 2ER9, 4ER4, 4APE, 1EPL, 1EPO														
1ER8	nat	ssi								$P2_1$	43.2 75.7 42.9	90 97.1 90	2.0	(99)
Also 2ER0, 2ER7, 3ER3, 4ER1, 4ER2, 5ER2, 3ER5, 1EPM, 1EPN, 1EPQ, 1EPR, 1EPP, 1EED														
Renin (E.C.3.4.23.15), *Homo sapiens*														
1BBS	rec		20	0.05M Cit	4.5	10% PEG 4	0.6M NaCl		hd	$P2_13$	143.1 143.1 143.1	90 90 90	2.8	(100)
Feline immunodeficiency virus protease (E.C.3.4.23.16), *Felis catus*														
1FIV	rec									$P3_121$	50.7 50.7 74.5	90 90 120	2.0	(101)
Penicillopepsin (E.C.3.4.23.20), *Penicillium janthinellum*														
1APT	nat	ssi								C2	97.6 46.5 66.4	90 116.2 901.8		(102)
Also 1APU, 1APV, 1APW, 1PPL, 1PPM, 1PPK, 3APP														
Pepsin (E.C.3.4.23.23), *Mucor pulsillus*														
1MPP	nat		11	0.1M PP	4.5	2M AS	1% acetone		b	$P2_12$	70.0 104.0 46.3	90 90 90	2.0	(103)
Rennilase (E.C.3.4.23.23), *Rhizomucor miehei*														
1ASI	nat									$P2_12_12_1$	41.7 51.2 173.3	90 90 90	2.8	(104)

Table 1. Continuous

pdb	nat	inhibitor	pc.	buffer	pH	precipi-tant	additive	T °C	met	spcgr	cell edges (Å)	angles (°)	res (Å)	lit
Human immunodeficiency virus type 1 (HIV-1), *HIV virus*														
3HVP	synth.		6	0.1M Imid	7.0	0.25M NaCl	10 mM DTT	4	hd	$P4_12_12$	50.2 50.2 107.1	90 90 90	2.7	(105)
Also 3PHV, 2HVP, 1HHP														
7HVP	synth. ssi		5	0.1M Ac	5.4	57% AS			hd	$P2_12_12_1$	51.2 58.8 62.0	90 90 90	2.4	(106)
Also 4HVP, 1HIH, 8HVP, 1HIV, 1HVC, 1HVL, 1HVK, 1HVI, 1HVJ, 1HVS														
9HVP	rec			0.015M Ac	4.5	5% AS	5 mM DTT	20	vd	$P6_1$	63.3 63.3 83.6	90 90 120	2.8	(107)
Also 1HPV, 1HOS, 1HTF, 1AAQ, 1HBV, 1HPS, 1SBG, 1HVR														
5HVP	rec	sp	6	0.01M MES	5.0	0.6M NaCl	1 mM DTT, 1 mM EDTA	21		$P2_12_12$	58.4 86.7 46.3	90 90 90	2.0	(108)
Also 1HTE, 1HTG, 4PHV														
1HEG	rec	ssi.								$P6_122$	64.1 64.1 84.2	90 90 120	2.2	(109)
Also 1HEF														
Human immunodeficiency virus type 2 (HIV-2), *HIV virus*														
1HII	rec	ssi	2	0.1M Ac	5.4	23% AS		4		$P2_12_12_1$	33.4 64.2 99.5	90 90 90	2.3	(110)
2HPE	rec	sp								$P2_1$	58.2 43.8 39.2	90 106.2 902.0		(111)
2HPF	rec	sp								$P6_5$	80.2 80.2 71.1	90 90 120	3.0	(111)
2MIP	rec	ssi								$P4_3$	55.1 55.1 138.9	90 90 90	2.2	(112)
1IDA	rec	ssi								$P4_32_12$	62.6 62.6 115.8	90 90 90	1.7	(112)
Also 1IDB														

Table 1. Continuous

pdb	nat	inhibitor	pc.	buffer	pH	precipi-tant	additive	T °C	met	spcgr	cell edges (Å)	angles (°)	res (Å)	lit
1IVP	mut	ssi								$P2_12_12_1$	33.2 45.0 135.7	90 90 90	2.5	(113)
							Also 1IVQ							
Siman immunodeficiency virus (SIV), *SIV virus*														
1SIV	rec	ssi								I222	46.3 101.5 118.8	90 90 90	2.5	(114)
2SAM	mut									$C222_1$	62.7 32.2 96.1	90 90 90	2.4	(115)
							Also 1SIP							
Myeloblastoma associated viral protease														
1MVP										$P3_121$	88.9 88.9 78.8	90 90 120	2.0	(116)
							Also 2MVP							
Rous sarcoma virus protease														
2RSP										$P3_121$	88.9 88.9 78.8	90 90 120	2.0	(117)
METALLOENDOPROTEINASES (E.C.3.4.24.x)														
Fibroblast collagenase (E.C.3.4.24.7), *Homo sapiens*, fibroblast														
1CGF	rec									$P2_1$	71.0 50.5 48.0	90 100 90	2.1	(118)
1HFC	rec		15	3mM Tris	7.5	AS	0.4 M NaCl 5 mM CaCl$_2$			$P2_12_12_1$	109.2 44.6 36.3	90 90 90	1.6	(119)
1CGE	rec			0.1M Tris	8.0	2M formate	1mM CaCl$_2$ 50µM ZnCl$_2$	4	vd	$P4_12_12$	72.6 72.6 75.1	90 90 90	1.9	(118)

Table 1. Continuous

pdb	nat	inhibitor	pc.	buffer	pH	precipitant	additive	T °C	met	spcgr	cell edges (Å)	angles (°)	res (Å)	lit
1CGL	rec		13	0.1M Tris	7.5	1.5M formate 0.2M NaAc 1% MPD	1mM CaCl$_2$ 50μM ZnCl$_2$		hd	P6$_4$	78.2 78.2 87.4	90 90 120	2.4	(120)
Neutrophil collagenase (E.C.3.4.24.7), *Homo sapiens*, neutrophil														
1MNC	nat									P2$_1$2$_1$2$_1$	34.9 61.3 68.3	90 90 90	2.1	(121)
Astacin (E.C.3.4.24.21), *Astacus astacus*														
1AST	nat				7.0	AS				P3$_1$21	62.0 62.0 98.5	90 90 120	1.8	(122)
Elastase (E.C.3.4.24.26), *Pseudomonas aeruginosa*														
1EZM	nat		10	0.06M MES	6.0	20% PEG8	2mM CaCl$_2$ 0.1M PA cross-linker			P2$_1$	41.9 90.0 40.8	90 113.8 901.5		(123)
Neutral proteinase (E.C.3.4.24.27), *Bacillus cereus*														
1NPC	nat			0.05M succ	6.0	18% PEG6	10 mM CaCl$_2$		hd	P6$_5$22	76.5 76.5 201.0	90 90 120	2.0	(124)
Thermolysin (E.C.3.4.24.27), *Bacillus thermoproteolyticus*														
6TMN	nat	ssi								P6$_1$22	94.1 94.1 131.4	90 90 120	1.6	(125)
Also 1TMN, 2TMN, 3TMN, 4TMN, 5TMN, 4TLN, 5TLN, 7TLN, 1TLP, 1THL, 1HYT, 1LNE, 1LNA, 1LNB, 1LNC, 8TLN, 1LND, 1LNF														
Alkaline protease, *Pseudomonas aeruginosa*														
1KAP													1.6	(126)

Table 1. Continuous

pdb	nat	inhibitor	pc.	buffer	pH	precipi-tant	additive	T °C	met	spcgr	cell edges (Å)	angles (°)	res (Å)	lit
Serratia protease (serralysin) (E.C.3.4.24.40), *Serratia marcescens*														
1SAT	nat		10	0.1M cit	6.3	30% PEG8	0.1M AS	4	sd	$P2_12_12_1$	151.0 109.2 42.6	90 90 90	1.8	(127)
1SRP	nat		15	0.		2% PEG6	27% sat. AS	24	md	$P2_12_12_1$	109.1 150.9 42.6	90 90 90	1.8	(128)
1SMP	nat	ECI	10	0.2M Tris	8-9	1.2M AS		20	sd	$P4_3$	108.8 108.8 87.7	90 90 90	2.0	(129)
Atrolysin C, *Crotalus atrox*														
1HTD	nat									$P6_5$	97.3 97.3 87.8	90 90 120	2.1	(130)
Adamalysin II (E.C.3.4.24.46), *Crotalus adamanteus*														
1IAG					5.0	1.8M AS				$P3_21 2$	73.6 73.6 96.4	90 90 120	2.0	(131)
THREONINE PROTEINASES (E.C.3.4.99.x)														
Proteasome (E.C.3.4.99.46), *Thermoplasma acidophilum*														
1PMA			7	5mM MPS	7.5	15% PEG1	0.1M PP	RT	vd	$P2_12_12_1$	311.9 209.0 117.2	90 90 90	3.4	(132)

Table 2. Abbreviations used

Headings	
pdb:	pdb-entry code
nat:	native
nat α:	nat α
nat β:	nat β
nat ε:	nat ε
nat γ:	nat γ
pro-:	zymogen
rec:	recombinant
mut:	mutant
mod:	chemical modified
anh.	anhydrous
pc.:	protein concentration in mg/ml
T:	temperature
met:	crystallization method
spcgr:	spacegroup
res:	resolution
lit:	literature
vd:	vapour diffusion
hd:	hanging drop
sd:	sitting drop
md:	microdialysis
d:	dialysis
tg:	temperature gradient
ssi:	small synthetic inhibitor
sp:	synthetic peptide inhibitor
Chemicals	
MPD:	methyl pentanediol
BA:	benzamidine

Table 2. Continuous

Headings	
cit.:	citrate
AS:	ammonium sulfate
PEG 6:	polyethylene glycol 6000
Imid.	imidazole
IM:	imidazolmalate
NaAc:	sodium acelate
PP:	phosphate buffer
MES:	2-morpholino-ethanesulfonic acid
MPS:	2-morpholino-propane-sulfonic acid
MeOH:	methanol
AE:	aminoethanol
PA:	phenanthroline
succ.:	succinate buffer
Proteins	
CP A:	carboxypeptidase A
CPDW-II:	serine carboxypeptidase II
ser-CP:	serine carboxypeptidase
OMTKY3:	turkey ovomucoid inhibitor third domain
PSTI3(rec):	human pancreatic secretory trypsin inhibitor variant 3 (E. coli)
PSTI4(rec):	human pancreatic secretory trypsin inhibitor variant 4 (E. coli)
PTI:	pancreatic trypsin inhibitor
str. inh:	Streptomyces subtilisin inhibitor
CMTI-I	Cucurbita maxima trypsin inhibitor
AB-I:	Phaseolus angularis inhibitor (Bowman-Birk)
MCTI-A:	Momordica charantia seed trypsin inhibitor
BPTI:	basic pancreatic trypsin inhibitor
APPI:	Homo sapiens amyloid beta-protein precursor inhibitor domain

Table 2. Continuous

Headings	
TFPI-II	Homo sapiens tissue factor pathway inhibitor domain II
PPSTI:	Sus scrofa pancreatic sectretory trypsin inhibitor
DFP:	Diisopropyl-fluorophosphofluoridate
PPACK:	D-Phe-Pro-Arg-chloromethylketone
TM:	Thrombomodulin
hir-der:	hirudin-derivative
PF2:	prothrombin fragment 2
DNA-fr.:	DNA-fragment
FP-a:	fibrinopeptide alpha
CI-2:	barley chymotrypsin inhibitor 2
Glu-sgp:	glutamic acid specific protease
RMCP:	rat mast cell protease
SVCP:	sindbis virus capsid protein
RI:	renin inhibitor
PCI1:	Russet Burbank potato inhibitor
FIV:	feline immunodeficiency virus
HIV:	human immunodeficiency virus
SIV:	simian immunodeficiency virus
MAV:	myeloblastosis associated virus
RSV:	Rous sarcoma virus
ECI:	Erwinia chrysanthemi inhibitor

References

(1) H. Kim, W.N. Lipscomb: X-ray crystallographic determination of the structure of bovine lens leucine aminopeptidase complexed with amastatin: Formulation of a catalytic mechanism featuring a gem-diolate transition state. Biochemistry 32, 8465, 1993.

(2) B. Chevrier, C. Schalk, H. d'Orchymont, J.M. Rondeau, D. Moras, C. Tarnus: Crystal structure of *Aeromonas proteolytica* aminopeptidase: A prototypical member of the co-catalytic zinc enzyme family. Structure 2, 283 1994.

      C. Schalk, J.-M. Remy, B. Chevrier, D. Moras, C. Tarnus, C: Rapid purification of the Aeromonas proteolytica aminopeptidase: crystallization and preliminary X-ray data. Arch. Biochem. Biophys. 294, 91-97, 1992.

  (3) A. Guasch, M. Coll, F.X. Aviles, R. Huber: Three dimensional structure of porcine pancreatic procarboxypeptidase A. A comparison of the A and B zymogens and their determinants for inhibition and activation. J. Mol. Biol. 224, 141 1992

  (4) M.F.Schmid, J.R.Herriott: Structure of carboxypeptidase B at 2.8Å resolution. J.Mol.Biol. 103, 175 1976

  (5) D.-I. Liao, K. Breddam, R.M. Sweet, T. Bullock, S.J. Remington: Refined atomic model of wheat serine carboxypeptidase II at 2.2Å resolution. Biochemistry 31, 9796 1992.
      K.P. Wilson, D.I. Liao, T. Bullock, S.J. Remington, K. Breddam: Crystallization of serine carboxypeptidases. J Mol Biol 211, 301-303,1990.

  (6) T.L.Bullock, S.J.Remington: Structure of the complex of L-benzylsuccinate with wheat serine carboxypeptidase II at 2.0Å resolution. Biochemistry 33, 11127-11134, 1994.

  (7) J.A. Kelly, J.R. Knox, C. Moews, G.J. Hite, J.B. Bartolone, H. Zhao, B. Joris, J.-M. Frere, J.-M. Ghuysen: 2.8Å structure of penicillin-sensitive D-Alanyl carboxypeptidase-transpeptidase from *Streptomyces r61* and complexes with beta-lactams. J. Biol. Chem. 260, 6449, 1985

  (8) S.B. Sorensen, M. Raaschou-Nielsen, U.H. Mortensen, S.J. Remington, K. Breddam: Site-directed mutagenesis on (serine) carboxypeptidase Y from yeast. The significance of Thr 60 and Met 398 in hydrolysis and aminolysis reactions. J. Am. Chem. Soc. 117, 5944, 1995.
      J.A. Endrizzi, K. Breddam, S.J. Remington: 2.8Å structure of yeast serine carboxypeptidase. Biochemistry 33, 11106-11120, 1994.

  (9) H. Kim, W.N. Lipscomb: Crystal structure of the complex of carboxypeptidase A with a strongly bound phosphonate in a new crystalline form: comparison with structures of other complexes. Biochemistry 29, 5546, 1990

(10) D.C. Rees, W.N. Lipscomb: Refined crystal structure of the potato inhibitor complex of carboxypeptidase A at 2.5Å resolution. J. Mol. Biol. 160, 475 1982
      D.C. Rees, W.N. Lipscomb: Structure of potato inhibitor complex of carboxypeptidase A at 5.5-A resolution. Proc Natl Acad Sci U S A 77, 277-280,1980.

(11) D.C. Rees, M. Lewis, W.N. Lipscomb: Refined crystal structure of carboxypeptidase A at 1.54Å resolution. J. Mol. Biol. 168, 367, 1983.

(12) A. Tulinsky, R.A. Blevins: Structure of a tetrahedral transition state complex of alpha-chymotrypsin at 1.8Å resolution. J. Biol. Chem. 262, 7737, 1987.

(13) F. Frigerio, A. Coda, L. Pugliese, C. Lionetti, E. Menegatti, G. Amiconi, H., Schnebli, Ascenzi, M. Bolognesi: Crystal and molecular structure of the bovine alpha-chymotrypsin-eglin C complex at 2.0Å resolution . J. Mol. Biol. 225, 107, 1992.
      L. Pugliesi, G. Gatti, M. Bolognesi, A. Coda, E. Menegatti, H.P. Schnebli, P. Ascenzi, G. Amiconi, G: Preliminary crystallographic data on the complex of bovine alpha-chymotrypsin with the recombinant proteinase inhibitor eglin c from Hirudo medicinalis. J. Mol. Biol. 208, 511-513, 1989.

(14) M. Fujinaga, A.R. Sielecki, R.J. Read, W. Ardelt, M. Laskowski junior, M.N.G. James: Crystal and molecular structures of the complex of alpha-chymotrypsin with its inhibitor turkey ovomucoid third domain at 1.8Å resolution. J. Mol. Biol. 195, 397 1987.

(15) H.J. Hecht, M. Szardenings, J. Collins, D. Schomburg: Three-dimensional structure of the complexes between bovine chymotrypsinogen A and two recombinant variants of human pancreatic secretory trypsin inhibitor (Kazal-type). J. Mol. Biol. 220, 711 1991

(16) D. Wang, W. Bode, R. Huber: Bovine chymotrypsinogen α. X-ray crystal structure, analysis and refinement of a new crystal form at 1.8Å resolution. J. Mol. Biol. 185, 595 1985.

(17) S.T. Freer, J.Kraut, J.D.Robertus, H.T.Wright, N.H.Xuong: Chymotrypsinogen, 2.5Å crystal structure, comparison with alpha-chymotrypsin, and implications for zymogen activation. Biochemistry 9, 1997 1970

(18) M. Harel, C.-T. Su, F. Frolow, I. Silman, J.L. Sussman: Gamma-chymotrypsin is a complex of alpha-chymotrypsin with its own autolysis products. Biochemistry 30, 5217 1991
G.H. Cohen, E.W. Silverton D.R. Davies: Refined crystal structure of gamma-chymotrypsin at 1.9 A resolution. Comparison with other pancreatic serine proteases. J Mol Biol 148, 449-479, 1981

(19) K. Kreutter, A.C.U. Steinmetz, T.-C. Liang, M. Prorok, R. Abeles, D. Ringe: Three-dimensional structure of chymotrypsin inactivated with (2s)n-acetyl-l-alanyl-l-phenylalanyl-chloroethyl-ketone: implications for the mechanism of inactivation of serine proteases by chloroketones. Biochemistry 33, 13792-13800, 1994.

(20) M. Marquart, J. Walter, J. Deisenhofer, W. Bode, R. Huber: The geometry of the reactive site and of the peptide groups in trypsin, trypsinogen and its complexes with inhibitors. Acta Crystallogr. B39, 480 1983

(21) J. Walter, W. Steigemann, T., Singh, H. Bartunik, W. Bode, R. Huber: On the disordered activation domain in trypsinogen. Chemical labelling and low- temperature crystallography. Acta Crystallogr. B38, 1462 1982

(22) A.Y. Lee, M. Hagihara, R. Karmacharya, M.W. Albers, S.L. Schreiber, J. Clardy: Atomic structure of the trypsin-cyclotheonamide A complex: Lessons for the design of serine protease inhibitors. J. Am. Chem. Soc. 115, 12619 1993

(23) W.F. Mangel, T. Singer, D.M. Cyr, T.C. Umland, D.L. Toledo, R.M. Stroud, J.W. Pflugrath, R.M. Sweet: Structure of an acyl-enzyme intermediate during catalysis: (guanidinobenzoyl) trypsin. Biochemistry 29, 8351 1990

(24) Y. Takeuchi, T. Nonaka, K.T. Nakamura, S. Kojima, K-I. Miura, Y. Mitsui: Crystal structure of an engineered subtilisin inhibitor complexed with bovine trypsin. Proc. Nat. Acad. Sci. USA 89, 4407 1992

(25) W. Bode, H.J. Greyling, R. Huber, J. Otlewski, T. Wilusz: The refined 2.0Å X-ray crystal structure of the complex formed between bovine beta-trypsin and CMTI-I, a trypsin inhibitor from squash seeds (*Cucurbita maxima*): topological similarity of the squash seed inhibitors with the carboxypeptidase A inhibitor from potatoes. FEBS Lett. 242, 285 1989

(26) Y. Tsunogae, I. Tanaka, T. Yamane, J.-I. Kikkawa, T. Ashida, C. Ishikawa, K. Watanabe, S. Nakamura, K. Takahashi: Structure of the trypsin-binding domain of Bowman-Birk type protease inhibitor and its interaction with trypsin. J. Biochem. (Tokyo) 100, 1637 1986

(27) H. Fehlhammer, W. Bode: The refined Crystal structure of bovine b-trypsin at 1.8 A resolution. J.Mol.Biol. 98, 683-692, 1975
R.M. Stroud, L.M. Kay, R.E. Dickerson: The structure of bovine trypsin: electron density maps of the inhibited enzyme at 5 Angstrom and at 2-7 Angstrom resolution. J Mol Biol 83, 185-208, 1974.

(28) M. Bolognesi, G. Gatti, E. Menegatti, M. Guarneri, M. Marquart, E. Papamokos, R. Huber: Three-dimensional structure of the complex between pancreatic secretory inhibitor (Kazal type) and trypsinogen at 1.8Å resolution. Structure solution, crystallographic refinement and preliminary structural interpretation. J. Mol. Biol. 162, 839 1982
M. Bolognesi, A. Coda, M. Guarneri, E. Menegatti: Preliminary crystallographic data for the structure of the complex between Kazal trypsin inhibitor and trypsinogen. J. Mol. Biol. 145, 603-605, 1981

(29) Q. Huang, S. Liu, Y. Tang: The refined 1.6Å resolution crystal structure of the complex formed between porcine beta-trypsin and MCTI-A, a trypsin inhibitor of squash family. J. Mol. Biol. 229, 1022 1993
Huang, Q., Liu, S., Tang, Y., Zeng, F. and Qian, R. Amino acid sequencing of a trypsin inhibitor by refined 1.6 A X-ray crystal structure of its complex with porcine b-trypsin. FEBS Lett. 297, 143-146 (1992)

(30) Q. Huang, Z. Wang, Y. Li, S. Liu, Y. Tang: Refined 1.8Å resolution crystal structure of porcine epsilon-trypsin. Biochim. Biophys. Acta 1209, 77 1994

(31) M.J.M. Burgering, L.P.M. Orbons, A. van der Doelen, J. Mulders, H.J.M. Theunissen, P.D.J. Grootenhuis, W. Bode, R. Huber, M.T. Stubbs: The second Kunitz domain of human tissue factor pathway inhibitor. Cloning, structure determination and interaction with factor Xa. J. Mol. Biol., 269, 395-407, 1997

(32) C. Gaboriaud, L. Serre, O. Guy-Crotte, E. Forest, J.-C. Fontecilla-Camps: Crystal structure of human trypsin 1: unexpected phosphorylation of tyrosine 151. J. Mol. Biol. 259, 995-1010, 1996.

(33) J.J. Perona, C.A. Tsu, C.S. Craik, R.J. Fletterick: Crystal structures of rat anionic trypsin complexed with the protein inhibitors APPI and BPTI. J. Mol. Biol. 230, 919 1993

(34) S. Sprang, T. Standing, R. J. Fletterick, R. M. Stroud, J. Finer-Moore, N.-H. Xuong, R. Hamlin, W. J. Rutter, C. S. Craik: The three-dimensional structure of Asn-102- mutant of trypsin. Role of Asp-102 in serine protease catalysis: Science 237, 905 1987

(35) J.J. Perona, C.A. Tsu, M.E. McGrath, C.S. Craik, R.J. Fletterick: A negative charge in the binding pocket of trypsin. J. Mol. Biol. 230, 934 1993

(36) G.I. Berglund, A.O. Smalas, A. Hordvik, N.P. Willassen: The crystal structure of anionic salmon trypsin in a second crystal form. To be published

(37) A.O. Smalas, E.S. Heimstad, A. Hordvik, N.P. Willassen, R. Male: Cold-adaption of enzymes: structural comparison between salmon and bovine trypsins. Proteins 20, 149-166, 1994.

(38) R.J. Read, M.N.G. James: Refined crystal structure of *Streptomyces griseus* trypsin at 1.7Å resolution. J. Mol. Biol. 200, 523 1988

(39) E. Skrzypczak-Jankun, E. Carperos, K.G. Ravichandran, A. Tulinsky, M. Westbrook, J.M. Maraganore: Structure of the hirugen and hirulog 1 complexes of alpha-thrombin. J. Mol. Biol. 221, 1379 1991

(40) I.I. Mathews, K.P. Padmanabhan, A. Tulinsky, J.E. Sadler: The structure of a nonadecapeptide of the fifth EGF domain of thrombomodulin complexed with thrombin. Biochemistry 33, 13547-13552, 1994.

(41) D.W. Banner, P. Hadvary: Crystallographic analysis at 3.0Å resolution of the binding to human thrombin of four active site-directed inhibitors. J. Biol. Chem. 266, 20085 1991

(42) W. Bode, I. Mayr, U. Baumann, R. Huber, S.R. Stone, J. Hofsteenge: The refined 1.9Å crystal structure of human alpha-thrombin: interaction with D-Phe-Pro-Arg-chloromethylketone and significance of the Tyr-Pro-Pro-Trp insertion segment. EMBO J. 8, 3467 1989

(43) R.K. Arni, K. Padmanabhan, K.P. Padmanabhan, T.-P. Wu, A. Tulinsky: Structures of the noncovalent complexes of human and bovine prothrombin fragment 2 with human 3 PPACK-thrombin. Biochemistry 32, 4727 1993

(44) T.J. Rydel, A. Tulinsky, W. Bode, R. Huber: The refined structure of the hirudin-thrombin complex. J. Mol. Biol. 221, 583 1991

(45) M.T. Stubbs, H. Oschkinat, I. Mayr, R. Huber, H. Angliker, S.R. Stone, W. Bode: The interaction of thrombin with fibrinogen - a structural basis for its specificity, Eur. J. Biochem. 206, 187-195, 1992.

(46) J. Priestle, J. Rahuel, H. Rink, M. Tones, M.G. Gruetter: Changes in interactions in complexes of hirudin derivatives and human alpha-thrombin due to different crystal forms. Protein Sci. 2, 1630 1993

(47) K. Padmanabhan, K.P. Padmanabhan, J.D. Ferrara, J.E. Sadler, A. Tulinsky: The structure of alpha-thrombin inhibited by a 15-mer single-stranded DNA aptamer. J. Biol. Chem. 268, 17651 1993

(48) I.I. Mathews, K.P. Padmanabhan, Ganesh, A. Tulinsky, M. Ishil, J. Chen, C.W. Turck, S.R. Coughlin: Crystallographic structures of thrombin complexed with thrombin receptor peptides: expected and novel binding modes. Biochemistry 33, 3266-3279, 1994.

(49) T.J. Rydel, M. Yin, K.P. Padmanabhan, D.T. Blankenship, A.D. Cardin, E. Correa, J.W. Fenton II, A. Tulinsky: Crystallographic structure of human gamma-thrombin. J. Biol. Chem. 269, 22000 1994

(50) J. Vitali, D. Martin, M.G. Malkowski, W.D. Robertson, J.B. Lazar, R.C. Winant, H. Johnson, B.F.P. Edwards: The structure of a complex of bovine alpha-thrombin and recombinant hirudin at 2.8Å resolution. J. Biol. Chem. 267, 17670, 1992

(51) P.D. Martin, W. Robertson, D. Turk, R. Huber, W. Bode, B.F.P. Edwards: The structure of residues 7-16 of the A alpha chain of human fibrinogen bound to bovine thrombin at 2.3Å resolution. J. Biol. Chem. 267, 7911 1992.

(52) H. Brandstetter, D. Turk, H.W. Hoeffken, D. Grosse, J. Stuerzebecher, D. Martin, B.F.P. Edwards, W. Bode: Refined 2.3Å x-ray crystal structure of bovine thrombin complexes formed with the benzamidine and arginine-based thrombin inhibitors NAPAP, 4-TAPAP and MQPA: A starting point for improving antithrombotics. J. Mol. Biol. 226, 1085 1992.

(53) A. van de Locht, D. Lamba, M. Bauer, R. Huber, T. Friedrich, B. Kröger, W. Höffken, W. Bode: Two heads are better than one: crystal structure of the insect derived double domain Kazal inhbitor Rhodniin in complex with thrombin. The EMBO Journal 14, 5149 1995

(54) K. Padmanabhan, K.P. Padmanabhan, A. Tulinsky, C.H. Park, W. Bode, R. Huber, D.T. Blankenship, A.D. Cardin, W. Kisiel: Structure of human des(1-45) factor Xa at 2.2Å resolution. J. Mol. Biol. 232, 947 1993.

(55) R. Bone, N.S. Sampson, A. Bartlett, D.A. Agard: Crystal structures of alpha-lytic protease complexes with irreversibly bound phosphonate esters. Biochemistry 30, 2263 1991

(56) Z. Dauter, C. Betzel, N. Genov, N. Pipon, K.S. Wilson: The complex between the subtilisin from a mesophilic bacterium and the leech inhibitor eglin-C. Acta Crystallogr. B47, 707 1991

(57) H. Brandstetter, M. Bauer, R. Huber, P. Lollar, W. Bode: X-ray structure of clotting factor IXa: Active site and module structure related to Xase activity and hemophilia B. Proc. Natl. Acad. Sci. USA 92, 9796 1995

(58) Z. Chen, W. Bode: Refined 2.5Å X-ray crystal structure of the complex formed by porcine kallikrein A and the bovine pancreatic trypsin inhibitor. Crystallization, Patterson search, structure determination, refinement, structure and comparison with its components and with the bovine trypsin-pancreatic trypsin inhibitor complex. J. Mol. Biol. 164, 283 1983

(59) W. Bode, Z. Chen, K. Bartels, C. Kutzbach, G. Schmidt-Kastner, H. Bartunik: Refined 2Å X-ray crystal structure of porcine panceratic kallikrein A, a specific trypsin-like serine protease. Crystallization, structure determination, crystallographic refinement, structure and its comparison with bovine trypsin. J. Mol. Biol. 164, 237 1983

(60) I. Li de la Sierra, E. Papamichael, C. Sakarelos, J.-L. Dimicoli, T. Prange: Interaction of the peptide CF3-Leu-Ala-NH-C6H4-CF3 (TFLA) with porcine pancreatic elastase. X-Ray studies at 1.8Å. J. Mol. Recog. 3, 36 1990

(61) C.A. Veale, R. Bernstein, C. Bryant, C. Ceccarelli, J.R. Damewood, R. Earley, S.W. Feeney, B. Gomes, B.J. Kosmider, G.B. Steelman, R.M. Thomas, E., Vacek, J.C. Williams, D.J. Wolanin, S. Woolson: Nonpeptidic inhibitors of human leukocyte elastase. 5. Design, synthesis, and X-ray crystallography of a series of orally active 5-amino-pyrimidin-6-one-containing trifluoromethylketones. J. Med. Chem. 38, 98-108, 1995.

(62) M.A. Navia, B.M. McKeever, J.P. Springer, T.-Y. Lin, H.R. Williams, E.M. Fluder, C.P. Dorn, K. Hoogsteen: Structure of human neutrophil elastase in complex with a peptide chloromethyl ketone inhibitor at 1.84Å resolution. Proc. Nat. Acad. Sci. USA 86, 7 1989
Williams, H.R., Lin, T.-Y., Navia, M.A., Springer, J.P., McKeever, B.M., Hoogsteen, K. and Dorn, C.P.Jr.: Crystallization of human neutrophil elastase. J. Biol. Chem. 262, 17178-17181, 1987.

(63) S. Tsunasawa, T. Masaki, M. Hirose, M. Soejima, F. Sakiyama: The primary structure and structural characteristics of *Achromobacter lyticus* protease I, a lysine-specific serine protease. J. Biol. Chem. 264, 3832 1989

(64) T. Yamane, T. Kani, T. Hatanaka, A. Suzuki, T. Ashida, T. Kobayashi, S. Ito, O. Yamashita: Crystal structure of a new alkaline serine protease (m-protease) from *Bacillus sp.* ksm-k16. To be published

(65) D.W. Goddette, C. Paech, S.S. Yang, J.R. Mielenz, C. Bystroff, M. Wilke, R.J. Fletterick: The crystal structure of the *Bacillus lentus* alkaline protease, subtilisin Bl, at 1.4Å resolution. J. Mol. Biol. 228, 580 1992

(66) P.A. Fitzpatrick, A.C.U. Steinmetz, D.Ringe, A.M. Klibanov: Enzyme crystal structure in a neat organic solvent. Proc. Nat. Acad. Sci. USA 90, 8653 1993
Petsko, G.A. and Tsernoglou, D: The structure of subtilopeptidase A. I. X-ray crystallographic data. J. Mol. Biol. 106, 4553-456, 1976.

(67) C.A. McPhalen, M.N.G. James: Structural comparison of two serine proteinase-protein inhibitor complexes: Eglin-C-subtilisin carlsberg and CI-2-subtilisin novo. Biochemistry 27, 6582 1988.
McPhalen, C.A., Schnebli, H.P. and James, M.N.G. Crystal and molecular structure of the inhibitor eglin from leeches in complex with subtilisin Carlsberg. FEBS Lett. 188, 55, 1985

(68) W. Bode, E. Papamokos, D. Musil: The high-resolution X-ray crystal strucuture of the comples formed between subtilisin Carlsberg and eglin-C, an elastase inhibitor from the leech *Hirudo medicinalis*. Strucutural analysis, subtilisin strucuture and interface geometry. Eur. J. Biochem. 166, 673 1987

(69) R. Syed, Z.-P. Wu, J.M. Hogle, D. Hilvert: Crystal structure of selenosubtilisin at 2.0Å resolution. Biochemistry 32, 6157 1993

(70) D.W. Heinz, J.P. Priestle, J. Rahuel, K.S. Wilson, M.G. Gruetter: Refined crystal structures of subtilisin novo in complex with wild-type and two mutant eglins: comparison with other serine proteinase inhibitor complexes. J. Mol. Biol. 217, 353 1991

(71) C. Betzel, G.P. Pal, W. Saenger: Synchrotron x-ray data collection and restrained least-squares refinement of the crystal structure of proteinase K at 1.5Å resolution. Acta Crystallogr. B44, 163 1988.

(72) A. Teplyakov, I., Kuranova, E.H. Harutyunyan, B.K. Vainshtein, C. Froemmel, W.E. Hoehne, K.S. Wilson: Crystal structure of thermitase at 1.4Å resolution. J. Mol. Biol. 214, 261
C. Frömmel, G. Hausdorf, W.E. Hoehne, U. Behnke, H. Ruttloff: Characterization of a protease from *Thermoactinomyces vulgaris* (thermitase). 2. Single-step fine purification and protein-chemical characterization. Acta Biol. Med. Ger. 37, 1193-1204 (1978)

(73) P. Gros, M. Fujinaga, B.W. Dijkstra, K.H. Kalk, W.G.J. Hol: Crystallographic refinement by incorporation of molecular dynamics: The thermostable serine protease thermitase complexed with eglin-C. Acta Crystallographica B45, 488-499, 1989.

(74) D. Lamba, M. Bauer, R. Huber, S. Fischer, R. Rudolph, U. Kohnert, W. Bode: The 2.3 A crystal structure of the catalytic domain of recombinant two-chain human tissue-type plasminogen activator. J. Mol. Biol. 258, 117 1996

(75) V.L. Nienaber, K. Breddam, J.J. Birktoft: A glutamic acid specific serine protease utilizes a novel histidine triad in substrate binding. Biochemistry 32, 11469 1993

(76) S.J. Remington, R.G. Woodbury, R.A. Reynolds, B.W. Matthews, H. Neurath: The structure of rat mast cell protease II at 1.9Å resolution. Biochemistry 27, 8097 1988

(77) L. Tong, G. Wengler, M.G. Rossmann: The refined structure of sindbis virus core protein in comparison with other chymotrypsin-like serine proteinase structures. J. Mol. Biol. 230, 228 1993
U. Boege, M. Cygler, G. Wengler, P. Dunas, J. Tsao, M. Luo, T.J. Smith, M.G. Rossmann: Sindbis virus core protein crystals. J. Mol. Biol. 208, 79-82

(78) M.N.G. James, A.R. Sielecki, G.D. Brayer, L.T.J. Delbaere, C.-A. Bauer: Structures of product and inhibitor complexes of *Streptomyces griseus* protease A at 1.8Å resolution. A model for serine protease catalysis. J. Mol. Biol. 144, 43 1980
G.D. Brayer, L.T.J. Delbaere, M.N.G. James: Molecular structure of crystalline Streptomyces griseus protease A at 2.8 A resolution. J. Mol. Biol. 124, 243-259, 1978.

(79) R.J. Read, M. Fujinaga, A.R. Sielecki, M.N.G. James: Structure of the complex of *Streptomyces griseus* protease B and the third domain of the turkey ovomucoid inhibitor at 1.8Å resolution. Biochemistry 22, 4420 1983
M. Fujinaga, R.J. Read, A. Sielecki, W. Ardelt, M. Laskowski Jr., M.N.G. James: Refined crystal structure of the molecular complex of Streptomyces griseus protease B, a serine protease, with the third domain of the ovomucoid inhibitor from turkey. PNAS USA 79, 4868-4872, 1982

(80) H.M. Greenblatt, C.A. Ryan, M.N.G. James: Structure of the complex of *Streptomyces griseus* proteinase B and polypeptide chymotrypsin inhibitor-1 from russet burbank potato tubers at 2.1Å resolution. J. Mol. Biol. 205, 201 1989

(80a) M. Fujinaga, M.N.G. James: Rat submaxillary gland serine protease, tonin. Structure solution and refinement at 1.8Å resolution. J. Mol. Biol. 195, 373 1987.
K. Hayakawa, J.A. Kelly, M.N.G. James: Crystal data for tonin, an enzyme involved in the formation of angiotensin II. J. Mol. Biol. 123, 107-111, 1978

(81) D. Musil, D. Zucic, D. Turk, R.A. Engh, I. Mayr, R. Huber, T. Popovic, V. Turk, T. Towatari, N. Katunuma, W. Bode: The refined 2.15Å x-ray crystal structure of human liver cathepsin B: The structural basis for its specificity. EMBO J. 10, 2321 1991

(82) Z. Jia, S. Hasnain, T. Hirama, X. Lee, J.S. Mort, R. To, C.P. Huber: Crystal structures of recombinant rat cathepsin B and a cathepsin B-inhibitor complex: implications for structure-based inhibitor design. J. Biol. Chem. 270, 5527 1995
X. Lee, F.R. Ahmed, T. Hirama, C.P. Huber, D.R. Rose, R. To, S. Hasnain, A. Tam, J.S. Mort: Crystallization of recombinant rat cathepsin B. J. Biol. Chem. 265, 5950-5951, 1990.

(83) D. Yamamoto, K. Matsumoto, H. Ohishi, T. Ishida, M. Inoue, K. Kitamura, H. Mizuno: Refined X-ray structure of papain-E-64-c complex at 2.1Å resolution. J. Biol. Chem. 266, 14771 1991
J. Drenth, J.N. Jansonius, R. Koekoek, B.G. Wolthers: The structure of papain. Adv. Prot. Chem. 25, 79-115, 1971.

(84) M.T. Stubbs, B. Laber, W. Bode, R. Huber, R. Jerala, B. Lenarcic, V. Turk: The refined 2.4Å X-ray crystal structure of recombinant human stefin B in complex with the cysteine proteinase papain: a novel type of proteinase inhibitor interaction. EMBO J. 9, 1939 1990

(85) E. Schroeder, C. Phillips, E. Garman, K. Harlos, C. Crawford: X-ray crystallographic structure of a papain-leupeptin complex. FEBS Lett. 315, 38 1993

(86) M.-J. Kim, D. Yamamoto, K. Matsumoto, M. Inoue, T. Ishida, H. Mizuno, S. Sumiya, K. Kitamura: Crystal structure of papain-E64-c complex. Binding diversity of E64-c to papain s-2- and s-3- subsites. Biochem. J. 287, 797 1992

(87) RW. Pickersgill, G.W. Harris, E. Garman: Structure of monoclinic papain at 1.60Å resolution. Acta Crystallogr. B48, 59 1992

(88) E.N. Baker, E.J. Dodson: Crystallographic refinement of the structure of actinidin at 1.7Å resolution by fast fourier least-squares methods. Acta Crystallogr. A36, 559 1980
Baker, E.N., Preliminary Crystallographic data for Actinidin, a thiol protease from Actinidia chinensis. J. Mol. Biol. 74, 411-412, 1973

(89) R.W. Pickersgill, P. Rizkallah, G.W. Harris, W. Goodenough: Determination of the structure of papaya protease omega. Acta Crystallogr. B47, 766 1991

(90) C. Abad-Zapatero, T.J. Rydel, J. Erickson: Revised 2.3Å structure of porcine pepsin. Evidence for a flexible subdomain. Proteins 8, 62-81, 1990.

(91) J.B. Cooper, G. Khan, G. Taylor, I.J. Tickle, T.L. Blundell: X-ray analyses of aspartic proteases. Three-dimensional structure of the hexagonal crystal form of porcine pepsin at 2.3Å resolution. J. Mol. Biol. 214, 199 1990

(92) L. Chen, J.W. Erickson, T.J. Rydel, C.H. Park, D. Neidhart, J. Luly, C. Abad-Zapatero: Structure of a pepsin/renin inhibitor complex reveals a novel crystal packing induced by minor chemical alterations in the inhibitor. Acta Crystallogr. B48, 476 1992

(93) A.R. Sielecki, M. Fujinaga, R.J. Read, M.N.G. James: Refined structure of porcine pepsinogen at 1.8Å resolution. J. Mol. Biol. 219, 671 1991
M.N.G. James, A.R.C. Sielecki: Molecular structure of an aspartic proteinase zymogen, porcine pepsinogen, at 1.8 A resolution.Nature 319, 33-38,1986.
S.N Rao, S.N. Koszelak, J.A. Hartsuck: Crystallization and preliminary crystal data of porcine pepsinogen. J. Biol. Chem. 252, 8728-8730, 1977

(94) S.A. Moore, A.R. Sielecki, M. Chernaia, N. Tarasova, M.N.G. James: Crystal and molecular structures of human progastricsin at 1.62Å resolution. J. Mol. Biol. 247, 466-485, 1995.

(95) P. Strop, J. Sedlacek, J. Stys, Z. Kaderabkova, I. Blaha, L. Pavlickova, J. Pohl, M. Fabry, Kostka, M. Newman, C. Frazao, A. Searer, I.J. Tickle, T.L. Blundell: Engineering enzyme sub-site specificity: Preparation, kinetic characterization and X-ray analysis at 2.0Å resolution of Val111-Phe site-mutated calf chymosin. Biochemistry 29, 9863 1990.

(96) E.T. Baldwin, T.N. Bhat, S. Gulnik, M. Hosur, R.C. Sowder II, R.E. Cachau, J. Collins, A.M. Silva, J.W. Erickson: Crystal structures of native and inhibited forms of human cathepsin D: implications for lysosomal targeting and drug design. Proc. Nat. Acad. Sci. USA 90, 6796 1993
S. Gulnick, E.T. Baldwin, N. Tarasova, N. J. Erickson: Human liver cathepsin D. Purification, crystallization and preliminary X-ray diffraction analysis of a lysosomal enzyme. J. Mol. Biol. 227, 265-270,1992.

(97) K. Suguna, E.A. Padlan, R. Bott, J. Boger, D.R. Davies: Structures of complexes of rhizopuspepsin with pepstatin and other statine-containing inhibitors. Proteins 13, 195-205, 1992.

(98) J.B. Cooper, S.I. Foundling, T.L. Blundell, R.J. Arrowsmith, C.J. Harris, J.N. Champness: A rational approach to the design of antihypertensives. X-ray studies of complexes between aspartic proteinases and aminoalcohol renin inhibitors. Editor, R. Leeming: Topics in Medicinal Chemistry 308 1988 Royal Society of Chemistry, London UK isbn 085186-706-5

(99) J.B. Cooper, S.I. Foundling, T.L. Blundell, J. Boger, R.A. Jupp, J. Kay: X-ray studies of aspartic proteinase-statine inhibitor complexes. Biochemistry 28, 8596 1989

(100) V. Dhanaraj, C. Dealwis, C. Frazao, M. Badasso, B. L. Sibanda, I.J. Tickle, J.B. Cooper, H.P.C. Driessen, M. Newman, C. Aguilar, S.P. Wood, T.L. Blundell, M. Hobart, K.F. Geoghegan, M.J. Ammirati, D.E. Danley, B.A. O'Connor, D.J. Hoover: X-ray analyses of peptide inhibitor complexes define the structural basis of specificity for human and mouse renins. Nature 357, 466 1992
M. Badasso, C. Frazo, B.L. Sibanda, V. Dhanaraj, C. DeAlwis, J.B. Cooper, S.P. Wood, T.L. Blundell, K. Murakami, H. Miyazaki, P.M. Hobart, K.F. Geoghegan, M.J. Ammirati, A.J. Lanzetti, D.E. Danley, B.A. O'Connor, D.J. Hoover, J. Sueiras-Diaz, D.M. Jones, M. Szelke: Crystallization and preliminary X-ray analysis of complexes of peptide inhibitors with human recombinant and mouse submandibular renins. J. Mol. Biol. 223, 447-453, 1992

(101) A. Wlodawer, A. Gustchina, L. Reshetnikova, J. Lubkowski, J.A. Zdanov, K.Y. Hui, E.L. Angleton, W.G. Farmerie, M.M. Goodenow, D. Bhatt, L. Zhang, B.M. Dunn: Structure of an inhibitor complex of protease from feline immunodeficiency virus. Nature Struct. Biol. 2, 480-488, 1995.

(102) M.N.G. James, A.R. Sielecki, J. Moult: Crystallographic analysis of a pepstatin analogue binding to the aspartyl proteinase penicillopepsin at 1.8Å resolution. Editor J. Hruby, D.H. Rich: Peptides: structure and function. V.1 521 1983 Proceedings of

the Eighth American Peptide Symposium Pierce Chemical Company, Rockford, IL ISBN 0-935940-02-2

(103) M. Newman, F. Watson, P. Roychowdhury, H. Jones, M. Badasso, A. Cleasby, S.P. Wood, I.J. Tickle, T.L. Blundell: X-ray analyses of aspartic proteinases. Structure and refinement at 2.0Å resolution of the aspartic proteinase from *Mucor pusillus*. J. Mol. Biol. 230, 260 1993

(104) Z. Jia, M. Vandonselaar, P. Schneider, J.W. Quail: Crystallization and preliminary X-ray structure solution of *Rhizomucor meihei* aspartic proteinase. To be published

(105) A. Wlodawer, M. Miller, M. Jaskolski, B.K. Snthyanarayana, E. Baldwin, I.T. Weber, L.M. Selk, L. Clawson, J. Schneider, S.B.H. Kent: Conserved folding in retroviral proteases. Crystal structure of a synthetic HIV-1 protease. Science 245, 616 1989
McKeever, B.M., Navia, M.A., Fitzgerald, P.M.D., Springer, J.P., Leu, C.-T., Heimbach, J.C., Herber, W.K., Sigal, I.S. and Darke, P.L Crystallization of the Aspartylprotease from the Human Immunodeficiency Virus, HIV-1. J. Biol. Chem. 264, 1919-1921, 1989.

(106) A.L. Swain, M.M. Miller, J. Green, D.H. Rich, J. Schneider, S.B.H. Kent, A. Wlodawer: X-ray crystallographic structure of a complex between a synthetic protease of human immunodeficiency virus 1 and a substrate-based hydroxyethylamine inhibitor. Proc. Nat. Acad. Sci. USA 87, 8805 1990.

(107) J. Erickson, D.J. Neidhart, J. van Drie, D.J. Kempf, X.C. Wang, D.W. Norbeck, J.J. Plattner, J.W. Rittenhouse, M. Turon, N. Wideburg, W.E. Kohlbrenner, R. Simmer, R. Helfrich, D.A. Paul, M. Knigge: Design, activity and 2.8Å crystal structure of a C-2- symmetric inhibitor complexed to HIV-1 protease. Science 249, 527 1990

(108) P.M.D. Fitzgerald, B.M. McKeever, J.F. van Middlesworth, J.P. Springer, J.C. Heimbach, C.-T. Leu, W.K. Herber, R.A.F. Dixon, L. Darke: Crystallographic analysis of a complex between human immunodeficiency virus type 1 protease acetyl-pepstatin at 2.0Å resolution. J. Biol. Chem. 265, 14209 1990

(109) H.M. Krishna Murthy, E.L. Winborne, M.D. Minnich, J.S. Culp, C. Debouck: The crystal structures at 2.2Å resolution of hydroxyethylene - based inhibitors bound to human immunodeficiency virus type 1 protease show that the inhibitors are present in two distinct orientations. J. Biol. Chem. 267, 22770 1992

(110) J.P. Priestle, A. Fassler, J. Rosel, M. Tintelnot-Blomley, P. Strop, M.G. Gruetter: Comparative analysis of the X-ray structures of HIV-1 and HIV-2 proteases in complex with CGP 53820, a novel pseudosymmetric inhibitor. Structure 3, 381, 1995

(111) A.M. Mulichak, J.O. Hui, A.G. Tomasselli, R.L. Heinrikson, K.A. Curry, C.-S. Tomich, S. Thaisrivongs, T.K. Sawyer, K.D. Watenpaugh: The crystallographic structure of the protease from human immunodeficiency virus type 2 with two synthetic peptidic transition state analog inhibitors. J. Biol. Chem. 168, 13103 1993

(112) L. Tong, S. Pav, C. Pargellis, F. Do, D. Lamarre, C. Anderson: Crystal structure of human immunodeficiency virus (HIV) type 2 protease in complex with a reduced amide inhibitor and comparision with HIV-protease structures. Proc. Nat. Acad. Sci. USA 90, 8387 1993

(113) A.M. Mulichak, J.O. Hui, A.G. Tomasselli, R.L. Heinrikson, K.A. Curry, C.-S. Tomich, S. Thaisrivongs, T.K. Sawyer, K.D. Watenpaugh: The crystallographic structure of the protease from the immunodeficiency virus type 2 with two synthetic peptidic transition state analog inhibitors. J. Biol. Chem. 268, 13103-13109, 1993.

(114) B. Zhao, E. Winborne, M.D. Minnich, J.S. Culp, C. Debouck, S. Abdel-Meguid: Three-dimensional structure of a SIV protease/inhibitor complex. Implications for the design of HIV-1 and HIV-2 protease inhibitors. Biochemistry 32, 13054-13060, 1993.

(115) A.F. Wilderspin, R.J. Sugrue: Alternative native flap conformation revealed by 2.3Å resolution structure of SIV proteinase. J. Mol. Biol. 239, 97 1994

(116) D.H. Ohlendorf, S.I. Foundling, J.J. Wendoloski, J. Sedlacek, P. Strop, F.R. Salemme: Structural studies of the retroviral proteinase from avian myeloblastosis associated virus. Proteins 14, 382-391, 1992.

(117) M. Jaskolski, M. Miller, J.K.M. Rao, J. Leis, A. Wlodawer: Structure of the aspartic protease from rous sarcoma retrovirus refined at 2Å resolution. Biochemistry 29, 5889-5898, 1990.

(118) B. Lovejoy, A.M. Hassell, M.A. Luther, D. Weigl, S.R. Jordan: Crystal structures of recombinant 19-kda human fibroblast collagenase complexed to itself. Biochemistry 33, 8207 1994

(119) J.C. Spurlino, A.M. Smallwood, D.D. Carlton, T.M. Banks, K.J. Vavra, J.S. Johnson, E.R. Cook, J. Falvo, R.C. Wahl, T.A. Pulvino, J.J. Wendoloski, D.L. Smith: 1.56 Å structure of mature truncated human fibroblast collagenase. Proteins. Struct. Funct. 19, 98 1994.

(120) B. Lovejoy, A. Cleasby, A.M. Hassell, K. Longley, M.A. Luther, D. Weigl, G. McGeehan, A.B. McElroy, D. Drewry, M.H. Lambert, S.R. Jordan: Structure of the catalytic domain of fibroblast collagenase complexed with an inhibitor. Science 263, 375 1994
A.M. Hassell, R.J. Anderegg, D. Weigl, M.V. Milburn, W. Burkhart, G.F. Smith, P. Graber, T.N. Wells, M.A. Luther, S.R. Jordan: Preliminary X-ray diffraction studies of recombinant 19 kDa human fibroblast collagenase. J. Mol. Biol. 236, 1410-1412, 1994.

(121) T. Stams, J.C. Spurlino, D.L. Smith, R.C. Wahl, T.F. Ho, M.W. Qoronfleh, T.M. Banks, B. Rubin: Structure of human neutrophil collagenase reveals large S1' specificity pocket. Nat. Struct. Biol. 1, 119 1994

(122) W. Bode, F.-X. Gomis-Rueth, R. Huber, R. Zwilling, W. Stoecker: Structure of astacin and implications for activation of astacins and zinc-ligation of collagenases. Nature 358, 164 1992.

(123) M.M. Thayer, K.M. Flaherty, D.B. McKay: Three-dimensional structure of the elastase of *Pseudomonas aeruginosa* at 1.5Å resolution. J. Biol. Chem. 266, 2864 1991.

(124) W. Stark, R.A. Pauptit, K.S. Wilson, J.N. Jansonius: The structure of neutral protease from *Bacillus cereus* at 0.2-nm resolution. Eur. J. Biochem. 207, 781 1992
R.A. Pauptit, R. Karlsson, D. Picot, J.A. Jenkins, A.-S. Niklaus-Reimer, J.N. Jansonius: Crystal structure of neutral protease from Bacillus cereus refined at 3.0A resolution and comparison with the homologous but more thermostable enzyme thermolysin. JMB 199 525-537, 1988

(125) D.E. Tronrud, H.M. Holden, B.W. Matthews: Structures of two thermolysin-inhibitor complexes that differ by a single hydrogen bond. Science 235, 571 1987

(126) U. Baumann, S. Wu, K.M. Flaherty, D.B. McKay: Three-dimensional structure of the alkaline protease of *Pseudomonas aeruginosa*: a two-domain protein with a calcium-binding parallel beta roll motif. EMBO J. 12, 3357 1993

(127) U. Baumann: Crystal structure of the 50 kda metallo protease from *S. marcescens*. J. Mol. Biol. 242, 244 1994

(128) K. Hamada, H. Hiramatsu, Y. Katsuya, Y. Hata, Y. Matsuura, Y. Katsube: Structural analysis of serratia protease. Acta Crystallogr. A49 153 1993
Y. Katsuya, K. Hamada, Y. Hata, N. Tanaka, M. Sato, Y. Katsube, K. Kakinchi, K. Miyata: Preliminary X-ray studies on Serratia Protease. J. Biochem. 98, 1139-1142, 1985.

(129) U. Baumann, M. Bauer, S. Létoffé, P. Delepelaire, C. Wandersman: Crystal structure of a complex between *Serratia marcescens* metallo-protease and an inhibitor from *Erwinia chrysanthemi*. J. Mol. Biol. 248, 653 1995

(130) D. Zhang, I. Botos, F.-X. Gomis-Rüth, R. Doll, C. Blood, F.G. Njoroge, J.W. Fox, W. Bode, E.F. Meyer: Structural interaction of natural and synthetic inhibitors with the venom metalloproteinase, atrolysin C (HT-D). Proc. Nat. Acad. Sci. USA 91, 8447 1994

(131) F.-X. Gomis-Rüth, L.F. Kress, W. Bode: First structure of a snake venom metalloproteinase: a prototype for matrix metalloproteinases/collagenases. EMBO J. 12, 4151 1993.

(132) J. Löwe, D. Stock, B. Jap, P. Zwickl, W. Baumeister, R. Huber: Crystal structure of the 20S proteasome from the archaeon *T. acidophilum* at 3.4Å resolution. Science 268, 533 1995.

MIX
Papier aus verantwortungsvollen Quellen
Paper from responsible sources
FSC® C105338

If you have any concerns about our products,
you can contact us on
ProductSafety@springernature.com

In case Publisher is established outside the EU,
the EU authorized representative is:
Springer Nature Customer Service Center GmbH
Europaplatz 3, 69115 Heidelberg, Germany

Printed by Libri Plureos GmbH
in Hamburg, Germany